Board Review Series

Cell Biology and Histology
2nd edition

PROJECT #2 — CREATE LESSON FROM   ENDOCRINE
                                   LYMPHOID
                                   CARDIOVASCULAR
                                   BLOOD

Board Review Series

Cell Biology and Histology

2nd edition

Leslie P. Gartner, Ph.D.
Associate Professor of Anatomy
Department of Anatomy
University of Maryland
Dental School
Baltimore, Maryland

James L. Hiatt, Ph.D.
Associate Professor of Anatomy
Department of Anatomy
University of Maryland
Dental School
Baltimore, Maryland

Judy M. Strum, Ph.D.
Professor of Anatomy
Department of Anatomy
University of Maryland
School of Medicine
Baltimore, Maryland

Harwal Publishing

Philadelphia • Baltimore • Hong Kong • London • Munich • Sydney • Tokyo

A Waverly Company

Harwal

Library of Congress Cataloging-in-Publication Data

Gartner, Leslie P., 1943–
    Histology/cell biology / Leslie P. Gartner, James L. Hiatt,
Judy M. Strum.—2nd ed.
      p.   cm.
    Rev. ed. of: Histology. c1988.
    Includes index.
    ISBN 0-683-03426-X (pbk. : alk paper)
    1. Histology—Outlines, syllabi, etc.   2. Cytology—Outlines,
syllabi, etc. I. Hiatt, James L., 1934–   .   II. Strum, Judy M.
(Judy May) III. Gartner, Leslie P., 1943–   Histology. IV. Title.
    [DNLM: 1. Histology—outlines. QS 18 G244h 1993]
QM551.G37 1993
611′.018′076—dc20
DNLM/DLC
for Library of Congress   93-6948
                                                  CIP

10 9 8 7 6 5 4 3

Contents

Preface to the Second Edition

We were pleased with the reception that the first edition of **Histology** was accorded by students who used it with success in preparing for National Boards examinations.

In order to prepare students better for the United States Medical Licensing Examination (USMLE Step 1), it was necessary to rewrite much of the material presented in the first edition. The emphasis that is now placed on cell and molecular biology in these examinations required that we weave a substantial amount of cell biology into the histology text, which is reflected in the title as well.

We have added four new chapters, "Plasma Membrane," "Nucleus," "Cytoplasm," and "Extracellular Matrix," and have revised the remaining chapters. In doing so, we have strived to present Board-driven material as succinctly as possible. A tremendous amount of subject matter, which comprises both cell biology and histology, is compressed into a concise but highly comprehensible presentation.

Many new illustrations prepared specifically for this edition will assist the student in speedily grasping essential concepts. In order to present the subject matter in a more clinically relevant fashion, the didactic component of each chapter concludes with a section entitled "Clinical Considerations."

The organization of the self-examinations at the conclusion of each chapter has remained unaltered; however, the format has been changed to coincide with that of USMLE Step 1. Finally, the last section of the book is dedicated to a comprehensive USMLE-type examination.

Leslie P. Gartner, Ph.D.
James L. Hiatt, Ph.D.
Judy M. Strum, Ph.D.

Acknowledgments

We gratefully acknowledge the excellent guidance we received in preparing this revision from Susan Kelly, Managing Editor of the Board Review Series, and to Ruth Steyn for her copyediting skills. For the excellent illustrations, we thank Tim Phelps, Rob Duckwall, Jerry Gadd, and also Iris Sadowsky, one of our graduate students. Additionally, we owe thanks to Kim Frank and Sharon Liu, both graduate students, for their help in completing the many tasks necessary to bringing this edition to fruition.

1

Plasma Membrane

I. Overview—The Plasma Membrane

- —is also known as the **plasmalemma** or **cell membrane**.
- —is about 7.5 nm thick.
- —is composed of a **lipid bilayer** and associated proteins (as are all membranes of the cell).
- —envelops the cell and aids in maintaining its structural and functional integrity.
- —functions as a **semipermeable** membrane between the cytoplasm and the external environment.
- —is a sensory device that permits the cell to recognize (and be recognized by) other cells and macromolecules.
- —is composed of an **inner leaflet** (facing the cytoplasm) and an **outer leaflet** (facing the extracellular environment).
- —exhibits a trilaminar structure (called the **unit membrane**) when thin sections are examined by transmission electron microscopy.

II. Fluid Mosaic Model of the Plasma Membrane

A. Lipid bilayer (Figure 1.1)
- —is freely permeable to small nonpolar lipid-soluble molecules but is impermeable to charged ions.

1. Molecular structure—the lipid bilayer
- —is composed of phospholipids, glycolipids, and cholesterol.

a. Phospholipids
- —are amphipathic, meaning that they possess a polar (**hydrophilic**) head and two nonpolar (**hydrophobic**) fatty acyl tails.
- **(1)** The **polar head** of each phospholipid molecule faces the membrane surface, whereas the double hydrophobic tail projects into the interior of the membrane.
- **(2)** The **tails** of the phospholipids in the inner and outer leaflets form weak bonds that attach the leaflets to one another.

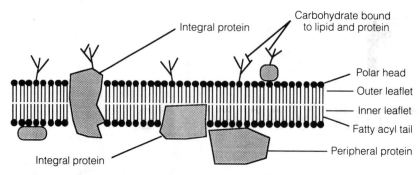

Figure 1.1. Diagrammatic representation of the plasma membrane showing the outer (*top*) and inner (*bottom*) leaflets of the unit membrane. The hydrophobic fatty acyl tails and the polar heads of the phospholipids constitute the lipid bilayer in which integral proteins are embedded. Peripheral proteins are located primarily on the cytoplasmic aspect of the inner leaflet.

b. Glycolipids

–are present in the outer leaflet only.

–have **polar carbohydrate residues** that extend from the outer leaflet into the extracellular space, forming part of the **glycocalyx**.

c. Cholesterol

–is located in both leaflets of the plasma membrane.

–constitutes approximately 2% of the plasmalemma lipids.

–assists in maintaining the structural integrity of the plasma membrane.

2. Fluidity of the lipid bilayer

–is crucial to such cellular activities as **exocytosis, endocytosis, membrane trafficking,** and **membrane biogenesis**.

–is **increased** by a rise in temperature and by greater unsaturation of the hydrocarbon (fatty acyl) tails.

–is **decreased** by an increase in the membrane's cholesterol content.

B. Membrane proteins (see Figure 1.1)

1. Integral proteins

–are dissolved in the lipid bilayer.

–are **amphipathic,** containing **hydrophilic** amino acids and **hydrophobic** amino acids, some of which interact with the hydrocarbon tails of the membrane phospholipids.

–may extend into both leaflets or into the inner leaflet only.

–include **transmembrane proteins,** which span the entire plasma membrane and function as membrane **receptors** and/or **transport proteins**. The latter employ various mechanisms for transferring polar molecules across the membrane.

–are often **glycoproteins,** most of which are transmembrane proteins.

–remain preferentially attached to the **P-face** (protoplasmic face), the external surface of the inner leaflet, and seldom remain attached to the **E-face,** the internal surface of the outer leaflet, in freeze-fractured membranes (Figure 1.2).

2. Peripheral proteins

–do **not** extend into the lipid bilayer.

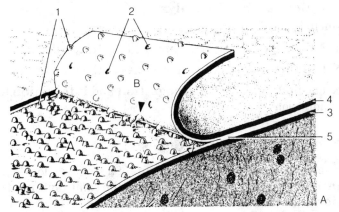

Figure 1.2. Freeze-fracturing cleaves the plasma membrane (*5*). The impressions (*2*) of the transmembrane proteins are evident on the E-face between the inner leaflet (*3*) and outer leaflet (*4*). The integral proteins (*1*) remain preferentially attached to the P-face (*A*), the external surface of the inner leaflet; fewer proteins remain associated with the E-face (*B*), the internal surface of the outer leaflet. The arrowhead indicates a transmembrane protein attached to both the E-face and P-face. (Reprinted with permission from Krstic RV: *Ultrastruktur der Säugertierzelle.* Berlin, Springer Verlag, 1976, p 177.)

—are located on the cytoplasmic aspect of the inner leaflet. In some cells (e.g., thymocytes), the outer leaflets possess covalently attached glycolipids to which peripheral proteins are anchored and thus project into the intercellular space.

—bond to the polar groups of membrane phospholipids or to integral membrane proteins.

—usually function as part of the **cytoskeleton** or of an **intracellular secondary messenger system**.

3. Functional characteristics of membrane proteins

    a. Ratio of lipid to protein in plasma membranes varies from 1:1 (by weight) in most cells to 4:1 in myelin.

    b. Some membrane proteins can **diffuse laterally** in the lipid bilayer, whereas others are **immobile,** being held in place by interactions with cytoskeletal constituents.

C. Glycocalyx

—is the **sugar coat** commonly associated with the extracytoplasmic aspect of the outer leaflet of the plasma membrane.

—may measure as much as 50 nm in thickness.

—is also known as the **cell coat**.

—varies in appearance in transmission electron micrographs (e.g., in thickness, fuzziness).

—may be associated with extracellular matrix components that adhere to the cell surface.

1. Composition—the glycocalyx

—consists of polar oligosaccharide side chains linked covalently to most protein and some lipid (glycolipid) constituents of the plasma membrane.

—contains cell-surface **proteoglycans,** which consist of membrane integral proteins to which are bound **glycosaminoglycans** (negatively charged polysaccharides).

—can be revealed by lectin binding to the component carbohydrates.

2. Functions of the glycocalyx

—include the following:

a. Aiding in **cellular attachment** to extracellular matrix components

b. **Binding** of antigens and enzymes to the cell surface

c. Facilitating **cell–cell recognition** and **interaction** (e.g., sperm–egg adhesion)

III. Membrane Transport Processes

—may involve the unidirectional transport of a single molecule (**uniport**) or the **cotransport** of two different molecules in the same (**symport**) or the opposite (**antiport**) direction.

A. Passive transport (Figure 1.3)

—includes **simple diffusion** and **facilitated diffusion,** neither of which requires energy.

—involves passage of molecules across the plasma membrane **down a concentration or electrochemical gradient**.

1. Simple diffusion

—transports small nonpolar molecules (e.g., O_2 and N_2) and small uncharged polar molecules (e.g., H_2O, CO_2, and glycerol).

—exhibits little specificity.

—occurs at a **rate proportional to the concentration gradient** of the transported molecule across the membrane.

2. Facilitated diffusion

—occurs via **channel** and/or **carrier proteins,** which exhibit specificity for the transported molecules.

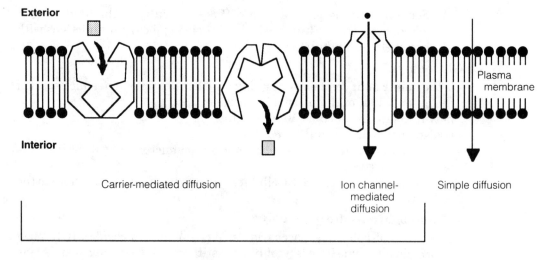

Figure 1.3. Passive transport of molecules across plasma membranes occurs by simple diffusion (*right*) and facilitated diffusion mediated by ion channel proteins (*center*) and carrier proteins (*left*).

−increases the transport rate over that obtainable by simple diffusion, in effect making the membrane permeable to molecules to which it would otherwise be almost impermeable (e.g., ions and large polar molecules).

a. Ion channel proteins

−are highly folded transmembrane proteins that form small aqueous pores across membranes through which specific small, water-soluble molecules and ions pass down an electrochemical gradient.

b. Carrier proteins

−are highly folded transmembrane proteins that undergo reversible conformational alterations, resulting in transport of specific molecules across the membrane.

−function in **both** passive transport and active transport.

B. Active transport

−is an **energy-requiring process** for transporting a molecule **against** an electrochemical gradient via carrier proteins.

1. Na^+–K^+ pump

−involves the **antiport** transport of Na^+ ions out of the cell and of K^+ ions into the cell mediated by Na^+–K^+ **ATPase,** a carrier protein.

a. Mechanism

(1) Three Na^+ ions are pumped **out** of the cell.

(2) Two K^+ ions are brought **into** the cell.

(3) To complete these two steps, a single ATP is hydrolyzed by the Na^+–K^+ ATPase.

(4) The binding site for K^+ is on the extracellular portion of the carrier protein, whereas the binding site for Na^+ is on the cytoplasmic side.

(5) **Oubain,** which has the same binding site as K^+, inhibits **both** ion transport and the Na^+–K^+ ATPase.

b. Function

(1) The pump acts primarily to **maintain constant cell volume** by reducing the ion concentration (and thus the osmotic pressure) intracellularly and by increasing it extracellularly, thus decreasing the flow of water into the cell.

(2) The pump also plays a minor role in the maintenance of a **potential difference** across the plasma membrane.

2. Glucose transport

−involves the **symport** movement of glucose across an epithelium (**transepithelial transport**).

−is frequently powered by an electrochemical Na^+ gradient, which drives carrier proteins that are located at specific regions of the cell surface.

−occurs by the following sequence of events:

a. Na^+ ions bind to the carrier proteins, inducing conformational alterations in them that unmask attachment sites for glucose on the **external** aspect of the plasma membrane.

b. Once glucose binds, another conformational change in the carrier proteins brings the Na^+ and glucose inside the cell.

    c. Elsewhere in the plasma membrane, a Na^+–K^+ pump drives Na^+ ions out of the cell, thus maintaining the Na^+ concentration gradient.

    d. At other sites, glucose exits the cell via glucose-specific carrier proteins **down** a chemical gradient.

C. Facilitated diffusion of ions

—is mediated by selective ion channel proteins, which permit only certain ions to traverse them.

—may also occur via **ionophores,** small lipid-soluble molecules that can insert themselves into the lipid bilayer and increase membrane permeability to specific ions.

1. K^+ leak channels

—are **ungated**.

—are the most common ion channel; thus K^+ is most responsible for establishing a potential difference across the plasma membrane.

2. Gated ion channels

—open only **transiently** in response to various stimuli.

—include the following types:

    a. Voltage-gated channels open upon alteration of the potential difference across the membrane.

    b. Mechanically gated channels open in response to a mechanical stimulus (e.g., the tactile response of the hair cells in the inner ear).

    c. Ligand-gated channels open upon binding of a **signaling molecule** or **ion** to the channel protein.

        (1) Neurotransmitter-gated channels open in response to binding of a neurotransmitter substance.

        (2) Nucleotide-gated channels respond to an increase in the intracellular concentration of nucleotides such as cAMP, which binds to the cytoplasmic aspect of the channel protein (e.g., in olfactory cells).

    d. G-protein–gated K^+ channels, which are present in cardiac muscle cells, open following binding of the neurotransmitter acetylcholine to the G-protein–linked **muscarinic acetylcholine receptor**.

3. Voltage-gated Na^+ channels

—function in the generation of **action potentials** in nerve cells (see Chapter 9 VIII B and Figure 9.2).

IV. Cell-to-Cell Communication

A. Signaling molecules

—are secreted by cells and function in communication with other cells.

—include **neurotransmitter substances,** which are released into the synaptic cleft (see Chapter 8 IV A; Chapter 9 IV D); **endocrine hormones,** which are carried in the bloodstream and act on distant target cells; and hormones released into the intercellular space, which act on nearby cells **(paracrine hormone)** or the releasing cell **(autocrine hormone)**.

1. Lipid-soluble signaling molecules

   –include certain hormones (e.g., testosterone).

   –penetrate the plasma membrane and bind to **receptors in the cytoplasm** or **nucleus,** activating intracellular messengers.

   –frequently influence gene transcription.

2. Hydrophilic signaling molecules

   –include neurotransmitters and numerous hormones (e.g., serotonin, thyroid-stimulating hormone, insulin).

   –bind to and activate **cell-surface receptors** (as do some lipid-soluble signaling molecules).

   –have diverse physiologic effects (see Chapter 13).

B. Membrane receptors

   –are mostly glycoproteins.

   –are located on the cell surface and bind specific signaling molecules.

1. Functions of membrane receptors

   a. Control plasmalemma permeability by regulating the conformation of ion channel proteins (e.g., the nicotinic acetylcholine receptor).

   b. Regulate the entry of molecules into the cell (e.g., the delivery of cholesterol via low-density lipoprotein receptors).

   c. Bind extracellular matrix molecules to the cytoskeleton; these receptors, called **integrins,** are essential for cell–cell contact and cell–matrix interactions.

   d. Act as transducers to transfer extracellular events intracellularly to secondary messenger systems.

   e. Permit pathogenic organisms that mimic normal ligands to enter cells.

2. Channel-linked receptors

   –include **neurotransmitter-gated** ion channels.

   –bind a signaling molecule that temporarily opens or closes the gate, permitting or inhibiting the movement of ions across the cell membrane.

   –include **nicotinic acetylcholine receptors** on the muscle-cell sarcolemma at the myoneural junction (see Chapter 8 IV A).

3. Catalytic receptors

   –are single-pass transmembrane proteins whose extracytoplasmic moiety functions as a receptor and whose **cytoplasmic aspect functions as a protein kinase**.

   –may lack an extracytoplasmic moiety and as a result be continuously activated; such defective receptors are coded for by some **oncogenes**.

   –include receptors for insulin, epidermal growth factor, and platelet-derived growth factor.

   a. Insulin binds to its receptor, which **autophosphorylates**. The insulin–receptor complex then is **endocytosed,** enabling the complex to perform its function within the cell.

   b. Growth factors bind to many of the catalytic receptors, inducing the cells to proliferate.

4. G-protein–linked receptors

—are transmembrane proteins **associated with an ion channel or an enzyme** that is bound to the cytoplasmic surface of the plasma membrane.

—interact with **G protein** (GTP-binding regulatory protein) after binding of a signaling molecule; this interaction results in the activation of **intracellular secondary messengers,** the most common of which are **cyclic AMP** (cAMP) and **Ca^{2+}**.

a. G proteins (Table 1.1)

—include stimulatory G protein (**G$_s$**), inhibitory G protein (**G$_i$**), phospholipase C activator G protein (**G$_p$**), and transducin (**G$_t$**).

—of all types, have the same general composition and **act as signal transducers** mediating the interaction between the receptor and a membrane-bound enzyme or ion channel.

b. Composition and mechanism of G$_s$ protein (Figure 1.4)

(1) Three subunits (α, β, γ), located in the inner leaflet of the plasma membrane, compose G$_s$ protein.

(2) The α subunit binds **GTP** when the receptor is activated by association with a signaling molecule; it binds **GDP** when the receptor is not activated.

(3) When GTP is bound, the α subunit dissociates from the other two and activates **adenylate cyclase**.

(4) The α subunit also catalyzes hydrolysis of GTP to GDP, thus controlling the duration of its own activation.

V. Plasma Membrane–Cytoskeleton Association

—is via **integrins,** whose extracytoplasmic domain binds to components of the extracellular matrix and whose intracellular domain binds to cytoskeletal components.

—stabilizes the plasma membrane and is important in determining and maintaining cell shape.

Table 1.1. Functions and Examples of G Proteins

Type	Function	Result	Examples
G$_s$	Activates adenylate cyclase leading to formation of cAMP	Activation of protein kinases	Binding of epinephrine to β-adrenergic receptors increases cAMP levels in cytosol
G$_i$	Inhibits adenylate cyclase preventing formation of cAMP	Protein kinases remain inactive	Binding of epinephrine to α_2-adrenergic receptors decreases cAMP levels in cytosol
G$_p$	Activates phospholipase C leading to formation of inositol triphosphate and diacylglycerol	Influx of Ca^{2+} into cytosol and activation of protein kinase C	Binding of antigen to membrane-bound IgE causes the release of histamine by mast cells
G$_t$	Activates cGMP phosphodiesterase in rod-cell membranes leading to hydrolysis of cGMP	Hyperpolarization of rod-cell membrane	Photon activation of rhodopsin causes rod cells to fire

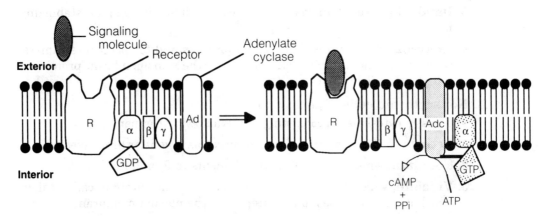

Figure 1.4. Functioning of G_s-protein–linked receptors. The signaling molecule binds to the receptor, which causes the α subunit of the G_s protein to bind GTP as well as to dissociate from the β and γ subunits. Activation of adenylate cyclase by the GTP–α-subunit complex stimulates synthesis of cAMP, one of the most common intracellular messengers.

A. Red blood cells (Figure 1.5A)

 −have integrins, which are referred to as **band 3 proteins,** in their plasma membrane.

 −possess a cytoskeleton consisting of the following components:

 1. Spectrin, a long flexible protein (about 110 nm long), which forms **tetramers** and provides a scaffolding for structural reinforcement

 2. Actin, which attaches to binding sites on the spectrin tetramers, thus forming a lattice-work of spectrin tetramers held together by actin

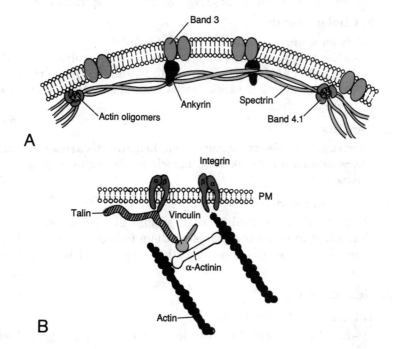

Figure 1.5. Plasmalemma–cytoskeleton association in red blood cells (A) and nonerythroid cells (B). (Reprinted with permission from Widnell CC and Pfenninger KH: *Essential Cell Biology.* Baltimore, Williams & Wilkins, 1990, p 82.)

3. **Band 4.1 protein,** which binds to the spectrin–actin complex, stabilizing it

4. **Ankyrin,** which is linked to band 3 proteins and to spectrin tetramers, thus affixing the spectrin–actin complex to the transmembrane proteins of the plasma membrane

B. **Nonerythroid cells** (Figure 1.5*B*)

  –possess a cytoskeleton consisting of the following components:

  1. **Actin** (and perhaps **fodrin,** which serves as a nonerythroid spectrin)

  2. **α-Actinin,** which crosslinks actin filaments to form a meshwork

  3. **Vinculin,** which binds to α-actinin and to another protein, called **talin,** which in turn attaches to the integrin in the plasma membrane

VI. Clinical Considerations

A. **Cystinurea**

  –is a hereditary condition caused by abnormal carrier proteins that are unable to remove cystine from the urine, resulting in the formation of kidney stones.

B. **Toxins**

  1. **Tetrodotoxin**

    –is produced by the puffer fish.
    –**inactivates Na$^+$ channels** by occupying the Na$^+$-binding sites, thus paralyzing the prey.
    –ingestion by humans quickly causes dizziness and tingling about the mouth, which may be followed by ataxia, respiratory paralysis, and death.

  2. **Cholera toxin**

    –is an exotoxin produced by *Vibrio cholera*.
    –**alters G$_s$ protein** so that it is unable to hydrolyze its GTP molecule. As a result, cAMP levels increase in the absorptive surface cells of the intestine, leading to excess loss of sodium ions and water, which causes severe diarrhea.

C. **Venoms**

  –such as those of some poisonous snakes, **inactivate acetylcholine receptors** located on the skeletal muscle sarcolemma at neuromuscular junctions.

D. **Autoimmune diseases**

  –may produce antibodies that specifically **bind to certain plasma–membrane receptors,** causing them to become activated.
  –include **Graves' disease** (hyperthyroidism), which is discussed in Chapter 13 VIII B.

E. **Genetic defects**

  1. **Defective G$_s$ proteins**

    –may lead to mental retardation and/or diminished growth and sexual development.
    –may result in decreased responses to certain hormones.

2. Hereditary spherocytosis

—is characterized by fragile, misshaped red blood cells (**spherocytes**).

—results from a **defective spectrin** that has decreased ability to bind to band 4.1 protein.

—is marked by anemia due to the increased destruction of spherocytes in the spleen.

Review Test

Directions: Each of the numbered items or incomplete statements in this section is followed by answers or by completions of the statement. Select the **one** lettered answer or completion that is **best** in each case.

1. A herpetologist, bitten by a poisonous snake, is brought into the emergency room with progressive muscle paralysis. The venom is probably incapacitating his

(A) Na^+ channels
(B) Ca^{2+} channels
(C) phospholipids
(D) acetylcholine receptors
(E) spectrin

2. Which one of the following statements concerning the lipid bilayer is FALSE?

(A) The presence of cholesterol decreases the fluidity of the lipid bilayer
(B) Glycolipids are located in the outer leaflet
(C) Each phospholipid possesses a polar head that projects into the center of the membrane
(D) Each lipid molecule is amphipathic
(E) The glycocalyx is formed largely by carbohydrate residues projecting into the extracellular space

3. Which of the following statements concerning integral membrane proteins is FALSE?

(A) They do *not* extend into the lipid bilayer
(B) They are amphipathic molecules
(C) They may be transmembrane proteins
(D) They may be integrins
(E) They may function as membrane receptors

4. Which one of the following statements concerning plasma–membrane components is FALSE?

(A) G proteins are composed of three subunits
(B) The glycocalyx is usually composed of phospholipids
(C) Channel proteins are not energy dependent
(D) Gated channels are open only transiently
(E) Ankyrin binds to integrins of the red blood cell plasma membrane

5. Which one of the following transport processes requires energy?

(A) Facilitated diffusion
(B) Passive transport
(C) Active transport
(D) Simple diffusion

6. Which one of the following substances is unable to traverse the plasma membrane by simple diffusion?

(A) O_2
(B) N_2
(C) Na^+
(D) Glycerol
(E) CO_2

7. Symport refers to the process of transporting

(A) a molecule into the cell
(B) a molecule out of the cell
(C) two different molecules in the opposite direction
(D) two different molecules in the same direction
(E) a molecule between the cytoplasm and the nucleus

8. One of the ways that cells communicate with each other is by secretion of various molecules. The secreted molecule is known as

(A) a receptor molecule
(B) a signaling molecule
(C) a spectrin tetramer
(D) an integrin
(E) an anticodon

9. The hormone ACTH travels through the bloodstream, enters connective tissue spaces, and attaches to specific sites on target-cell membranes. These sites are

(A) peripheral proteins
(B) signaling molecules
(C) G proteins
(D) G-protein–linked receptors
(E) ribophorins

10. Examination of the blood smear of a young patient displays misshapen red blood cells, and the pathology report indicates hereditary spherocytosis. Defects in which one of the following proteins causes this condition?

(A) Signaling molecules
(B) G proteins
(C) Spectrin
(D) Hemoglobin
(E) Ankyrin

Directions: Each group of items in this section consists of lettered options followed by a set of numbered items. For each item, select the **one** lettered option that is most closely associated with it. Each lettered option may be selected once, more than once, or not at all.

Questions 11–15

Match each characteristic below with the pump or channel that fits best.

(A) Na^+–K^+ pump
(B) Glucose carrier protein driven by a Na^+ gradient
(C) K^+ leak channels
(D) Voltage-gated Na^+ channels
(E) Ligand-gated channels

11. Maintenance of constant cell volume

12. Establishment of a potential difference across the cell membrane

13. Initiation of muscle contraction

14. Formation of action potentials

15. A type of symport

Questions 16–20

Match each characteristic below with the G protein that fits best.

(A) G_s
(B) G_i
(C) G_p
(D) G_t

16. Formation of inositol triphosphate

17. Hyperpolarization of rod cells

18. Synthesis of cAMP

19. Prevention of cAMP synthesis

20. Mental retardation due to decreased responses to certain hormones

Answers and Explanations

1–D. Snake venom usually blocks acetylcholine receptors, thus preventing depolarization of the muscle cell. The Na^+ and Ca^{2+} channels are not incapacitated by snake venoms.

2–C. Each phospholipid molecule is composed of a polar head and two nonpolar fatty acyl tails. The polar head faces the membrane surface; it is the double hydrophobic tails that project towards the center of the membrane.

3–A. Integral membrane proteins are of two types, those that extend through the entire thickness of the lipid bilayer (transmembrane proteins) and those that extend half way into the lipid bilayer. It is the peripheral proteins that do not extend into the lipid bilayer.

4–B. The glycocalyx—the sugar coat on the membrane surface—is composed mostly of polar carbohydrate residues.

5–C. Active transport requires energy. Facilitated diffusion, which is mediated by membrane proteins, and simple diffusion, which involves passage of material directly across the lipid bilayer, are types of passive transport.

6–C. Na^+ and other ions require channel (carrier) proteins for their transport across the plasma membrane. The other substances, which are small nonpolar molecules and small uncharged polar molecules, can traverse the plasma membrane by simple diffusion.

7–D. The coupled transport of two different molecules in the same direction is termed symport.

8–B. Cells can communicate with each other by releasing signaling molecules that attach to receptor molecules on target cells.

9–D. G-protein–linked receptors are sites where ACTH and some other signaling molecules attach. Binding of ACTH to its receptor causes G_s protein to activate adenylate cyclase, setting in motion the specific response elicited by the hormone.

10–C. Hereditary spherocytosis is caused by a defect in spectrin that renders the protein incapable of binding to band 4.1 protein, thus destabilizing the spectrin–actin complex of the cytoskeleton. Although defects in hemoglobin—the respiratory protein of erythrocytes—also cause red blood cell anomalies, hereditary spherocytosis is not one of them.

11–A. The main function of the Na^+–K^+ pump is to maintain constant cell volume by reducing the ion concentration intracellularly and increasing it extracellularly, thus decreasing the flow of water into the cell.

12–C. K^+ leak channels are primarily responsible for establishing the potential difference across the plasma membrane.

13–E. Acetylcholine receptors, which are ligand-gated channels, initiate muscle contraction.

14–D. Voltage-gated Na^+ channels function in the formation of action potentials.

15–B. Glucose transport into epithelial cells is mediated by Na^+-powered glucose symports.

16–C. Inositol triphosphate is one of the byproducts of the hydrolysis of the membrane lipid phosphatidylinositolbiphosphate. The enzyme responsible for this hydrolysis is phospholipase C, which is activated by G_p protein.

17–D. Absorption of a photon by rhodopsin causes it to form a bond with G_t, resulting in activation of cGMP phosphodiesterase, the enzyme that reduces cGMP levels in the rod cytosol. As a result, the rod plasma membrane becomes hyperpolarized, and the rod is no longer able to inhibit the postsynaptic cell.

18–A. G_s protein activates the enzyme adenylate cyclase, facilitating the formation of cAMP.

19–B. G_i protein inhibits adenylate cyclase, preventing the formation of cAMP.

20–A. Hereditary defects in G_s protein can cause decreased responsiveness to certain hormones, resulting in mental retardation, diminished growth, and diminished sexual development.

2

Nucleus

I. Overview—The Nucleus (Figure 2.1)

–includes the **nuclear envelope, nucleolus, nucleoplasm,** and **chromatin**.
–contains genetic apparatus encoded in the **deoxyribonucleic acid** (DNA) of chromosomes.
–directs protein synthesis in the cytoplasm via **ribosomal ribonucleic acid** (rRNA), **messenger RNA** (mRNA), and **transfer RNA** (tRNA), which are synthesized in the nucleus.

II. Nuclear Envelope (Figure 2.2)

–surrounds the nuclear material and consists of **two parallel membranes** separated from each other by a narrow **perinuclear cisterna**.
–is perforated at intervals by openings called **nuclear pores**.

A. Outer nuclear membrane

–is about 6 nm thick.
–faces the cytoplasm and is continuous at certain sites with the rough endoplasmic reticulum (RER).
–is surrounded on its cytoplasmic aspect by a loosely arranged mesh of intermediate filaments (**vimentin**).
–is studded with **ribosomes** on its cytoplasmic surface (see Figure 2.1), which can synthesize proteins that enter into the perinuclear cisterna.

B. Inner nuclear membrane

–is about 6 nm thick.
–faces the nuclear material but is separated from it and is supported on its inner surface by the **nuclear lamina,** a fibrous lamina that is 80–300 nm thick and composed primarily of **lamins A, B,** and **C**—a class of intermediate filament proteins that help to organize the nuclear envelope and perinuclear chromatin.

C. Perinuclear cisterna

–is 20–40 nm wide.
–is located **between** the inner and outer nuclear membranes.

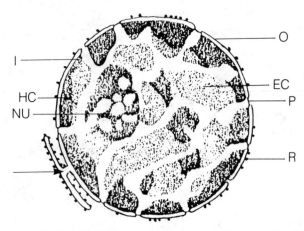

Figure 2.1. Diagram of the cell nucleus. The nuclear envelope, formed by the inner (*I*) and outer (*O*) nuclear membranes, is interrupted with nuclear pores (*P*). Ribosomes (*R*) are located on the outer nuclear membrane, and the envelope is continuous with the rough endoplasmic reticulum (*arrow*). The inactive heterochromatin (*HC*) is dense and mostly confined to the periphery of the nucleus. Euchromatin (*EC*), the active form, is less dense and dispersed throughout. The nucleolus (*NU*) contains fibrillar and granular portions.

      —is continuous with the cisternae of the RER.
      —is interrupted at various locations by nuclear pores.

D. Nuclear pores
      —average 80 nm in diameter.
      —permit passage of certain molecules in either direction between the nucleus and cytoplasm.
      —are formed by fusion of the inner and outer nuclear membranes.

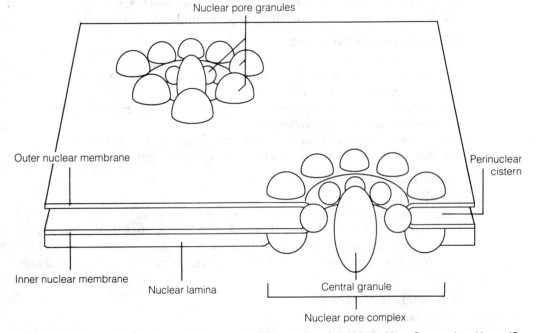

Figure 2.2. Diagram of the nuclear membrane and nuclear pore complex. (Adapted from Stevens A and Lowe JS: *Histology*. London, Gower Medical Publishing, Inc., 1992, p 12.)

—are associated with the nuclear pore complex, whose cylindrical aqueous channel opening is about 9 nm in diameter and permits the passage of small water-soluble molecules.

—vary in number and distribution depending on cell type and state of differentiation.

E. Nuclear pore complex (see Figure 2.2)

—is an arrangement of proteins around each nuclear pore.

—is composed of **eight sets of large protein granules** arranged in an octagonal fashion embedded in the margin of the pore and attached to the inner and outer nuclear membranes.

—has a **large central granule** that often plugs the pore and probably represents preribosomal subunits (or other particles) in transit between the nucleus and cytoplasm.

—selectively controls movement of various proteins into and out of the nucleus, probably by altering the diameter of the pore lumen.

—possesses nuclear pore **receptor proteins,** which recognize **nuclear import signals** on certain nuclear proteins (e.g., **nucleoplasmin**); once recognized, these proteins are actively transported into the nucleus.

—exports newly formed RNA molecules and ribosomal subunits from the nucleus to the cytosol (via active transport).

III. Nucleolus

—is a well-defined, but **not** membrane-bounded, nuclear inclusion (sometimes more than one) present in cells that are actively synthesizing proteins.

—generally is detectable only when the cell is in **interphase**.

—is involved in the synthesis of **rRNA** and its assembly into precursors of ribosomes.

—contains **nucleolar organizers**—regions of some chromosomes (in humans, chromosomes 13, 14, 15, 21, and 22) where rRNA genes are located; these regions are involved in reconstituting the nucleolus during the G_1 phase of the cell cycle.

—contains the following distinct regions:

A. Fibrillar centers

—are spherical areas of **inactive ribosomal DNA** (rDNA) surrounded by a dense fibrillar region.

B. Fibrillar regions

—are composed of fibrils (5 nm in diameter) around and between the fibrillar centers.

—contain **transcriptionally active rDNA**.

—represent early stages in the formation of rRNA precursors.

C. Granular regions

—are composed of 15-nm particles representing **maturing ribosomal precursors**.

IV. Nucleoplasm

—is the portion of the protoplasm that is surrounded by the nuclear envelope.

—consists of a matrix and various types of particles.

A. Nuclear matrix

—acts as a scaffold that aids in organizing the nucleoplasm.

—contains the following **structural components:** fibrillar elements, the nuclear pore–nuclear lamina complex, residual nucleoli, and a residual ribonucleoprotein (RNP) network.

—contains the following **functional components,** which are involved in the transcription and processing of mRNA and rRNA: steroid receptor–binding sites, carcinogen-binding sites, heat-shock proteins, DNA viruses, and viral proteins (T antigen).

B. Nuclear particles

1. Interchromatin granules

—are clusters of irregularly distributed particles (20–25 nm in diameter) that contain ribonucleoprotein.

—contain various enzymes, including ATPase, GTPase, NAD–pyrophosphatase, and β-glycerophosphate, but their function is not yet clear.

2. Perichromatin granules (see Figure 2.1)

—are single dense granules (30–50 nm in diameter) with a less dense **halo** surrounding them.

—are located at the periphery of heterochromatin.

—exhibit a substructure of 3-nm packed fibrils.

—contain **4.7S RNA** and two peptides similar to those found in heterogeneous nuclear ribonucleoprotein particles.

—may represent **messenger ribonucleoprotein particles** (mRNPs).

—increase in number in liver cells exposed to carcinogens or temperatures above 37° C.

3. Heterogeneous nuclear ribonucleoprotein particles (hnRNPs)

—are complexes of **precursor mRNA** (pre-mRNA) with proteins.

—are involved in processing of pre-mRNA.

4. Small nuclear ribonucleoprotein particles (snRNPs)

—are complexes of proteins and **small RNAs**.

—are involved in hnRNP splicing or in cleavage reactions.

V. Chromatin (see Figure 2.1)

—is double-stranded DNA complexed with **histones** and **acidic proteins**.

—is responsible for **RNA synthesis**.

—resides within the nucleus in two forms: heterochromatin and euchromatin.

A. Heterochromatin

—appears in the light microscope as basophilic clumps of nucleoprotein.

—appears in the transmission electron microscope (EM) as dense granular clumps of nucleoprotein.

—is concentrated at the periphery of the nucleus and around the nucleolus, as well as being scattered throughout the nucleoplasm.

—corresponding to **one of the two X chromosomes,** is present in nearly all somatic cells of female mammals. During interphase, the inactive X chromosome is visible as a dark-staining evagination protruding from the nucleus. This structure is called the **Barr body,** or **sex chromatin**.

—is **transcriptionally inactive**.

B. Euchromatin

–appears in the light microscope as a lightly stained, dispersed region of the nucleus.

–appears in the transmission electron microscope as electron-lucent regions among heterochromatin.

–is **transcriptionally active**.

VI. Chromosomes

–become visible during mitosis and meiosis when their constituent chromatin is condensed.

A. Structure—chromosomes

–consist of chromatin extensively folded into loops and maintained in its conformation by DNA-binding proteins (Figure 2.3).

–contain a single long DNA molecule and associated proteins, which are assembled into **nucleosomes,** the structural unit of chromatin packing.

1. Extended chromatin

–contains two copies each of **histones H2A, H2B, H3,** and **H4,** which form the nucleosome core around which the DNA double helix is wrapped two full turns. Nucleosomes are spaced at intervals of 200 base pairs.

–resembles "beads on a string" at the transmission EM level with the beads being **nucleosomes** and the string being **linker DNA**.

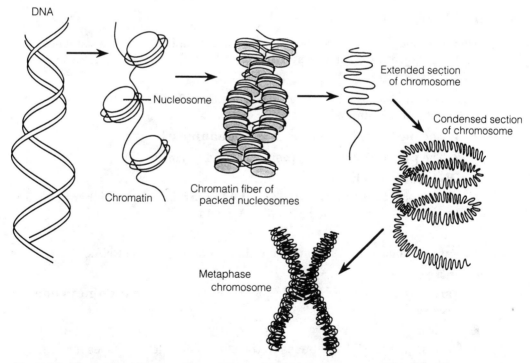

Figure 2.3. Diagram showing the packing of chromatin into the condensed metaphase chromosome. Nucleosomes contain two copies each of histones H2A, H2B, H3, and H4 in extended chromatin. An additional histone, H1, is present in condensed chromatin. (Adapted from Widnell CC and Pfenninger KH: *Essential Cell Biology.* Baltimore, Williams & Wilkins, 1990, p 47.)

2. Condensed chromatin

—contains an additional histone, **H1,** which wraps around groups of nucleosomes forming 30-nm-diameter fibers, the structural unit of the chromosome.

B. G-banding

—is observed in chromosomes during mitosis after staining with Giemsa, which is specific for DNA sequences rich in **adenine** (A) and **thymine** (T).

—is thought to represent DNA loops highly folded on themselves.

—is used in identifying chromosomes and is characteristic for each species.

C. Karyotype

—refers to the **number and morphology of chromosomes** and is characteristic for each species.

1. Haploid number (n) is the number of chromosomes in germ cells (23 in humans).

2. Diploid number ($2n$) is the number of chromosomes in somatic cells (46 in humans).

D. Genome

—is the **total genetic complement** of an individual stored in its chromosomes.

—is represented in humans by the 22 pairs of **autosomes** and the 1 pair of **sex chromosomes** (either **XX** or **XY**) totaling 46 chromosomes.

VII. DNA

—is a very long linear molecule composed of multiple nucleotide sequences.

—acts as a **template for the synthesis of RNA**.

A. Nucleotides

—are composed of a base (purine or pyrimidine), a deoxyribose sugar, and a phosphate group.

1. The **purines** are **adenine** (A) and **guanine** (G).

2. The **pyrimidines** are **cytosine** (C) and **thymine** (T).

B. DNA double helix

—consists of **two complementary DNA strands** bound together by hydrogen bonds between the base pairs A-T and G-C.

C. Exons

—are regions of the DNA molecule that **code** for specific RNAs.

D. Introns

—are regions of the DNA molecule, between exons, that **do not code** for RNAs.

E. Codon

—is a sequence of **three bases** in the DNA molecule that codes for a **single amino acid**.

F. Gene

—is a segment of the DNA molecule that is responsible for the formation of a **single RNA molecule**.

VIII. RNA

—is a linear molecule similar to DNA but containing **ribose** instead of deoxyribose and **uracil** (U) instead of thymine.

—is synthesized by **transcription** of DNA, which is catalyzed by three **RNA polymerases:** I for rRNA, II for mRNA, and III for tRNA.

A. Messenger RNA

—carries the genetic code to the cytoplasm for directing **protein synthesis** (Figure 2.4).

—is a single-stranded molecule consisting of hundreds to thousands of nucleotides.

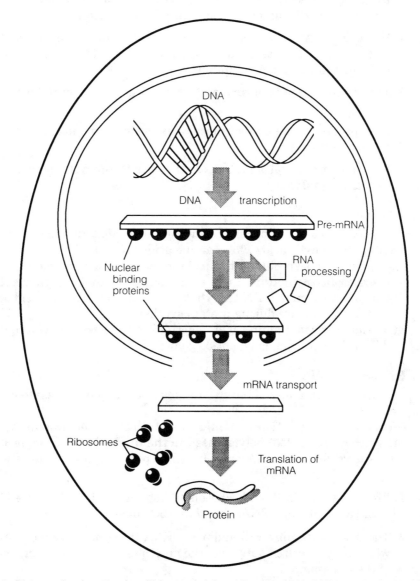

Figure 2.4. Diagram showing steps by which genetic information encoded in DNA is transcribed into mRNA and ultimately converted into proteins in the cytoplasm. (Adapted from Alberts B et al: *Biology of the Cell,* 2nd ed. New York, Garland Publishing, Inc., 1989, p 482.)

—contains codons that are **complementary** to the DNA codons from which it was transcribed, including one codon (**AUG**) for **initiating** protein synthesis and one of three codons (**UAA, UAG,** or **UGA**) for **terminating** protein synthesis.

—is synthesized in the following series of steps:

1. **RNA polymerase II** recognizes a **promoter** on a single strand of the DNA molecule and binds tightly to it.

2. The DNA helix unwinds for about two turns, separating the DNA strands and exposing the **codons** that act as the template for synthesis of the complementary RNA molecule.

3. RNA polymerase II moves along the DNA strand and promotes base pairing between DNA and complementary RNA nucleotides.

4. When the RNA polymerase II recognizes a **chain-terminator** on the DNA molecule, it terminates its association with the DNA and is released to repeat the process of transcription.

5. The primary transcript, **pre-mRNA,** associates with proteins to form an **hnRNP.**

6. RNA processing then takes place to excise introns and splice exons, producing an **mRNP.**

7. Subsequent removal of proteins as the mRNP enters the cytoplasm yields **functional mRNA.**

B. **Transfer RNA**

—is folded into a cloverleaf shape and contains about 80 nucleotides, terminating in adenylic acid (where amino acids attach).

—combines with a specific amino acid that has been activated by an enzyme.

—possesses an **anticodon,** a triplet of nucleotides that recognizes the complementary codon in mRNA, so that insertion of each amino acid occurs in the proper sequence during protein synthesis.

—transfers activated amino acid molecules to the ribosome where they are added to the growing polypeptide chain.

C. **Ribosomal RNA**

—associates with many different proteins (including enzymes) to constitute **ribosomes.**

—functions in association with mRNA and tRNA during protein synthesis.

—is synthesized by RNA polymerase I **in the nucleolus** as a single **45S precursor rRNA** (pre-rRNA), which is then **processed to form ribosomes** as follows (Figure 2.5):

1. Pre-rRNA associates with ribosomal proteins and is then cleaved into the 28S, 18S, and 5.8S rRNAs found in ribosomes.

2. The RNP containing 28S and 5.8S rRNA then combines with 5S rRNA, which is synthesized outside of the nucleolus, to form the **large subunit** of the ribosome.

3. The RNP containing 18S rRNA forms the **small subunit** of the ribosome.

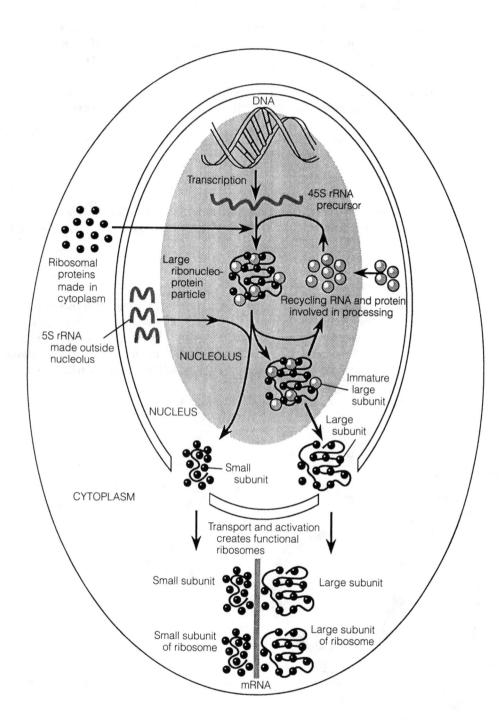

Figure 2.5. Diagram showing formation of rRNA and its processing into ribosomal subunits, which occurs in the nucleolus. (Adapted from Alberts B et al.: *Biology of the Cell,* 2nd ed. New York, Garland Publishing, Inc., 1989, p 542.)

IX. **Cell Cycle** (Figure 2.6)

–consists of two major periods: **interphase** (the interval between cell divisions) and **mitosis** (the period of cell division), also known as the **M phase**.

–varies greatly in length in different types of cells but is repeated each time a cell divides.

–is **temporarily suspended** in nondividing resting cells (e.g., peripheral lymphocytes), which are in the G_0 **state**. Such cells may reenter the cycle and begin to divide again.

–is **permanently interrupted** in differentiated cells that do not divide (e.g., cardiac muscle cells and neurons).

A. Interphase

–is considerably longer than the M phase.

–is the period during which **the cell doubles in size and DNA content**.

–is divided into three separate phases (G_1, **S, and** G_2) during which specific cellular functions occur.

1. Phases

a. G_1 phase

–is the **gap phase** just after mitosis during which cell growth and protein synthesis occur, restoring the daughter cells to normal volume and size.

–is when certain "trigger proteins" are synthesized, enabling the cell to reach a threshold (**restriction point**) and proceed to the S phase. Cells that fail to reach the restriction point become resting cells and enter the G_0 state.

–lasts from a few hours to several days.

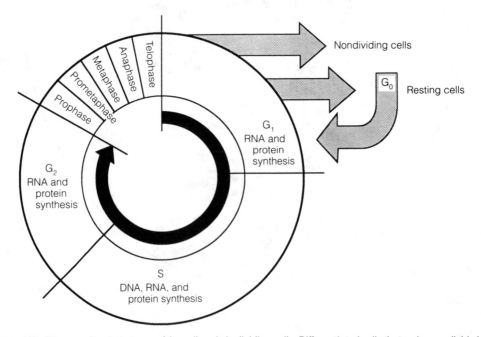

Figure 2.6. Diagram showing stages of the cell cycle in dividing cells. Differentiated cells that no longer divide have left the cycle, whereas resting cells in the G_0 state may reenter the cycle and begin dividing again. (Adapted from Widnell CC and Pfenninger KH: *Essential Cell Biology.* Baltimore, Williams & Wilkins, 1990, p 58.)

b. S phase

 - is the **synthetic phase** during which DNA replication and protein synthesis occur, resulting in **duplication of the chromosomes**.
 - is the period when centrioles are self-duplicated.
 - lasts 8–12 hours in most cells.

c. G_2 phase

 - is the **gap phase,** which follows the S phase and extends to mitosis.
 - is when the cell prepares to divide; the centrioles grow to maturity; energy required for the completion of mitosis is stored; and RNA and proteins necessary for mitosis are synthesized.
 - lasts 2–4 hours.

2. Control factors

a. S-phase activator triggers the initiation of DNA synthesis.

b. M-phase delaying factor, which may be identical with S-phase activator, inhibits formation of the M-phase promoting factor and thus prevents the cell from proceeding into the M phase until all of the DNA has been replicated.

c. M-phase promoting factor induces the cell to enter the M phase.

B. Mitosis (Figure 2.7)

 - follows the G_2 phase and completes the cell cycle.
 - involves division of the nucleus (**karyokinesis**) and division of the cytoplasm (**cytokinesis**), resulting in the production of two **identical** daughter cells.
 - lasts 1–3 hours.
 - includes five major stages: prophase, prometaphase, metaphase, anaphase, and telophase.

1. Prophase

 - begins when the chromosomes condense and become rod-like. Each chromosome consists of two parallel **sister chromatids** attached to one another at the **centromere,** a constricted region along the chromosome.
 - is when the nucleolus and nuclear envelope disappear.

a. Centrosome contains **centrioles** and is the principal **microtubule-organizing center** (MTOC) of the cell.

b. Centrioles migrate to opposite poles of the cell and give rise to the **spindle fibers** and **astral rays** of the mitotic spindle.

c. Kinetochores begin to develop at the centromere region and function as MTOCs.

2. Prometaphase

 - is the stage during which the kinetochore completes development and attaches to specific spindle microtubules, forming **kinetochore microtubules**. Spindle microtubules that *do not* attach to kinetochores are called **polar microtubules.**

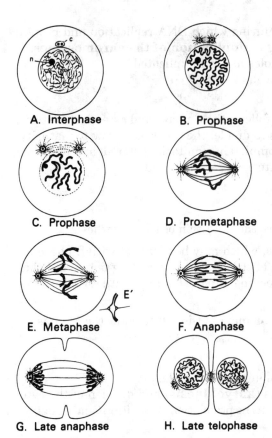

A. Interphase
B. Prophase
C. Prophase
D. Prometaphase
E. Metaphase
F. Anaphase
G. Late anaphase
H. Late telophase

Figure 2.7. Diagram illustrating events in various phases of mitosis. *c* = centrioles; *n* = nucleus; *E'* = enlargement of one metaphase chromosome being pulled apart. (Reprinted from Kelly DE et al: *Bailey's Textbook of Microscopic Anatomy,* 18th ed. Baltimore, Williams & Wilkins, 1984, p 89.)

3. Metaphase

—is the period when the condensed chromosomes are aligned at the equatorial plate of the mitotic spindle.

4. Anaphase

—begins as the chromatids separate (at the centromere) and daughter chromosomes move to opposite poles of the cell.

—is associated with elongation of the spindle.

—is also characterized (in its later stages) by a **cleavage furrow** that begins to form around the cell due to contraction of a band of actin filaments called the **contractile ring**.

5. Telophase

—is characterized by a deepening of the cleavage furrow, which leaves the **midbody** (containing overlapping polar microtubules) between the newly forming daughter cells.

—is associated with a depolymerization of microtubules in the midbody, facilitating the **completion of cytokinesis** and formation of two identical daughter cells.

—includes **reformation of the nuclear envelope** around the condensed chromosomes in the daughter cells.

    –is associated with the **reappearance of nucleoli,** which arise from specific nucleolar organizer regions (called secondary constriction sites) on five chromosomes.

    –is completed as the daughter nuclei gradually enlarge and the dense chromosomes disperse to form the typical interphase nucleus with heterochromatin and euchromatin.

X. Meiosis

–is a special form of cell division in which the **chromosome number is reduced** from diploid ($2n$) to haploid (n).

–occurs in developing germ cells (spermatozoa and oocytes) in preparation for sexual reproduction. Subsequent fertilization results in **diploid zygotes.**

–involves a doubling of the DNA content of the original diploid cell in the S phase followed by two successive cell divisions that give rise to **four haploid cells.**

–is accompanied by **recombination** of maternal and paternal genes by crossing over and random assortment, yielding the unique haploid genome of the gamete.

–in men and women involves the same nuclear events but different cytoplasmic events (Figure 2.8).

–is divided into the following stages:

A. Reductional division (meiosis I)

–occurs following interphase during which the 46 chromosomes are duplicated, giving the cell a **4CDNA content** (considered to be the total DNA content of the cell).

1. Prophase I

–is divided into the following five stages:

a. Leptotene, during which the chromatin condenses into the visible chromosomes, each of which contains two chromatids joined at the centromere

b. Zygotene, during which homologous maternal and paternal chromosomes pair and make physical contact (**synapsis**) via the **synaptonemal complex,** forming a **tetrad**

c. Pachytene, during which **chiasmata** are formed and **crossing over** (random exchanging of genes between segments of homologous chromosomes) occurs—an event that is crucial for **increasing genetic diversity**

d. Diplotene, during which the chromosomes continue to condense and chiasmata can be observed, indicating sites where crossing over has taken place

e. Diakinesis, during which the nucleolus disappears, chromosomes are condensed maximally, and the nuclear envelope disappears

2. Metaphase I

–includes alignment of homologous sets of chromosomes on the equatorial plate of the meiotic spindle in a random arrangement, thus facilitating genetic mixing.

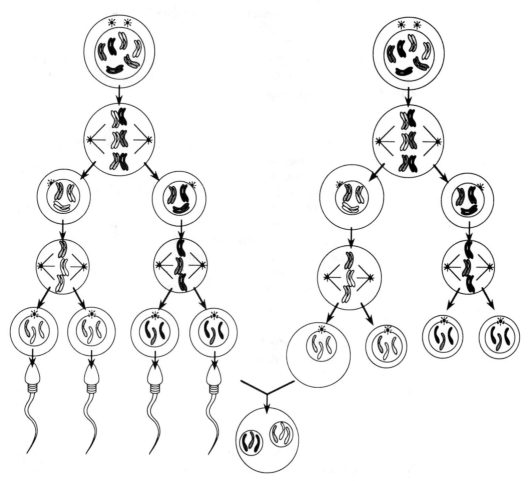

Figure 2.8. Diagram illustrating meiosis in the male and female. Spermatogenesis in the male gives rise to sperm, each containing the haploid number of chromosomes. Oogenesis in the female gives rise to an ovum with the haploid number of chromosomes. Fertilization reconstitutes the diploid number of chromosomes in the resulting zygote. (Adapted from Widnell CC and Pfenninger KH: *Essential Cell Biology.* Baltimore, Williams & Wilkins, 1990, p 69.)

   —includes attachment of spindle fibers from either pole to the **kineto-chore** of any one of the chromosome pairs, thus assuring that genetic mixing takes place.

3. Anaphase I

   —is similar to anaphase in mitosis except that each chromosome consists of **two chromatids that remain held together**.

   —involves migration of chromosomes to the poles.

4. Telophase I

   —is similar to telophase in mitosis.

   —includes reformation of the nucleus and cytokinesis, forming two daughter cells.

   a. Each daughter cell now contains 23 chromosomes (the haploid (n) number) but has a **2CDNA content** (the diploid amount).

    b. Each chromosome is composed of **two sister chromatids,** which are similar but **not genetically identical**.

B. Equatorial division (meiosis II)

    —begins soon after the completion of meiosis I, following a very brief interphase **without DNA replication**.

    —involves separation of sister chromatids in the two daughter cells formed in meiosis I and their distribution as chromosomes into **four daughter cells,** each containing its own **unique recombined genetic material (1CDNA; *n*)**.

    —involves events similar to those in mitosis; thus the stages are named similarly (prophase II, metaphase II, anaphase II, and telophase II).

    —occurs more rapidly than mitosis.

XI. Clinical Considerations

A. Euchromatin–heterochromatin ratio

    —generally is higher in rapidly proliferating cells (e.g., malignant cells) than in less active cells.

B. Aneuploidy

    —refers to any deviation in the normal number of chromosomes.

    —can be detected by karyotyping.

    —includes **trisomy** (the presence of a third chromosome of one type) and **monosomy** (the absence of one member of a chromosome pair).

1. Down's syndrome

    —is associated with an extra chromosome 21 (**trisomy 21**).

    —is characterized by mental retardation, short stature, stubby appendages, congenital heart malformations, and other defects.

2. Klinefelter's syndrome

    —is generally associated with aneuploidy of the sex chromosomes (**XXY**).

    —is characterized by infertility, variable degrees of masculinization, and small testes.

3. Turner's syndrome

    —is generally associated with **monosomy** of the sex chromosomes (**XO**).

    —is characterized by short stature, sterility, and variable other abnormalities.

    —is compatible with life in contrast to other types of monosomy, which are lethal.

C. Transformed cells

    —have lost their ability to respond to regulatory signals controlling the cell cycle and may undergo cell division indefinitely, thus becoming cancerous.

    —may be arrested in mitosis by the administration of **Vinca alkaloids;** other drugs block purine and pyrimidine synthesis, thus arresting cells in the S phase of the cell cycle.

D. Oncogenes

—represent **mutations of certain regulatory genes,** called **proto-oncogenes,** which normally stimulate or inhibit cell proliferation and development.

—may result from random genetic accidents or viruses.

—dominate the normal alleles (proto-oncogenes), causing a **deregulation** of cell division, which leads to a cancerous state.

—are responsible for bladder cancer and acute myelogenous leukemia.

Review Test

Directions: Each of the numbered items or incomplete statements in this section is followed by answers or by completions of the statement. Select the **one** lettered answer or completion that is **best** in each case.

1. Which of the following statements regarding the nuclear pore complex is FALSE?

(A) It restricts communication between the nucleus and the cytoplasm
(B) It is bridged by a unit membrane
(C) It is composed of eight large protein granules arranged in an octagon about the pore
(E) It often possesses a dense central granule

2. Which one of the following nucleotides is NOT present in DNA?

(A) Thymine
(B) Adenine
(C) Uracil
(D) Cytosine
(E) Guanine

3. The nuclear envelope includes all of the following EXCEPT

(A) the outer nuclear membrane
(B) the inner nuclear membrane
(C) nuclear pores
(D) the nuclear lamina
(E) nuclear pore complexes

4. Which one of the following statements concerning the nucleolus is FALSE?

(A) It is involved in the synthesis of rRNA and its packaging into precursors of ribosomes
(B) It is surrounded by a membrane
(C) It is generally detected only at interphase
(D) It possesses regions that are transcriptionally active

5. Which one of the following statements about heterochromatin is TRUE?

(A) It is observable only by transmission electron microscopy
(B) It is observable in the light microscope as a light-staining, dispersed material
(C) It is concentrated near the periphery of the nucleus
(D) It is transcriptionally active

6. Which one of the following statements concerning DNA is FALSE?

(A) It is located mostly in the nucleus
(B) It is a complex molecule consisting of ribose, phosphate, and a base
(C) It is made up of two complementary strands bound together in a double helix by bonded base pairs
(D) It is the genetic template upon which RNAs are synthesized

7. Which one of the following statements concerning transcription is FALSE?

(A) A gene is transcribed into an RNA molecule by RNA polymerase
(B) Both exons and introns are transcribed
(C) The DNA molecule unwinds at the end of transcription
(D) The codon always represents a single amino acid
(E) The chain terminator on the DNA molecule stops the transcription process

8. Anticodons are located in

(A) mRNA
(B) rRNA
(C) tRNA
(D) snRNP
(E) hnRNP

9. DNA is duplicated in the cell cycle during the

(A) G_2 phase
(B) S phase
(C) M phase
(D) G_1 phase
(E) G_0 phase

10. During mitosis, sister chromatids begin their migration to opposite poles during

(A) telophase
(B) metaphase
(C) anaphase
(D) prophase

11. Which one of the following statements concerning meiosis I is FALSE?

(A) Prophase I is divided into five stages: leptotene, zygotene, pachytene, diplotene, and diakinesis
(B) During metaphase I, homologous chromosomes line up at the equator in random fashion
(C) The chromosome number is reduced from the diploid to the haploid number
(D) During telophase I, each chromosome consists of one chromatid and a centromere
(E) Diakinesis occurs at the end of prophase I

12. An individual who exhibits the characteristics of Down's syndrome is

(A) aneuploid for chromosome 21
(B) aneuploid for the X chromosome
(C) monoploid for chromosome 21
(D) monoploid for the Y chromosome

13. Transformed cells become malignant because they

(A) have a reduced amount of RNA polymerase
(B) are unable to respond to regulatory signals in the cell cycle
(C) cease mitotic activity
(D) contain extra chromosomes
(E) contain transcriptionally inactive heterochromatin

14. Microscopic observations of the proportions of the two types of chromatin in the nuclei of tissue sections can often be used to

(A) determine the karyotype
(B) identify the genome
(C) diagnose Turner's syndrome
(D) identify DNA viruses
(E) distinguish benign from malignant cells

15. Which one of the following statements concerning the human karyotype is FALSE?

(A) A normal human karyotype consists of 46 chromosomes
(B) There are 22 pairs of autosomes in humans
(C) There is one pair of sex chromosomes in women
(D) Aneuploidy is considered to be a normal karyotype

Directions: The group of items in this section consists of lettered options followed by a set of numbered items. For each item, select the **one** lettered option that is most closely associated with it. Each lettered option may be selected once, more than once, or not at all.

Questions 16–20

Match each description below with the corresponding lettered structure in the micrograph of a nucleus.

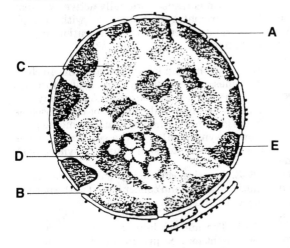

16. Transcriptionally inactive chromatin

17. A fibrous layer containing lamins A, B, and C

18. Site of rRNA synthesis

19. Transcriptionally active chromatin

20. Passageway for transfer of materials

Answers and Explanations

1–B. The nuclear pore complex contains a central aqueous channel, which permits passage of small water-soluble molecules. However, movement of proteins in and out of the nucleus is selectively controlled by the nuclear pore complex, which can regulate the diameter of the pore lumen.

2–C. DNA contains the purines adenine and guanine and the pyrimidines cytosine and thymine. In RNA, uracil, a pyrimidine, replaces thymine.

3–D. The nuclear lamina is located just inside the inner nuclear membrane and thus is not part of the nuclear envelope. The nuclear lamina contains lamins A, B, and C, which are intermediate filament proteins that help to organize the nuclear envelope and perinuclear chromatin.

4–B. The nucleolus is a non-membrane–bounded inclusion within the nucleus. It is observed at interphase and it disappears during mitosis.

5–C. Heterochromatin is observed in the light microscope as dark basophilic clumps of nucleoprotein concentrated at the periphery of the nucleus. It is transcriptionally inactive.

6–B. DNA is a long linear molecule consisting of the sugar deoxyribose in addition to phosphate and a base. Ribose is the sugar present in RNA.

7–C. The DNA molecule unwinds at the beginning of transcription after RNA polymerase has bound to the DNA near a promoter. Although both exons and introns are transcribed into RNA, the introns are removed during RNA processing.

8–C. Each tRNA possesses a triplet of nucleotides, called an anticodon, which recognizes the complementary codon in mRNA.

9–B. The S (synthesis) phase of the cell cycle is the period when DNA replication and histone synthesis occur, resulting in duplication of the chromosomes. At the end of the S phase, each chromosome consists of two identical chromatids attached to one another at the centromere.

10–C. Anaphase begins when the sister chromatids separate and begin their migration to opposite poles of the cell.

11–D. During metaphase I of meiosis, homologous chromosomes line up at the equator of the spindle. The homologous chromosomes dissociate from each other at anaphase I. At telophase I the chromosomes, each consisting of two chromatids, arrive at the pole. Though the chromosomes have been reduced to the haploid number, the DNA content is 2CDNA (diploid amount).

12–A. Aneuploid is the term applied to an abnormal number of chromosomes in the genome. In Down's syndrome, the individual has three number 21 chromosomes (trisomy 21), rather than the normal two.

13–B. Cells that lose their ability to respond to regulatory signals operating in the cell cycle are said to be transformed and ultimately become malignant.

14–E. The ratio of euchromatin to heterochromatin may be helpful in distinguishing benign from malignant cells, which are characterized by rapid, uncontrolled proliferation. Since euchromatin is transcriptionally active, whereas heterochromatin is not, cells with a high euchromatin–heterochromatin ratio may be malignant.

15–D. The human karyotype contains 46 chromosomes, represented by 22 pairs of autosomes and 1 pair of sex chromosomes. Aneuploidy represents an abnormal number of chromosomes in the genome—either of the autosomes (e.g., in Down's syndrome) or of the sex chromosomes (e.g., in Klinefelter's and Turner's syndromes).

16–E. Inactive chromatin, called heterochromatin, is dense and concentrated at the periphery of the nucleus.

17–A. The nuclear lamina is the cytoskeletal component located just inside of the inner nuclear membrane.

18–D. Synthesis of rRNA and assembly of the ribosomal subunits occurs in the nucleolus.

19–C. Euchromatin, the transcriptionally active form of chromatin, appears less dense than heterochromatin in electron micrographs.

20–B. The nuclear envelope is interrupted by nuclear pores through which materials can be transported in and out of the nucleus.

3

Cytoplasm

I. Overview—The Cytoplasm

—contains three main structural components: **organelles, inclusions,** and the **cytoskeleton**. The fluid component is called the **cytosol**.

—exhibits dynamic functional interactions among certain organelles that result in the uptake and release of material by the cell, protein synthesis, and intracellular digestion.

II. Structural Components

A. Organelles (Figure 3.1)

—are metabolically active units of living matter, which usually are limited by a membrane.

—are described in this section as viewed by transmission electron microscopy.

1. Plasma membrane

—is discussed in Chapter 1.

2. Ribosomes

—are 12-nm wide and 25-nm long.

—consist of a **small** and **large** subunit, which are composed of several types of rRNA and numerous proteins (Table 3.1; see Figure 2.5).

—may exist free in the cytosol or bound to membranes of the rough endoplasmic reticulum or outer nuclear membrane. Whether free or membrane bound, the ribosomes constitute a single interchangeable population.

—cluster along a single strand of mRNA to form a **polyribosome (polysome),** which exists in a spiral configuration.

—are the sites where **mRNA is translated into protein,** as follows:

a. **Small ribosomal subunit** binds mRNA and activated tRNAs; the **codons** of the mRNA then **base-pair** with the corresponding **anticodons** of the tRNA.

b. **Large ribosomal subunit** then binds to the complex, and **peptidyl**

35

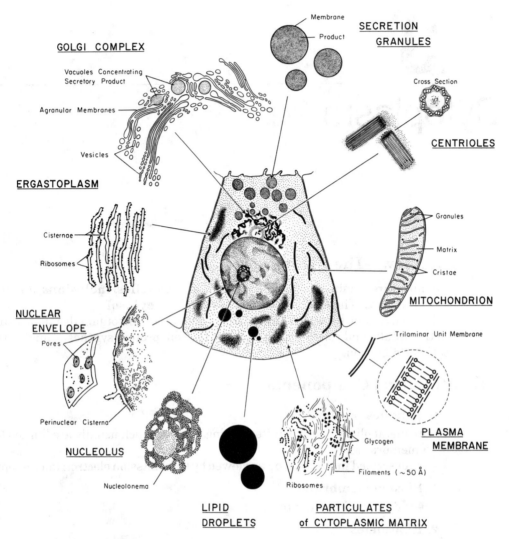

Figure 3.1. Diagram of a eukaryotic cell and its major organelles and inclusions. (Adapted from Fawcett DW: *Bloom and Fawcett's Textbook of Histology*, 11th ed. Philadelphia, WB Saunders, 1986, p 2.)

transferase in the large subunit catalyzes peptide-bond formation, resulting in addition of amino acids to the growing polypeptide chain.

  c. **Chain-terminating codon** (**UAA, UAG,** or **UGA**) causes release of the nascent polypeptide from the ribosome, and the subunits dissociate from the mRNA.

3. Rough endoplasmic reticulum (RER) [see Figure 3.1]

  —is the site where **noncytosolic proteins** are synthesized; these include secretory, plasma-membrane, and lysosomal proteins.

  —is a system of sacs, or cavities, bounded by membranes whose outer surface is studded with ribosomes (thus appearing rough) and whose interior region is called a **cisterna**.

Table 3.1. Ribosome Composition

Subunit	rRNA types	Number of proteins
Large (60S)	5S	
	5.8S	
	28S	49
Small (40S)	18S	33

–has a membrane that is **continuous** with the outer nuclear membrane, thus bringing the perinuclear cisterna into continuity with the cisternae of the RER.

–is abundant in cells synthesizing **secretory proteins;** in such cells, the RER is organized into many parallel arrays.

–has receptors (**ribophorins**) in its membrane to which the large ribosomal subunit binds.

4. Smooth endoplasmic reticulum (SER) [see Figure 3.1]

–is an irregular network of membrane-bounded channels that lacks ribosomes on its surface (thus appearing smooth).

–usually appears as branching anastomosing **tubules,** or **vesicles,** whose membranes do **not** contain ribophorins.

–is less common than RER but predominates in cells synthesizing steroids, triglycerides, and cholesterol.

–serves different **functions** in different cell types, as follows:

a. Steroid hormone synthesis

–occurs in Leydig cells of the testis, which make testosterone; in zona fasciculata cells of the adrenal cortex; and in other cells rich in SER.

b. Drug detoxification

–occurs in hepatocytes following proliferation of the SER in response to phenobarbital, since the SER contains the mixed function oxidases that metabolize this drug.

c. Muscle contraction and relaxation

–involves the release and recapture of calcium ions by the sarcoplasmic reticulum in the skeletal muscle (see Chapter 8 II F).

5. Annulate lamellae

–are parallel stacks of membranes (usually six to ten) that generally are located near the nucleus.

–resemble the nuclear envelope, including its pore complexes.

–are often arranged with their **annuli** (pores) in register.

–are **continuous** with the RER.

–are found in rapidly growing cells (e.g., germ cells, embryonic cells, and tumor cells), but their function and significance remain unknown.

6. Mitochondria (see Figure 3.1)

a. Morphology—mitochondria

–are rod-shaped organelles (0.2 μm wide and up to 7 μm long).

–possess an outer membrane, which bounds the organelle, and an inner membrane, which invaginates to form **cristae.**

—are subdivided into an **intermembrane compartment,** which is between the two membranes, and an **inner matrix compartment**.

—contain granules within the matrix that bind the divalent cations Mg^{2+} and Ca^{2+}.

b. Enzymes and genetic apparatus—mitochondria

—contain all the enzymes of the **Krebs (TCA) cycle** in the matrix, except for succinate dehydrogenase, which is located on the inner mitochondrial membrane.

—contain **elementary particles** (visible on negatively stained cristae) that contain **ATP synthase,** a special enzyme involved in **coupling oxidation to phosphorylation** of ADP to form ATP.

—possess (in the matrix) their own **genetic apparatus** composed of DNA (circular), mRNA, tRNA, and rRNA (with a limited coding capacity). However, most mitochondrial proteins are encoded by nuclear DNA.

c. Origin and reproduction—mitochondria

—might have **originated as symbionts** (intracellular parasites). According to this theory, anaerobic eukaryotic cells endocytosed aerobic microorganisms that evolved into mitochondria, which function in oxidative processes.

—proliferate by division (fission) of preexisting mitochondria, and typically have a 10-day lifespan.

d. Mitochondrial ATP synthesis

—occurs via the Krebs cycle, which traps chemical energy and produces ATP by **oxidation** of fatty acids, amino acids, and glucose.

—also occurs via a **chemiosmotic coupling mechanism** involving enzyme complexes of the **electron transport chain** and **elementary particles** present in the cristae (Figure 3.2).

e. Condensed mitochondria

—result from a **conformational change** in the **orthodox** form (typical morphology), which occurs in response to an uncoupling of oxidation from phosphorylation.

—are characterized by a decrease in the size of the inner compartment (accompanied by an increase in matrix density) and an enlargement of the intermembrane compartment.

—are present in **brown fat cells,** which produce **heat** rather than ATP due to a special transport protein in their inner membrane that **uncouples** respiration from ATP synthesis.

—**swell** in response to calcium, phosphate, and thyroxine, which induce an increase in water uptake and an uncoupling of phosphorylation; ATP reverses the swelling.

7. Golgi complex (apparatus) [see Figure 3.1]

—consists of several disk-shaped **cisternae (saccules)** arranged in a **stack**. Cisternae are slightly curved, with flat centers and dilated rims, but they vary in shape across the stack.

—reveals a distinct **polarity** across the stack.

—varies in size and development in different cell types.

Cytoplasm

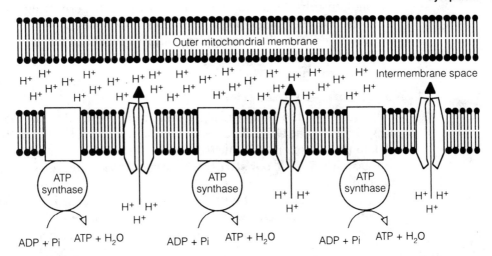

Figure 3.2. Chemiosmotic coupling mechanism for generating ATP in mitochondria. During electron transport, H$^+$ ions (protons) are pumped from the inner matrix compartment across the inner mitochondrial membrane into the intermembrane compartment. The electrochemical proton gradient thus created drives ATP synthase to catalyze the conversion of ADP and P$_i$ to ATP.

–has two major **functions:** processing of noncytosolic proteins synthesized in the RER (see Figure 3.6), and membrane retrieval, recycling, and redistribution (see Figure 3.4).

–includes the following regions:

a. Cis (entry) face of the Golgi comprises the forming (outer, convex) cisternae, which are located at the side of the stack facing the RER, and small **transfer vesicles**.

b. Trans (exit) face of the Golgi comprises the maturing (inner, concave) cisternae, which are located at the side of the stack facing vacuoles and secretory granules.

c. Medial compartment of the Golgi comprises a few cisternae lying between the cis and trans faces.

d. Trans Golgi network (TGN) lies apart from the last cisterna at the trans face and is separated from the Golgi stack. It corresponds to a tubular reticulum formerly called **GERL** (Golgi-associated endoplasmic reticulum from which **lysosomes** originate).

8. Coated vesicles

–are characterized by a visible cytoplasmic surface coat.

–exist in at least two varieties: clathrin-coated and non-clathrin–coated vesicles.

a. Clathrin-coated vesicles

–are coated with clathrin, which consists of three large and three small polypeptide chains that form a **triskelion** (three-legged structure). Thirty-six clathrin triskelions associate to form a polyhedral cage-like network around the vesicle.

—are formed during **receptor-mediated uptake** (endocytosis) of specific molecules by the cell, but lose their coat quickly, permitting clathrin to recycle back to the plasma membrane (Figure 3.3).

—are also associated with the regulated **signal-directed transport** of proteins from the trans Golgi network ultimately to lysosomes or to secretory granules.

b. Non-clathrin–coated vesicles

—are involved in the transport of proteins from the RER to the Golgi complex, from one Golgi cisterna to another, and from the Golgi complex to the plasma membrane (see Figure 3.6).

—are associated with **constitutive (unregulated) protein transport** (bulk flow).

—have coats that are not yet fully characterized but primarily consist of a protein called β-COP, which does **not** form a cage-like network around vesicles and has a different appearance than clathrin.

9. Lysosomes

—are dense, membrane-bounded organelles of diverse morphology and size.

—are formed when sequestered material fuses with a **late endosome** and digestion begins.

—may be identified in sections by cytochemical staining for **acid hydrolase**.

—possess special membrane proteins and more than 40 hydrolases, which are synthesized in the RER.

—possess ATP-powered proton pumps in their membrane that maintain an **acid pH** (≈ 5).

a. Formation of lysosomes via the lysosomal pathway (see Figure 3.3)

—involves the following intermediates:

(1) Early endosomes

—are irregular, peripherally located vesicles that contain receptor–ligand complexes.

—are formed by fusion of endocytic vesicles with an uncoupling vesicle to form the compartment of uncoupling of receptor and ligand (**CURL**).

—have acidic interiors (pH ≈ 6) maintained by a Na^+–K^+ ATPase, which is internalized from the plasma membrane. The acidity aids in the uncoupling of receptors and ligand; the receptors then return to the plasma membrane and the ligands are transferred to a late endosome.

(2) Late endosomes

—are also known as **endolysosomes** or the **intermediate compartment**.

—are irregular vesicles (pH ≈ 5.5) that can receive ligands from early endosomes (either by vesicle transport or endosome "maturation").

—contain **both lysosomal hydrolases and lysosomal membrane proteins;** these are formed in the RER, transported to the

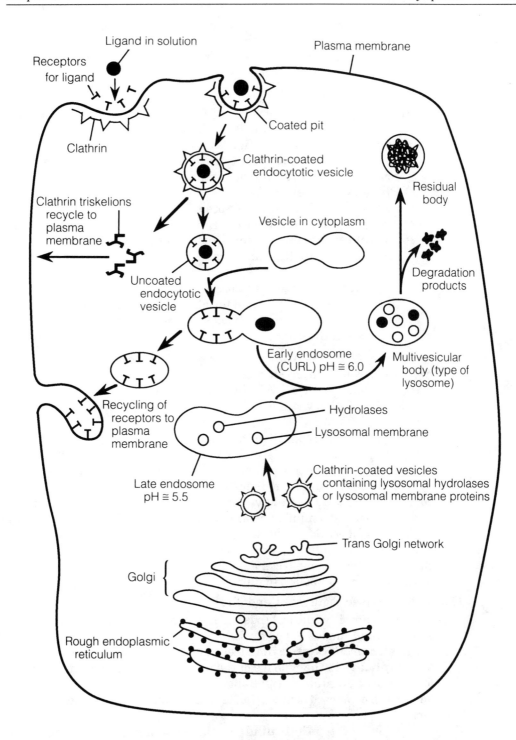

Figure 3.3. Diagram illustrating receptor-mediated endocytosis of a ligand (e.g., low-density lipoproteins) and the lysosomal degradative pathway. Clathrin triskelions quickly recycle back to the plasma membrane. The receptors and ligands then uncouple in the early endosome (CURL), followed by recycling of receptors back to the plasma membrane. The late endosome is the primary intermediate in the formation of lysosomes (e.g., multivesicular bodies). Material that is phagocytosed or organelles that undergo autophagy do not use the early endosomal pathway.

Golgi complex for processing, and delivered in **separate vesicles** to late endosomes.*

—fuse with early endosomes to form **multivesicular bodies,** which are a type of lysosome.

b. Types of lysosomes

—are named based on their content of recognizable material; otherwise the general term **lysosome** is used.

—include the following (see Figure 3.7):

(1) Multivesicular bodies are formed by fusion of an early endosome containing endocytic vesicles with a late endosome.

(2) Phagolysosomes are formed by fusion of a **phagocytic vacuole** with a late endosome or a lysosome.

(3) Autophagolysosomes are formed by fusion of an **autophagic vacuole** (containing cell constituents targeted for destruction that are enveloped by RER membranes) with a late endosome or lysosome.

(4) Residual bodies are lysosomes of any type that have expended their capacity to degrade material. They contain **undigested material** (e.g., lipofuscin and hemosiderin) and eventually may be excreted from the cell.

10. Peroxisomes (see Figure 3.1)

—are also known as **microbodies**.

—are membrane-bounded, typically ovoid organelles.

—may be identified in virtually all cells by a cytochemical reaction for **catalase**.

—appear in stained preparations as small organelles (0.15–0.25 μm in diameter); they may be larger in hepatocytes.

—originate from preexisting peroxisomes, which grow by importing specific cytosolic proteins and then divide by fission.

—contain three **oxidative enzymes** (D-amino acid oxidase, urate oxidase, and catalase) and a number of other enzymes whose functions vary from oxidation of fatty acids to the detoxification of substances such as ethanol.

—may contain a **nucleoid,** a crystalline core consisting of urate oxidase; the human peroxisome lacks a nucleoid.

—have a lifespan of about 5–6 days.

11. Centrioles (see Figures 3.1 and 2.7)

—exist as a pair of cylindrical rods (each 0.2 μm wide and 0.5 μm long) oriented at right angles to one another.

—are located in the **centrosome** (cell center).

—**self-duplicate** in the S phase of the cell cycle, as each parent centriole forms a **procentriole** at right angles to itself.

—have a wall composed of nine triplets of microtubules (9 + 0 axoneme pattern).

—are associated with **microtubule-organizing centers** (MTOCs), which are dense pericentriolar bodies that give rise to microtubules.

---

*The terms **primary** and **virgin lysosomes,** formerly used for tiny vesicles believed to be lysosomes that had not yet engaged in digestive activity, are no longer used.

      –form the poles of the mitotic spindle where microtubules originate or converge.

      –form **basal bodies,** which give rise to the axonemes of cilia and flagella (see Figure 5.3).

B. Inclusions

–are "lifeless" accumulations of material that are **not metabolically active** and usually are present in the cytosol only **temporarily**.

1. Glycogen

–appears as small clusters or (in hepatocytes) as larger aggregates (known as **rosettes**) of electron-dense, 20- to 30-nm beta particles, which are similar in appearance to, but larger than, ribosomes.

–is not bounded by a membrane but frequently lies close to the SER.

–serves as a **stored energy source** that can be degraded to glucose, which enters the bloodstream to elevate blood sugar levels.

2. Lipid droplets

–vary markedly in size and appearance depending on the method of fixation.

–are not bounded by a membrane.

–are storage forms of **triglycerides** (an energy source) and **cholesterol** (used in the synthesis of steroids).

3. Lipofuscin

–appears as membrane-bounded, electron-dense granular material varying greatly in size and often containing lipid droplets.

–represents a residue of undigested material present in residual bodies.

–increases with age and thus is called **age pigment**.

–is most common in nondividing cells (e.g., cardiac muscle cells, neurons) and in hepatocytes.

C. Cytoskeleton

–is the structural framework within the cytosol.

–functions in maintaining cell shape, stabilizing cell attachments, facilitating endocytosis and exocytosis, and promoting cell motility.

–includes the following components:

1. Microtubules

–are straight, hollow tubules 25 nm in diameter and several micrometers long.

–have a rigid wall composed of 13 protofilaments, each of which consists of a linear arrangement of **αβ-tubulin dimers.**

–are **polar,** with polymerization (assembly) and depolymerization (disassembly) occurring preferentially at one end (+ end).

–contain **microtubule-associated proteins** (MAPs), which stabilize them and bind them to other cytoskeletal components and organelles.

–are associated with **kinesin,** a force-generating protein, which serves as a "motor" for vesicle or organelle movement.

–have the following **functions**:

    a. Maintain cell shape

    b. Aid in the transport of macromolecules within the cytosol

  c. Promote the movement of cilia, flagella, and chromosomes

2. Microfilaments

 −are also known as **F actin** or **actin filaments**.
 −are 6 nm in diameter.
 −are composed of globular actin monomers (**G actin**) linked into a **double helix** having a 36-nm repeat.
 −display **polarity** like microtubules; that is, their polymerization and depolymerization occur preferentially at the (+) end.
 −are more stable than microtubules.
 −are abundant at the periphery of the cell where they are anchored to the plasma membrane via one or more intermediary proteins (e.g., α-actinin, vinculin, talin).
 −are involved in the following **cellular processes**:

 a. Sol–gel transformation of the cytosol

 b. Endocytosis and exocytosis

 c. Locomotion of nonmuscle cells

3. Intermediate filaments

 −are 8–11 nm in diameter.
 −constitute a population of heterogeneous filaments that includes keratin, vimentin, desmin, glial fibrillary acid protein (GFAP), lamins, and neurofilaments (Table 3.2).
 −in general, **provide mechanical strength** to cells.

Table 3.2. Intermediate Filaments

Protein	Location	Function
Keratin 19 distinct forms (acidic, neutral, and basic)	Epithelial cells	Provides structural support or tension-bearing role; markers for tumors of epithelial origin
Tonofilaments		Desmosome–hemidesmosome association
Desmin	Skeletal muscle Cardiac muscle Smooth muscle	Forms a framework linking myofibrils/myofilaments; marker for tumors of muscle origin
Vimentin	Fibroblasts Endothelial cells Vascular smooth muscle Chondroblasts Macrophages Mesenchymal cells	Is associated with nuclear envelope and pores; marker for connective tissue tumors
Glial fibrillary acidic protein (GFAP)	Astrocytes Oligodendrocytes Schwann cells	Provides structural support; marker for glial tumors
Neurofilaments	Neurons	Provide support for axons and dendrites; facilitate gel state of the cytosol
Lamins A, B, and C	Nuclear lamina of all cells	Organize nuclear envelope and perinuclear chromatin

4. Microtrabecular lattice

—is a three-dimensional meshwork of slender strands in the cytosol, which is observed only by high-voltage electron microscopy.
—is not universally accepted as an authentic structure.
—may compartmentalize metabolic activities in the cytosol, influence the movement of organelles, and affect the viscosity of the cytosol.

III. Interactions Among Organelles

—are involved in several important cellular processes.
—are the basis for a functional approach to the dynamics of cell biology.

A. Uptake and release of material by cells

1. Endocytosis

—is the **uptake (internalization) of material by cells**.
—includes pinocytosis, receptor-mediated endocytosis, and phagocytosis.

a. Pinocytosis (cell drinking)

—is the **nonspecific (random) uptake** of extracellular fluid and material in solution into pinocytic vesicles.

b. Receptor-mediated endocytosis

—is the **specific uptake** of a substance (e.g., low-density lipoproteins and protein hormones) by a cell with plasma-membrane receptors for that substance, termed a **ligand**.
—involves the following sequence of events (see Figure 3.3):
(1) A ligand **binds specifically** to its receptor on the cell surface.
(2) Ligand–receptor complexes cluster into a **clathrin-coated** pit, which invaginates and gives rise to a clathrin-coated vesicle containing the ligand.
(3) The clathrin coat is rapidly **lost** in an ATP-requiring step, leaving an **uncoated endocytic vesicle** containing the ligand.

c. Phagocytosis (cell eating)

—is the uptake of microorganisms, other cells, and foreign particles by a cell in a process that may or may not involve cell-surface receptors.
—is characteristic of cells—in particular **macrophages**—that degrade proteins and cellular debris.
—involves the following sequence of events:
(1) A macrophage binds, via its **Fc receptors,** to an antibody-coated (IgG-coated) bacterium.
(2) Binding progresses until the plasma membrane totally envelops the bacterium, forming a **phagocytic vacuole**.

2. Exocytosis

—is the **release of material** from the cell via fusion of a secretory granule membrane with the plasma membrane.
—requires interaction of receptors in **both** the granule and plasma membrane, as well as the **coalescence** (adherence and joining) of the two phospholipid membrane bilayers.
—includes both regulated and constitutive secretion.

a. Regulated secretion involves the release, in response to an **extracellular signal,** of proteins and other materials that are **stored** in the cell.

b. Constitutive secretion involves the more-or-less **continuous** release of material (e.g., collagen and serum proteins) without any intermediate storage step. An extracellular signal is **not** required for constitutive secretion.

3. Membrane recycling (Figure 3.4)

—maintains a relatively constant plasma-membrane surface area.

—occurs because the secretory granule membrane added to the plasma-membrane surface during exocytosis is **retrieved** during endocytosis via clathrin-coated vesicles. This vesicular membrane eventually is returned to the TGN and trans Golgi complex where it is recycled.

B. Protein synthesis

1. Synthesis of noncytosolic proteins

—involves translation of mRNAs encoding **secretory, membrane,** and **lysosomal proteins** on ribosomes at the surface of the RER, transport of the growing polypeptide chain across the RER membrane, and its processing within the RER.

a. Transport of nascent peptide into the RER

—is thought to occur by a mechanism described by the **signal hypothesis** as follows (Figure 3.5):

(1) mRNAs for secretory, membrane, and lysosomal proteins contain codons that encode a **signal sequence.**

(2) When the signal sequence is formed on the ribosome, a **signal recognition particle** (SRP) present in the cytosol binds to it.

(3) Synthesis of the growing chain stops until the SRP facilitates the relocation of the polysome to SRP receptors in the RER membrane.

(4) The large subunits of the ribosomes interact with **ribophorins** (RER integral membrane proteins), enabling the nascent polypeptide to enter the RER cisterna; the SRP detaches and protein synthesis resumes with the amino-terminal end inside the RER.

b. Posttranslational modification in the RER

—includes the following sequence of events:

(1) After the nascent polypeptide enters the cisterna, a **signal peptidase** cleaves the signal sequence from it.

(2) The polypeptide is glycosylated.

(3) Disulfide bonds form, converting the linear polypeptide into a globular form.

c. Protein transport from the RER to the Golgi complex

—includes the following sequence of events:

(1) Transitional elements of the RER give rise to **transfer vesicles** (not clathrin coated) containing newly synthesized protein.

(2) Transfer vesicles move to the cis face of the Golgi complex where they deliver the protein by coalescing (an ATP-requiring process) with the outermost Golgi cisterna.

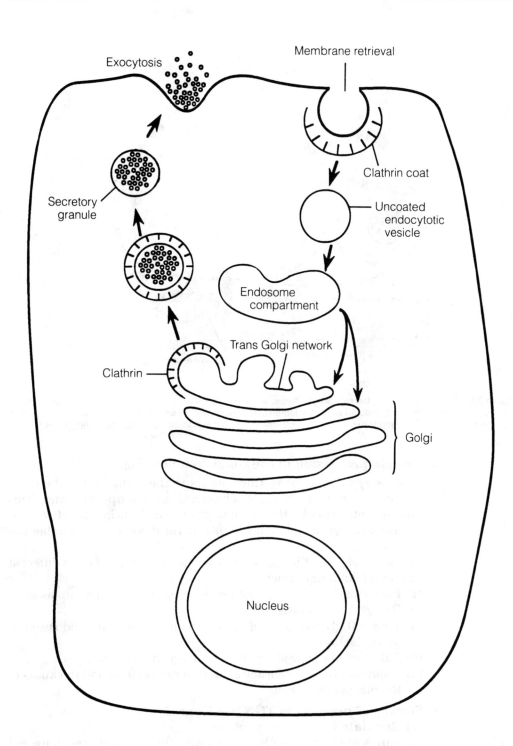

Figure 3.4. Recycling of secretory granule membrane after exocytosis. Membrane is retrieved via clathrin-coated vesicles, moves through endosomes, and then moves either to a trans Golgi cisterna or the trans Golgi network for reuse. (Adapted from Alberts B et al: *Molecular Biology of the Cell,* 2nd ed. New York, Garland Publishing, Inc., 1989, p 467.)

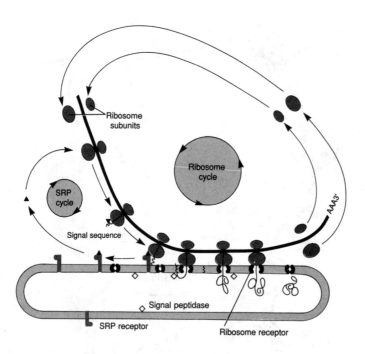

Figure 3.5. Diagram illustrating the signal hypothesis, which proposes a mechanism whereby nascent noncytosolic polypeptides become segregated within the lumen of the RER. *SRP* = signal recognition particle. (Reprinted from Widnell CC and Pfenninger KH: *Essential Cell Biology*. Baltimore, Williams & Wilkins, 1990, p 108.)

d. Protein processing in the Golgi complex (Figure 3.6)

—occurs as proteins move from the cis to the trans face of the Golgi complex through **distinct cisternal subcompartments**. This movement occurs by the budding-off of non-clathrin–coated vesicles from one cisterna and their fusion with the dilated rim of another cisterna.

—may include the following events, each of which occurs in a different cisternal subcompartment:

(1) Phosphorylation of mannose residues on lysosomal hydrolases
(2) Removal of mannose
(3) Terminal glycosylation of some proteins with sialic acid residues and galactose
(4) Sulfation and phosphorylation of amino acid residues
(5) Acquisition of a membrane similar in composition and thickness to the plasma membrane

e. Sorting of proteins in TGN (see Figure 3.6)

(1) Regulated secretory proteins

—are sorted from membrane and lysosomal proteins and delivered via clathrin-coated vesicles to condensing vacuoles; removal of water, via ionic exchanges, yields **secretory granules**.

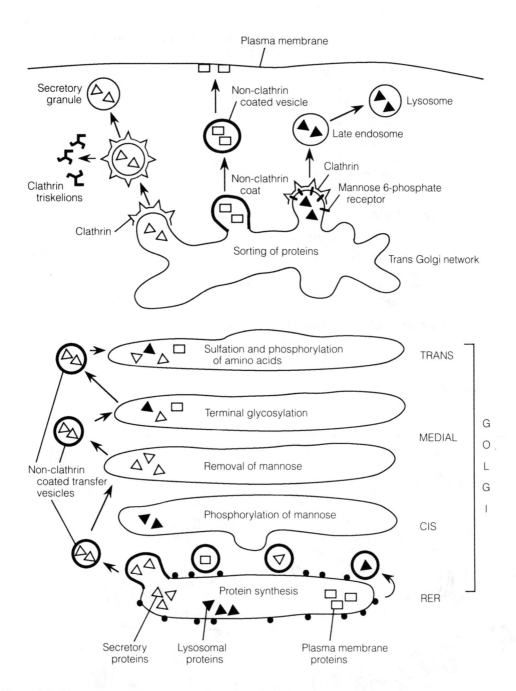

Figure 3.6. Diagram illustrating the modification of proteins in separate compartments of the Golgi complex and their sorting for movement to other cellular sites. Proteins synthesized in the rough endoplasmic reticulum (*RER*), which include secretory (△), membrane (▭), and lysosomal (▲) proteins, pass through the Golgi complex, whereas cytosolic proteins do not. Movement of a secretory protein is illustrated in the diagram. The heavy black line indicates portions of the RER and Golgi membrane that are pinched off to form non-clathrin–coated transfer vesicles that shuttle protein molecules from one site to another. All proteins do not undergo all of the chemical modifications (e.g., only lysosomal proteins undergo mannose phosphorylation).

(2) Lysosomal proteins

—are sorted into clathrin-coated regions of the TGN that have receptors for mannose 6-phosphate and are delivered to late endosomes via clathrin-coated vesicles.

(3) Plasma-membrane proteins

—are sorted into non-clathrin–coated regions of the TGN and delivered to the plasma membrane in non-clathrin–coated vesicles.

2. Synthesis of cytosolic proteins

—takes place on polyribosomes **free** in the cytosol.

—is directed by mRNAs that **lack** signal codons.

—yields proteins that are released directly into the cytosol (e.g., protein kinase, hemoglobin).

C. Intracellular digestion

1. Nonlysosomal digestion

—is the degradation of cytosolic constituents by mechanisms outside of the vacuolar lysosomal pathway.

—involves the turnover of short-lived proteins by different classes of nonlysosomal proteases.

2. Lysosomal digestion (Figure 3.7)

—is the degradation of material within various types of lysosomes by lysosomal enzymes.

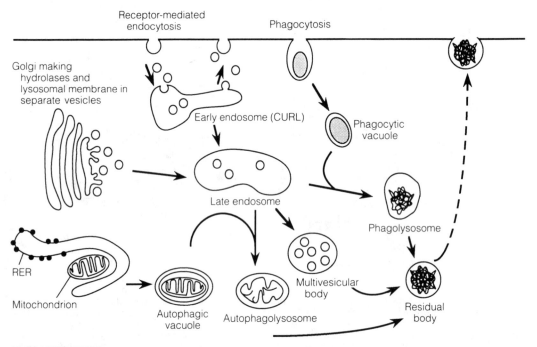

Figure 3.7. Diagram illustrating the different pathways of intracellular lysosomal digestion and the types of lysosomes involved in each.

–occurs by different pathways depending on the origin of the material to be degraded.

a. Heterophagy

–is the ingestion and degradation of **foreign material** taken into the cell by receptor-mediated endocytosis or phagocytosis.

(1) Digestion of **endocytosed** ligands occurs in **multivesicular bodies** (see Figure 3.3).

(2) Digestion of **phagocytosed** microorganisms and foreign particles begins and may be completed in **phagolysosomes**.

b. Autophagy

–is the segregation of an organelle or other cell constituent within membranes from the RER to form an **autophagic vacuole,** which is subsequently digested in an **autophagolysosome**.

c. Crinophagy

–is the fusion of **hormone secretory granules** with lysosomes and their subsequent digestion.

–is a means by which **excess numbers** of secretory granules are removed from the cell.

IV. Clinical Considerations

A. Lysosomal storage diseases

–are hereditary conditions in which the synthesis of specific lysosomal acid hydrolases is impaired.

–are characterized by the **inability of lysosomes to degrade certain compounds,** which accumulate and interfere with cell functioning.

–include the following:

1. Tay-Sachs disease, in which glycolipids accumulate in the nervous system

2. Glycogen storage disease, in which glycogen is abundant in the liver and muscle

3. Hurler's syndrome, in which glycosaminoglycans accumulate in many tissues and organs

B. Familial hypercholesterolemia

–is associated with a **decreased ability of cells to take in cholesterol,** which normally is ingested by receptor-mediated endocytosis of **low-density lipoproteins** (LDLs).

–is caused by an inherited genetic defect that results in an inability to synthesize LDL receptors or in the synthesis of defective ones unable to bind either to LDLs or to clathrin-coated pits.

–is characterized by an elevated level of cholesterol in the bloodstream, which facilitates early development of **atherosclerosis.** For this reason, afflicted individuals often die at an early age from coronary artery disease.

C. Tumor diagnosis

–often can be based on immunocytochemical identification of the intermediate filaments in tumor cells, since the type of intermediate filament present identifies the tissue from which the metastatic cancer cells originated.

Review Test

Directions: Each of the numbered items or incomplete statements in this section is followed by answers or by completions of the statement. Select the **one** lettered answer or completion that is **best** in each case.

1. Which one of the following statements concerning the cytoskeleton is FALSE?

(A) Keratin filaments are associated with the plaques of desmosomes
(B) Intermediate filaments provide mechanical strength to cells
(C) Desmin filaments are present in all three types of muscle cells
(D) Actin microfilaments are associated with hemidesmosome plaques
(E) Microtubules are associated with a molecular motor called kinesin

2. Which of the following vesicles does NOT contain a full complement of acid hydrolases?

(A) Multivesicular body
(B) Early endosome
(C) Autophagolysosome
(D) Lysosome
(E) Phagolysosome

3. Which of the following organelles divides by fission?

(A) Golgi complex
(B) Rough endoplasmic reticulum
(C) Peroxisome
(D) Smooth endoplasmic reticulum
(E) Centriole

4. A 30-year-old man with very high blood cholesterol levels (290) has been diagnosed with premature atherosclerosis. His father died of a heart attack at age 45, and his mother, age 44, has coronary artery disease. Which of the following is the most likely explanation of his condition?

(A) He has a lysosomal storage disease and cannot digest cholesterol
(B) He suffers from a peroxisomal disorder and produces low levels of hydrogen peroxide
(C) The smooth endoplasmic reticulum (SER) in his hepatocytes has proliferated and produced excessive amounts of cholesterol
(D) He has a genetic disorder and synthesizes defective LDL receptors
(E) He is unable to manufacture endosomes

Directions: Each group of items in this section consists of lettered options followed by a set of numbered items. For each item, select the **one** lettered option that is most closely associated with it. Each lettered option may be selected once, more than once, or not at all.

Questions 5–9

Match each process below with the option that fits best.

(A) Clathrin-coated vesicle
(B) Non-clathrin–coated vesicle
(C) Both
(D) Neither

5. Movement of protein from the RER to the cis face of the Golgi complex

6. Retrieval of secretion-granule membrane after exocytosis

7. Movement of glycoprotein from cis to medial Golgi cisternae

8. Uncoupling of endocytosed ligands from receptors

9. Movement of acid hydrolases from the trans Golgi network to a lysosome

Questions 10–13

Match each description below with the cell component that it best describes.

(A) Microfilament
(B) Intermediate filament
(C) Microtubule
(D) Microtrabecular lattice

10. Is associated with kinesin

11. Consists of globular actin monomers linked into a double helix

12. Has a rigid wall composed of 13 protofilament strands

13. Provides structural support to astrocytes

Questions 14–17

Match each description below with the corresponding lettered structure in the electron micrograph.

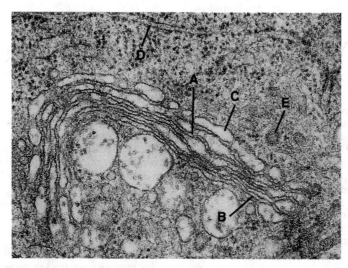

(A) Site where sorting of proteins occurs
(B) Site that receives transport vesicles from the RER
(C) Site where terminal glycosylation of noncytosolic proteins occurs
(D) Site where ribophorin I and II are located

Questions 18–20

Match each description below with the corresponding lettered structure in the electron micrograph.

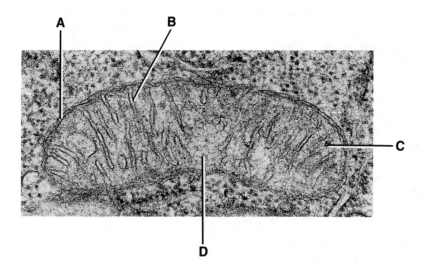

18. Site of the elementary particles associated with chemiosmotic coupling and ATP production

19. Site of enzymes associated with fatty acid oxidation

20. Site where divalent cations are bound

Answers and Explanations

1–D. Actin filaments are involved in endocytosis, exocytosis, and other activities associated with the cell surface. They are also associated with the sol–gel transformation of the cytosol and with contraction. Keratin filaments are associated with plaques of desmosomes and hemidesmosomes.

2–B. The early endosome is a mildly acidic compartment near the periphery of the cell in which the uncoupling of ligands and receptors takes place. It does not contain acid hydrolases.

3–C. A peroxisome originates from preexisting peroxisomes. It imports specific cytosolic proteins and then undergoes fission. The other organelle that divides by fission is the mitochondrion.

4–D. Cells import cholesterol by the receptor-mediated uptake of low-density lipoproteins (LDLs) in coated vesicles. Certain individuals inherit defective genes and cannot make LDL receptors, or they make defective ones that cannot bind to clathrin-coated pits. The result is an inability to internalize LDLs, which build to high levels in the bloodstream, predisposing the individual to premature atherosclerosis and the risk of heart attacks.

5–B. Transport of protein from the RER to the Golgi complex occurs via non–clathrin-coated vesicles, which fuse with the cis face.

6–A. Membrane recycling after exocytosis of the contents of a secretion granule occurs via clathrin-coated vesicles.

7–B. Transfer of material among the cisternae of the Golgi complex takes place via non-clathrin–coated vesicles.

8–D. The uncoupling of ligands and receptors internalized by receptor-mediated endocytosis occurs in the early endosome.

9–A. Proteins targeted for lysosomes leave the trans Golgi network in clathrin-coated vesicles.

10–C. Kinesin is a force-generating protein associated with microtubules. It serves as a molecular motor for transport of organelles and vesicles.

11–A. Globular actin monomers (G actin) polymerize into a double helix of filamentous actin (F actin)—or microfilament—in response to the regulatory influence of a number of actin-binding proteins.

12–C. A microtubule consists of $\alpha\beta$-tubulin dimers polymerized into a spiral around a hollow lumen to form a fairly rigid tubule. When cross-sectioned, the microtubule reveals 13 protofilament strands, which represent the tubulin dimers present in one complete turn of the spiral.

13–B. Glial filaments are a type of intermediate filament, composed of glial fibrillary acidic protein, present in fibrous astrocytes. These filaments are supportive, but they may also play additional roles in both normal and pathologic processes in the central nervous system.

14–B. The trans Golgi network is the site where sorting of protein occurs.

15–C. The cis cisterna of the Golgi complex is the site receiving transfer from the RER.

16–A. The medial Golgi cisterna, which lies between the cis and trans face of the Golgi complex, is the site where terminal glycosylation of noncytosolic proteins occurs.

17–D. The RER membrane is the site where ribophorin I and II are located. Ribophorins are integral membrane proteins that facilitate translocation of nascent polypeptides into the RER cisternae.

18–B. The inner mitochondrial membrane (which infolds to form cristae) is where the globular elementary particles associated with ATP synthetase are located. This enzyme converts ADP and P_i to ATP, a reaction driven by energy from the electrochemical proton gradient existing across the inner mitochondrial membrane.

19–D. The mitochondrial matrix is the site where enzymes associated with fatty acid oxidation are located.

20–C. Mitochondrial matrix granules bind divalent cations such as Ca^{2+} and Mg^{2+}.

4

Extracellular Matrix

I. Overview—The Extracellular Matrix

—is an **organized meshwork of macromolecules** surrounding and underlying cells.

—exhibits different functions in different tissues.

—alters the cells in contact with it by affecting their metabolic activities as well as influencing their shape, migration, division, and differentiation.

—varies in composition but consists of an amorphous **ground substance**— containing primarily glycosaminoglycans, proteoglycans, and glycoproteins— and **fibers** (Figure 4.1).

—along with water and other small molecules (e.g., nutrients, ions), constitutes the **extracellular enviroment**.

II. Glycosaminoglycans (GAGs)

—are long, unbranched polysaccharides composed of **repeating disaccharide units**.

—always possess an **amino sugar** (either N-acetylglucosamine or N-acetylgalactosamine) as one of their repeating disaccharides.

—are commonly **sulfated** and usually possess a **uronic acid sugar** (which has a carboxyl group) in the repeating disaccharide unit; as a result they have a **high negative charge**.

—generally are linked to a **core protein**.

—attract osmotically active cations (e.g., Na^+), resulting in a heavily hydrated matrix that strongly **resists compression**.

—form extended random coils occupying large volumes of space because of their inability to fold compactly.

—can be classified into four main groups based on their chemical structure (Table 4.1): **hyaluronic acid,** a huge, nonsulfated molecule (up to 20 μm in length) that is not attached to a core protein; **chondroitin sulfate** and **dermatan sulfate; heparin** and **heparan sulfate;** and **keratan sulfate**.

III. Proteoglycans

—consist of a **core protein from which many glycosaminoglycans extend**.

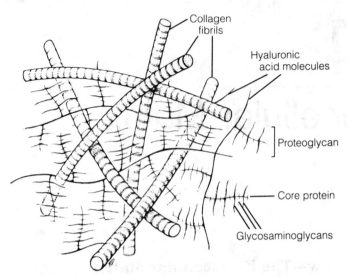

Figure 4.1. Illustration showing the major components of the extracellular matrix. (Adapted from Fawcett DW: *Bloom and Fawcett's A Textbook of Histology,* 11th ed. Philadelphia, WB Saunders Company, 1986, p 148.)

—are very large molecules shaped like a bottle brush (see Figure 4.1).

—may attach to hyaluronic acid, via their core protein, to form complex, gigantic aggregates.

—show marked heterogeneity in their core-protein content, molecular size, and the number and kinds of GAGs they contain.

—act as sites where **growth factors** (e.g., fibroblast growth factor) **and other signaling molecules may bind**.

—confer unique attributes (e.g., **selective permeability**) to the extracellular matrix in certain locations.

IV. Fibers

A. Collagen

—is the most abundant structural protein of the extracellular matrix.

—exists in at least 12 molecular types, which vary in the composition of the 3 α chains (Table 4.2).

Table 4.1. Classification of Glycosaminoglycans (GAGs)

Type	Linked to core protein	Sulfated	Major locations in body
Hyaluronic acid	No	No	Synovial fluid, vitreous humor, cartilage, skin, most connective tissues
Chondroitin sulfate	Yes	Yes	Cornea, cartilage, bone, adventitia of arteries
Dermatan sulfate	Yes	Yes	Skin, blood vessels, heart valves
Heparin	Yes	Yes	Lung, skin, liver, mast cells
Heparan sulfate	Yes	Yes	Basal laminae, lung, arteries, cell surfaces
Keratan sulfate	Yes	Yes	Cornea, cartilage, nucleus pulposus of intervertebral disk

Table 4.2. Characteristics of Common Collagen Types*

Molecular type	Cells synthesizing	Major location(s) in body	Function
I	Fibroblast Osteoblast Odontoblast	Dermis of skin, bone, tendons, ligaments, fibrocartilage	Resists tension
II	Chondroblast	Hyaline cartilage	Resists intermittent pressure
III	Fibroblast Reticular cell Smooth muscle Schwann cell Hepatocyte	Lymphatic system, cardiovascular system, liver, lung, spleen, intestine, uterus, endoneurium	Forms structural framework in expandable organs
IV	Endothelial cell Epithelial cell	Basal lamina	Provides support and filtration
	Muscle cell Schwann cell	External lamina	Acts as scaffold for cell migration
V	Mesenchymal cell	Placenta	Unknown
VII	Keratinocyte	Dermal–epidermal junction	Forms anchoring fibrils that secure lamina densa to underlying connective tissue

*The six **most abundant** collagen types are included in this table.

–is synthesized and assembled into **fibrils** by a series of **intracellular** and **extracellular** events (Figure 4.2).

1. Intracellular events in collagen synthesis

–occur in the following sequence:

a. Preprocollagen synthesis

–occurs at the surface of the rough endoplasmic reticulum (RER).
–is directed by mRNAs that encode the different types of α chains to be synthesized.

b. Hydroxylation

–of specific proline and lysine residues within the forming polypeptide chain occurs within the RER.
–is catalyzed by specific **hydroxylases,** which require vitamin C as a cofactor.

c. Attachment of sugars (glycosylation)

–to specific hydroxylysine residues also occurs in the RER.

d. Procollagen (triple-helix) formation

–is precisely regulated by **propeptides** (extra nonhelical amino acid sequences), which are located at both ends of each α chain.
–involves the three α chains aligning and coiling into a triple helix.
–occurs within the RER.

e. Addition of carbohydrates

–occurs in the Golgi complex to which procollagen is transported via transfer vesicles.
–results in completion of oligosaccharide side chains.

INTRACELLULAR EVENTS

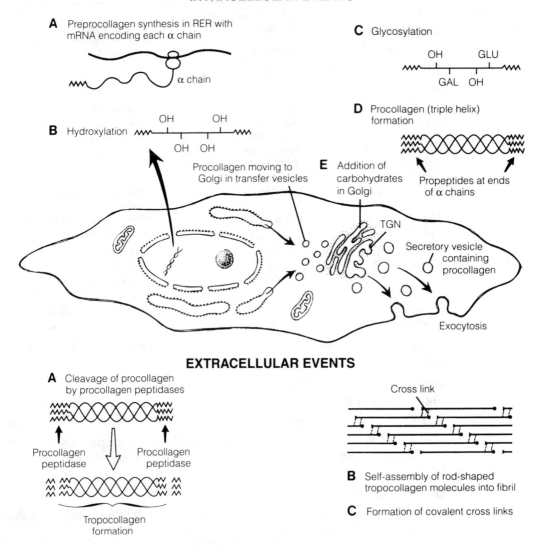

A Preprocollagen synthesis in RER with mRNA encoding each α chain

α chain

B Hydroxylation

OH   OH

OH   OH

C Glycosylation

OH   GLU

GAL   OH

D Procollagen (triple helix) formation

Procollagen moving to Golgi in transfer vesicles

E Addition of carbohydrates in Golgi

Propeptides at ends of α chains

TGN

Secretory vesicle containing procollagen

Exocytosis

EXTRACELLULAR EVENTS

A Cleavage of procollagen by procollagen peptidases

Procollagen peptidase

Procollagen peptidase

Tropocollagen formation

Cross link

B Self-assembly of rod-shaped tropocollagen molecules into fibril

C Formation of covalent cross links

Figure 4.2. Diagram showing the intracellular and extracellular steps involved in the synthesis of a collagen fibril. Adapted from Junqueira LC et al: *Basic Histology,* 7th ed. Norwalk, CT, Appleton & Lange, 1992, p 103.)

f. Secretion of procollagen

—by exocytosis occurs after secretory vesicles from the trans Golgi network are guided to the cell surface by microtubules and microfilaments.

2. Extracellular events in collagen synthesis

—occur in the following sequence:

a. Cleavage of procollagen

—is catalyzed by procollagen **peptidases,** which remove most of the propeptide sequences at the ends of each α chain, yielding **tropocollagen**.

b. Self-assembly of tropocollagen

–occurs as insoluble tropocollagen molecules aggregate near the cell surface.

–produces **fibrils** characteristic of types I, II, III, V, and VII collagen. These fibrils have a transverse banding periodicity of 67 nm in types I, II, and III collagen (Figure 4.3); the periodicity varies in other types of collagen.

c. Covalent-bond formation (cross-linking)

–occurs between adjacent tropocollagen molecules.

–involves formation of lysine- and hydroxylysine-derived aldehydes.

–provides great tensile strength to collagen fibrils.

3. Type IV collagen

–is unique in that it assembles into a **meshwork,** rather than fibrils.

–constitutes most of the **lamina densa** of basal laminae (and external laminae).

–**differs** from other collagen types as follows:

a. The propeptide sequences are not removed from the ends of its procollagen molecules.

b. Its triple-stranded helical structure is interrupted in many regions.

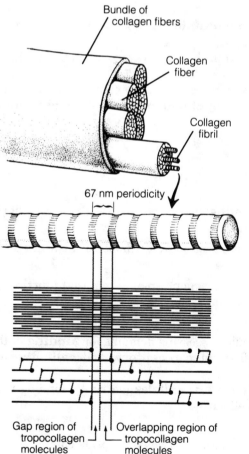

Figure 4.3. Illustration depicting the levels of organization in collagen fibers. As seen by light microscopy, collagen fibers consist of collagen fibrils (which typically reveal a 67-nm cross-banding, when observed by electron microscopy). The periodicity along the collagen fibril is due to the precise arrangement of tropocollagen molecules, which overlap each other, producing gap regions where electron-dense stains penetrate and produce a transverse banding across the fibril. (Adapted from Junqueira LC et al: *Basic Histology,* 7th ed. Norwalk, CT, Appleton & Lange, 1992, p 100.)

c. It forms head-to-head dimers, which interact to form lateral associations creating a sheet-like meshwork.

B. Elastic fibers

1. Components

a. Elastin

—is an amorphous structural protein.

—imparts remarkable elasticity to the extracellular matrix due to the presence of two unusual amino acids, **desmosine** and **isodesmosine,** in its primary structure.

—constitutes about 90% of elastic fibers or sheets.

—consists of molecules that are highly **cross-linked via lysine residues** to form an extensive network.

—exists in a variety of random-coil conformations, forming an elastic network that returns to its original shape after being stretched (similar to a rubber band).

b. Fibrillin

—is the main component of the **peripheral microfibrils** that are part of elastic fibers.

—is a glycoprotein that organizes elastin into fibers.

2. Synthesis of elastic fibers

—begins with the elaboration of **microfibrils,** which appear near the surface of the cell.

—continues as **elastin** forms within the spaces surrounded by bundles of microfibrils.

—is carried out by **fibroblasts** (in tendons and ligaments), **smooth muscle cells** (in large arteries), and by **chondrocytes** and **chondroblasts** (in elastic cartilage).

V. Glycoproteins

A. Fibronectin

1. Types and location

a. Matrix fibronectin is an **adhesive glycoprotein** that forms fibrils in the extracellular matrix.

b. Cell-surface fibronectin is a protein that transiently attaches to the surface of cells.

c. Plasma fibronectin is a circulating plasma protein that functions in blood clotting, wound healing, and phagocytosis.

2. Function—fibronectin

—is a **multifunctional** molecule possessing domains for **binding collagen, heparin,** various **cell-surface receptors,** and **cell-adhesion molecules.**

—mediates **cell adhesion** to the extracellular matrix by binding to fibronectin receptors on the cell surface.

B. Fibronectin receptor

—is a transmembrane protein, consisting of two polypeptide chains.

—belongs to the **integrin family of receptors,** and is known as a **cell-adhesion molecule** (CAM), since it enables cells to adhere to the extracellular matrix.

—binds to fibronectin via a specific tripeptide sequence (Arg-Gly-Asp); other extracellular adhesive proteins also contain this sequence.

—**functions** to link fibronectin outside the cell with cytoskeletal components (e.g., actin) inside the cell (Figure 4.4).

C. Laminin

—is located in the basal (external) lamina and is synthesized by adjacent cells.

—is a large **cross-shaped glycoprotein,** whose arms possess binding sites for cell-surface receptors (integrins), heparan sulfate, type IV collagen, and entactin.

—**functions** to mediate interaction between epithelial cells and the extracellular matrix by anchoring the cell surface to the basal lamina.

D. Entactin

—is a component of all basal (external) laminae.

—is a sulfated adhesive glycoprotein that **binds laminin**.

—functions to link laminin with type IV collagen of the lamina densa.

E. Tenascin

—is an adhesive glycoprotein, most abundant in embryonic tissues.

—is secreted by glial cells in the developing nervous system.

—**functions** in cell migration by promoting cell–matrix adhesion.

F. Chondronectin

—is a glycoprotein in cartilage that attaches chondrocytes to type II collagen.

—is a **multifunctional** molecule with binding sites for collagen, proteoglycans, and cell-surface receptors.

—**functions** in the development and maintenance of cartilage by influencing the composition of its extracellular matrix.

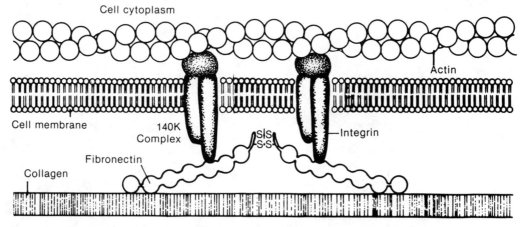

Figure 4.4. Diagram showing integrin receptors, such as the fibronectin receptor, linking molecules outside the cell with components inside the cell. Integrin receptors serve as transmembrane linkers that mediate reciprocal interactions between the cytoskeleton and the extracellular matrix. (Reprinted from Johnson KE: *Histology and Cell Biology,* 2nd ed. Baltimore, Williams & Wilkins, 1991, p 72.)

G. Osteonectin

–is an extracellular matrix glycoprotein in bone.

–**functions** to link minerals to type I collagen and to influence calcification by inhibiting crystal growth.

VI. Clinical Considerations

A. Scurvy

–is associated with a **deficiency of vitamin C**.

–results from the synthesis of poorly hydroxylated tropocollagen, which is unable to form either a stable triple helix or collagen fibrils.

–manifests itself in bleeding gums and eventual tooth loss.

–is reversed by administration of vitamin C.

B. Wound healing

–in adults involves the formation of **fibronectin "tracks"** along which cells migrate to their destinations.

1. In **connective tissue,** wound healing is often characterized by migration of fibroblasts across blood clots where they adhere to fibronectin.

2. In **epithelia** involves re-epithelialization, which depends on the **basal lamina serving as a scaffolding** for cell migration to cover the denuded area; then epithelial cell proliferation and replacement occurs.

C. Ehlers-Danlos type IV syndrome

–is a genetic defect in transcription of DNA or translation of mRNA encoding **type III collagen,** the major component of **reticular fibers**.

–presents often as a rupture of the bowel and/or large arteries, where reticular fibers normally ensheath the smooth muscle cells.

Review Test

Directions: Each of the numbered items or incomplete statements in this section is followed by answers or by completions of the statement. Select the **one** lettered answer or completion that is **best** in each case.

1. Which one of the following statements about the fibronectin receptor is FALSE?

(A) It binds to fibronectin via a specific tripeptide sequence (Arg-Gly-Asp)
(B) It mediates the linkage of molecules outside the cell with cytoskeletal components inside the cell
(C) It is located exclusively in the basal lamina
(D) It belongs to the integrin family of receptors
(E) It extends across the plasma membrane

2. A medical student presented at the emergency room and was diagnosed with a ruptured bowel, the result of a genetic condition called Ehlers-Danlos type IV syndrome. Which one of the following statements about this patients condition is FALSE?

(A) He has a defect in the synthesis of mRNA encoding type III collagen
(B) He has a defect in the translation of the mRNA encoding type IV collagen
(C) He has defective reticular fiber collagen
(D) He is at risk for rupturing his large arteries
(E) He has a defect in the translation of mRNA for type III collagen

3. Which one of the following statements about hyaluronic acid is FALSE?

(A) It does not possess a core protein
(B) It is a glycosaminoglycan
(C) It is a proteoglycan having a shape resembling a bottle brush
(D) It is not sulfated
(E) It is a very large molecule

4. Which one of the following statements about osteonectin is FALSE?

(A) It is present in bone matrix
(B) It is a glycoprotein
(C) It binds to type II collagen
(D) It influences calcification of bone

5. Which one of the following statements about scurvy is FALSE?

(A) One of its symptoms is bleeding gums
(B) It is a condition caused by poorly hydroxylated collagen
(C) It arises because of a deficiency of vitamin A
(D) It is often associated with a defective periodontal ligament
(E) It is alleviated by eating citrus fruits

6. All of the following events in collagen synthesis occur within the cell EXCEPT

(A) synthesis of preprocollagen
(B) hydroxylation of lysine residues
(C) triple-helix formation
(D) carbohydrate addition of procollagen
(E) tropocollagen self-assembly to form fibrils

Directions: Each group of items in this section consists of lettered options followed by a set of numbered items. For each item, select the **one** lettered option that is most closely associated with it. Each lettered option may be selected once, more than once, or not at all.

Questions 7–12

Match each description below with the protein that it describes best.

(A) Fibrillin
(B) Fibronectin
(C) Elastin
(D) Entactin
(E) Laminin
(F) Tenascin

7. A glycoprotein that organizes elastin into a fiber

8. An amorphous structural protein that stretches and recoils in a fashion similar to a rubber band

9. A glycoprotein across which fibroblasts migrate during wound healing

10. A cross-shaped glycoprotein present in the basal lamina

11. An adhesive glycoprotein that links type IV collagen with laminin in the lamina densa

12. The main component of peripheral microfibrils in an elastic fiber

Questions 13–18

Match each cell type below with the type of collagen it synthesizes.

(A) Type I collagen
(B) Type II collagen
(C) Type III collagen
(D) Type IV collagen
(E) Type V collagen

13. Reticular cell

14. Chondroblast

15. Odontoblast

16. Osteoblast

17. Epithelial cell

18. Endothelial cell

Questions 19–20

Match each description below with the protein that it describes best.

(A) Chondronectin
(B) Osteonectin
(C) Tenascin
(D) Entactin

19. A glycoprotein involved in bone crystallization

20. A multifunctional glycoprotein that binds chondrocytes to type II collagen

Answers and Explanations

1–C. The fibronectin receptor is a cell-surface receptor that is a transmembrane protein.

2–B. Ehlers-Danlos type IV syndrome is associated with a defect in the synthesis and translation of mRNA for type III (reticular) collagen.

3–C. Hyaluronic acid is a glycosaminoglycan, not a proteoglycan. The core protein of proteoglycans can attach to hyaluronic acid, forming very large aggregates.

4–C. Osteonectin binds to type I collagen in the bone matrix. Type II collagen is found in cartilage.

5–C. Scurvy is caused by a deficiency of vitamin C, which is a necessary cofactor in the hydroxylation of preprocollagen.

6–E. The formation and self-assembly of tropocollagen occurs in the extracellular space.

7–A. Fibrillin functions to organize elastin into fibers.

8–C. Elastin—an amorphous protein that forms an extensive network—has the ability to stretch and recoil.

9–B. Fibronectin forms "tracks" along which cells migrate. During wound healing in connective tissue, fibroblasts adhere to fibronectin in blood clots, facilitating the healing process.

10–E. Laminin is a large cross-shaped glycoprotein in basal (external) laminae. Its shape is related to its binding sites for integrins, heparan sulfate, type IV collagen, and entactin.

11–D. Entactin is a sulfated adhesive glycoprotein in basal (external) laminae that binds both type IV collagen and laminin.

12–A. Fibrillin is the major component of the peripheral microfibrils of elastic fibers.

13–C. Type III collagen is synthesized by reticular cells, as well as by Schwann cells, smooth muscle cells, hepatocytes, and fibroblasts.

14–B. Only chondroblasts synthesize type II collagen.

15–A. Odontoblasts produce type I collagen, which is a major organic component of dentin in teeth.

16–A. Osteoblasts produce type I collagen, which constitutes about 95% of the organic matrix of bone.

17–D. Epithelial cells manufacture type IV collagen, which forms the lamina densa that underlies them.

18–D. Endothelial cells produce type IV collagen, which forms the lamina densa, a part of the basal lamina beneath them.

19–B. Osteonectin—a bone-matrix glycoprotein—has an inhibitory effect on the formation of bone crystals.

20–A. Chondronectin—an extracellular matrix protein in cartilage—binds chondrocytes to type II collagen and influences the differentiation and maintenance of cartilage.

5

Epithelium and Glands

I. General Structure—Epithelia

—are **specialized layers** that line the internal and cover the external surfaces of the body.

—consist of a **sheet of cells** lying close to one another, with little intercellular space. These cells have distinct biochemical, functional, and structural domains that confer **polarity,** or sidedness, on epithelia.

—are separated from the underlying connective tissue and blood vessels by the **basement membrane**.

—are **avascular,** receiving nourishment by diffusion of molecules through the **basal lamina**.

—are classified into various types based on the **number of cell layers** (one is **simple** and more than one is **stratified**) and the **shape of the superficial cells** (Table 5.1).

II. Functions of Epithelia

A. Transcellular transport

—of molecules from one epithelial surface to another occurs by various processes including the following:

1. **Diffusion of oxygen and carbon dioxide** across the epithelial cells of lung alveoli and capillaries

2. **Carrier protein–mediated transport** of amino acids and glucose across intestinal epithelia

3. **Vesicle–mediated transport** of IgA and other molecules

B. Absorption

—occurs via **endocytosis** or **pinocytosis** (see Chapter 3 III A) in various organs (e.g., the proximal convoluted tubule of the kidney).

C. Secretion

—of various molecules (e.g., hormones, mucus, proteins) occurs by **exocytosis**.

Table 5.1. Classification of Epithelia

Type	Shape of Superficial Cell Layer	Typical Locations
One cell layer		
Simple squamous	Flattened	Endothelium (lining of blood vessels); mesothelium (lining of peritoneum and pleura)
Simple cuboidal	Cuboidal	Lining of distal tubule in kidney and ducts in some glands; surface of ovary
Simple columnar	Columnar	Lining of intestine, stomach, and excretory ducts in some glands
Pseudostratified	All cells rest on basal lamina, but not all reach the lumen, thus the epithelium appears "falsely stratified"	Lining of trachea, primary bronchi, nasal cavity, and excretory ducts in parotid gland
More than one cell layer		
Stratified squamous (nonkeratinized)	Flattened (nucleated)	Lining of esophagus, vagina, mouth, and true vocal cords
Stratified squamous (keratinized)	Flattened (and without nuclei)	Epidermis of skin
Stratified cuboidal	Cuboidal	Lining of ducts in sweat gland
Stratified columnar	Columnar	Lining of large excretory ducts in some glands and cavernous urethra
Transitional	Dome-shaped (when relaxed); flattened (when stretched)	Lining of urinary passages from renal calyces to the urethra

D. Selective permeability
—results from the presence of **tight junctions** between epithelial cells.
—permits fluids having different compositions and tonicity to exist on separate sides of an epithelial layer (e.g., intestinal epithelium).

E. Protection
—from abrasion and injury is provided by the **epidermis,** the epithelial layer of the skin.

III. Lateral Epithelial Surfaces
—contain specialized **junctions** that function in adhesion of or communication between cells and in restricting movement of materials into and out of lumina.

A. Junctional complex
—is an intricate arrangement of membrane-associated structures that function in cell-to-cell attachment of columnar epithelial cells.
—corresponds to the "terminal bar" observed in epithelia by light microscopy.
—consists of three distinct components—the zonula occludens, zonula adherens, and macula adherens—visible by electron microscopy.

1. Zonula occludens
—is also called a **tight junction**.

–is a **zone around the entire apical perimeter** of adjacent cells formed by **fusion of the outer leaflets** of their plasma membranes.

–is visible as a branching anastomosing network of intramembrane **strands** on the P-face and **grooves** on the corresponding E-face in freeze-fracture preparations of this zone of fusion. The strands consist of transmembrane proteins of each cell **attached directly to one another,** thus sealing off the intercellular space.

–**prevents movement of substances into the intercellular space** from the lumen. Its ability to do so (its "tightness") is directly related to the number and complexity of the strands.

–is analogous to the **fascia occludens,** a ribbon-like area of fusion between transmembrane proteins on adjacent **endothelial cells** lining capillaries.

2. Zonula adherens

–is also called a **belt desmosome** or **intermediate junction**.

–extends **completely around the perimeter** of epithelial cells, just basal to the zonula occludens.

–is characterized by a 10- to 20-nm separation between the adjacent plasma membranes, with an amorphous or filamentous material occupying the intercellular space.

–is associated on its cytoplasmic surfaces with a mat of **actin** filaments linked, via α-actinin and vinculin, to the transmembrane glycoprotein **E-cadherin**. This protein is markedly dependent upon calcium ions for promoting **adhesion** at this structurally supportive junction.

–is analogous to the **fascia adherens,** a ribbon-like adhesion zone present in the **intercalated disks** of cardiac muscle.

3. Macula adherens

–is also called a **desmosome**.

–is a small, discrete, disk-shaped **adhesive site**.

–is also commonly found at sites other than the junctional complex attaching epithelial cells together.

–is characterized by a **dense plaque** of **intracellular** attachment proteins, called **desmoplakins,** on the cytoplasmic surface of each opposing cell.

–has intermediate **keratin** filaments (**tonofilaments**) looping into and out of the dense plaque from the cytoplasm.

–also contains transmembrane linker glycoproteins.

B. Gap junction

–is also called a **nexus** or **communicating junction**.

–is common in certain tissues other than epithelia (e.g., central nervous system, cardiac muscle, and smooth muscle).

–**couples adjacent cells metabolically and electrically.**

–is a plaque-like entity composed of an **ordered array of subunits** called connexons, which extend beyond the cell surface and keep the opposing plasma membranes about 2 nm apart.

1. Connexons consist of six cylindrical subunits (composed of the protein **connexin**), which are arranged radially around a central channel with a diameter of 1.5 nm.

2. Precise **alignment of connexons** on adjacent cells produces a junction having a central **cell-to-cell channel** that permits passage of ions and small molecules with a molecular weight (MW) of less than 1200.

C. Lateral interdigitations

—are irregular, finger-like projections that **interlock** adjacent epithelial cells.

IV. Basal Epithelial Surfaces

A. Basal lamina

—is an **extracellular supportive structure** (20–100 nm thick) that is visible only by electron microscopy.
—is produced by the epithelium resting upon it.
—is composed mainly of **type IV collagen, laminin, entactin,** and **proteoglycans** (mostly heparan sulfate).
—consists of two zones: the **lamina rara** (or **lamina lucida**), which lies next to the plasma membrane, and the **lamina densa,** a denser meshwork of material that lies adjacent to the reticular lamina of the deeper connective tissue.
—and the underlying reticular lamina constitute the **basement membrane** observable by light microscopy.

B. Hemidesmosomes

—are specialized junctions that resemble one-half of a macula adherens (desmosome).
—mediate **adhesion** of epithelial cells to the underlying extracellular matrix.
—are present on the basal surface of **basal cells** in certain epithelia (e.g., tracheal epithelium and stratified squamous epithelium) and on **myoepithelial cells** where they lie adjacent to the basal lamina.
—consist of a dense cytoplasmic plaque (composed of desmoplakin and other proteins), which is linked via transmembrane receptor proteins (**integrins**) to laminins in the basal lamina. Anchoring filaments from the basal lamina extend into the connective tissue.
—**link the cytoskeleton with the extracellular matrix,** since keratin filaments in the cell terminate in the plaque.

C. Basal plasma-membrane infoldings

—are common in **ion-transporting epithelia** (e.g., distal convoluted tubule of the kidney, ducts in salivary glands).
—form deep invaginations that **compartmentalize mitochondria.**
—**function** to bring ion pumps (Na$^+$–K$^+$ ATPase) in the plasma membrane close to their energy supply (ATP produced in mitochondria).

V. Apical Epithelial Surfaces

—may possess specializations such as microvilli, stereocilia, and cilia.

A. Microvilli

—are finger-like **projections of epithelia** (approximately 1 μm in length) that **extend into a lumen,** increasing the area of the luminal surface.
—are characterized by a **glycocalyx** (sugar coat) on their exterior surface (see Chapter 1 II C).

—contain a bundle of about 30 **actin filaments,** which run longitudinally through the core of a microvillus and extend into the **terminal web,** a zone of intersecting filaments in the apical cytoplasm.

—constitute the **brush border** of kidney proximal tubule cells and the **striated border** of intestinal absorptive cells.

B. Stereocilia

—are very **long microvilli** (not cilia).

—are located in the **epididymis** and **vas deferens** of the male reproductive tract.

C. Cilia

—are **actively motile** processes (5–10 μm in length) extending from certain epithelia (e.g., tracheobronchial and oviduct epithelium).

—**propel substances** along epithelial surfaces.

—contain a **core of longitudinally arranged microtubules** (axoneme), which arises from a basal body during ciliogenesis.

1. Axoneme (Figure 5.1A)

—consists of nine doublet microtubules uniformly spaced around two central microtubules (**9 + 2 configuration**).

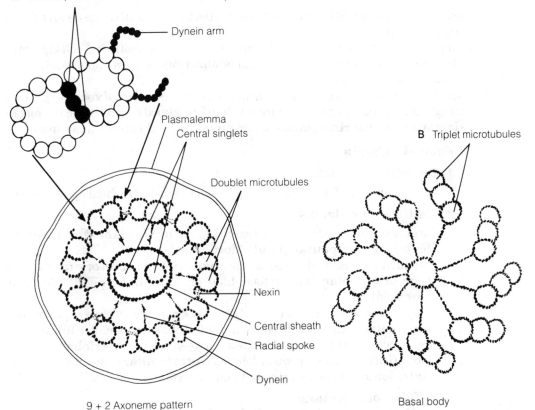

Figure 5.1. Cross-sectional diagrams of a cilium and basal body. (Part A adapted from Junqueira LC et al: *Basic Histology,* 7th ed. Norwalk, CT, Appleton & Lange, 1992, p 46. Part B adapted from West JB: *Best and Taylor's Physiological Basis of Medical Practice,* 12th ed. Baltimore, Williams & Wilkins, 1991, p 12.)

—also contains the following components:

a. Dynein arms extend unidirectionally from one member of each doublet microtubule and interact with adjacent doublets, so that they slide past one another. These arms consist of **dynein ATPase,** which splits ATP to liberate the energy necessary for active movement of a cilium.

b. Radial spokes extend from each of the nine outer doublets toward the central sheath.

c. Central sheath surrounds the two central microtubules. Together with the radial spokes, it regulates the ciliary beat.

d. Nexin is an elastic protein that connects adjacent doublet microtubules and helps maintain the shape of cilia.

2. Basal body (see Figure 5.1*B*)

—is a cylindrical structure located at the **base of each cilium**.

—consists of nine triplet microtubules arranged radially like a pinwheel (**9 + 0 configuration**).

—resembles a centriole (see Figure 3.1) but has a less complex central organization.

VI. Glands

—originate from an epithelium that penetrates the connective tissue and forms secretory units.

—consists of a functional portion of secretory and ductal epithelial cells (**parenchyma**), which is separated by a basal lamina from supporting connective tissue elements (**stroma**).

—are classified into three types based on the site of secretion: **Exocrine glands** secrete into a duct or onto a surface; **endocrine glands** secrete into the bloodstream; and **paracrine glands** secrete into the local extracellular space.

A. Exocrine glands

1. Unicellular glands

—are composed of a single cell (e.g., goblet cells in tracheal epithelium).

2. Multicellular glands

—are classified based on **duct branching** as **simple glands** (duct does not branch) or **compound glands** (duct branches).

—are further classified based on the **shape of the secretory unit** as **acinar** or **alveolar** (sac or flask-like) or **tubular** (straight, coiled, or branched).

—may be surrounded by a connective tissue capsule or have septa of connective tissue that divide the gland into **lobes** and smaller **lobules**.

—may have **ducts** that are located between lobes (**interlobar**), within lobes (**intralobar**), between lobules (**interlobular**), or within lobules (**intralobular**) such as striated and intercalated ducts.

a. Types of secretions

(1) **Mucus** is a viscous material, which usually protects or lubricates cell surfaces.

(2) **Serous secretions** are watery and often rich in enzymes.

(3) Mixed secretions contain both mucus and serous substances. Only multicellular glands can have mixed secretions.

   b. Mechanisms of secretion

   (1) In **merocrine** glands (e.g., parotid gland), the secretory cells release their contents by exocytosis.

   (2) In **apocrine** glands (e.g., lactating mammary gland), part of the apical cytoplasm of the secretory cells is released along with the contents.

   (3) In **holocrine** glands (e.g., sebaceous gland), the entire secretory cell along with its contents is released.

B. Endocrine glands

   —may be **unicellular** (e.g., individual endocrine cells in gastrointestinal and respiratory epithelia) or **multicellular** (e.g., adrenal gland).

   —**lack a duct system** and release secretory material into capillaries, which are abundant and may indent but do not penetrate the basal lamina of the glandular epithelium.

VII. Clinical Considerations

A. Immotile cilia syndrome

   —results from a genetic defect that causes an abnormal ciliary beat or the absence of any beat.

   —is characterized by cilia with axonemes that lack dynein arms and have other abnormalities.

   —is associated with recurrent respiratory infections, reduced fertility in women, and sterility in men.

   —may be associated with **situs inversus,** in which case the condition is known as **Kartagener's syndrome.**

B. Epithelial cell tumors

   —originate when cells fail to respond to normal growth regulatory mechanisms.

   —are **benign** when they remain **localized**.

   —are **malignant** when they **metastasize** to other parts of the body.

   1. Carcinomas are malignant tumors that arise from **surface epithelia**.

   2. Adenocarcinomas are malignant tumors that arise from **glands**.

C. Bullous pemphigoid

   —is an **autoimmune disease** in which antibodies against hemidesmosomes are produced.

   —is characterized by chronic, generalized **blisters** in the skin.

   —results in separation of the epithelium from the underlying substratum.

Review Test

Directions: Each of the numbered items or incomplete statements in this section is followed by answers or by completions of the statement. Select the **one** lettered answer or completion that is **best** in each case.

1. Which one of the following statements about epithelia is FALSE?

(A) They are polarized
(B) They are vascular
(C) They are separated from underlying connective tissue by a basal lamina
(D) They contain only a small amount of intercellular substance
(E) They line the lumen of blood vessels

2. Which one of the following statements about microvilli is FALSE?

(A) They include stereocilia
(B) They form the brush border in the proximal tubule of the kidney
(C) They facilitate absorption
(D) They form the striated border along intestinal absorptive cells
(E) They contain a core of keratin filaments

3. Which one of the following statements about cilia is FALSE?

(A) They contain dynein arms, which have ATPase activity
(B) They contain an axoneme
(C) They possess a 9 + 2 configuration of microtubules
(D) They are nearly identical to basal bodies

4. Which one of the following statements about the gap junction is FALSE?

(A) It is sometimes called a nexus
(B) It permits the passage of large proteins from one cell to an adjacent cell
(C) It is a plaque made up of many connexons
(D) It facilitates metabolic coupling between adjacent cells

5. Which one of the following statements about stratified squamous epithelium is FALSE?

(A) The surface layer of cells may be keratinized
(B) The cells in its most superficial layer are cuboidal
(C) It is supported by a basal lamina
(D) Its cells possess desmosomes

6. Which one of the following statements about the desmosome is FALSE?

(A) It is a disk-shaped adhesive junction between cells
(B) It possesses dense plaques composed in part of desmoplakins
(C) It permits the passage of ions from one cell to an adjacent cell
(D) Its adhesion is dependent upon calcium ions
(E) It possesses transmembrane linker glycoproteins

7. Simple squamous epithelium lines all of the following structures EXCEPT

(A) lung alveoli
(B) Bowman's capsule
(C) lymphatic capillaries
(D) the distal convoluted tubule
(E) the thin limb of the loop of Henle

8. Cells in exocrine glands use several mechanisms of secretion. Which one of the following terms does NOT refer to a mechanism of secretion?

(A) Apocrine
(B) Paracrine
(C) Merocrine
(D) Holocrine

9. Which one of the following statements does NOT apply to the basal lamina?

(A) It contains type IV collagen
(B) It includes the reticular lamina
(C) It consists of a lamina densa and a lamina rara (lucida)
(D) It contains proteoglycans (mostly heparan sulfate)

10. A medical student suffering with chronic respiratory infections seeks the advice of an ear, nose, and throat specialist. A biopsy of the student's respiratory epithelium reveals alterations in certain epithelial structures. This patient is most likely to have abnormal

(A) microvilli
(B) desmosomes
(C) cilia
(D) hemidesmosomes
(E) basal plasmalemma infoldings

Directions: Each group of items in this section consists of lettered options followed by a set of numbered items. For each item, select the **one** lettered option that is most closely associated with it. Each lettered option may be selected once, more than once, or not at all.

Questions 11–14

Match each description below with the corresponding lettered structure shown in the electron micrograph.

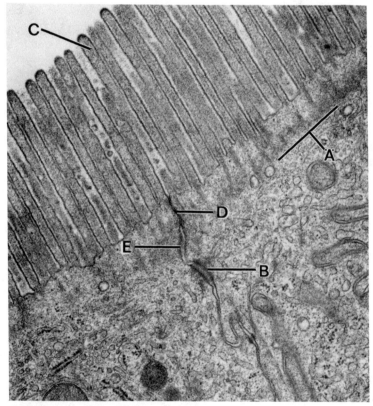

11. Contains a core of actin filaments

12. Is directly associated with keratin filaments (tonofilaments)

13. A zone in the apical cytoplasm containing more than one type of filament

14. Prevents movement of materials in the lumen into the intercellular space

Questions 15 and 16

Match the structure below with the word or phrase that most closely relates to it.

(A) Cilium
(B) A core of actin filaments
(C) Stereocilium
(D) Basal body

15. Axoneme

16. Centriole

Questions 17–19

Match each description below with the disease that is related to it.

(A) Immotile cilia syndrome
(B) Carcinoma
(C) Bullous pemphigoid
(D) Adenocarcinoma

17. An autoimmune disease

18. A hereditary disease that may be associated with infertility

19. A tumor arising from glandular epithelium

Answers and Explanations

1–B. Epithelia are avascular tissues, since blood vessels do not penetrate the basal lamina but remain within the connective tissue compartment.

2–E. Microvilli contain a core of actin (not keratin) filaments, which extend into the terminal web region of the epithelial cell.

3–D. Centrioles (not cilia) are nearly identical in structure to basal bodies (which give rise to cilia and flagella).

4–B. The gap junction "channel," which regulates the passage of ions and small molecules from cell to cell, excludes those having a molecular weight greater than 1200. Therefore, large proteins clearly cannot pass through gap junctions.

5–B. Stratified squamous epithelium is characterized by flattened (not cuboidal) cells in its superficial layer.

6–C. It is the gap junction (not the desmosome) that permits the passage of ions between adjacent cells.

7–D. The distal convoluted tubule is lined by simple cuboidal epithelium (not simple squamous epithelium).

8–B. Paracrine refers to the release of material from enteroendocrine cells into the local extracellular space.

9–B. The reticular lamina is not part of the basal lamina. It is the reticular lamina plus the basal lamina that constitutes the basement membrane observed by light microscopy.

10–C. Individuals with abnormal respiratory cilia often have recurrent respiratory infections if the cilia are unable to clear the respiratory epithelium of microorganisms, debris, and so forth. The student may have immotile cilia syndrome, which is caused by a genetic defect resulting in cilia whose axonemes lack dynein arms and thus are unable to beat.

11–C. Microvilli, which are located on the apical surface of epithelial cells, contain a bundle of about 30 actin filaments.

12–B. Desmosomes, located on the lateral surface of epithelial cells, are composed of a dense plaque of desmoplakins with keratin filaments looping into and out of the plaque.

13–A. The terminal web is an area in the apical cytoplasm where actin filaments from microvilli are anchored. The terminal web usually is free of organelles but contains myosin with which the actin filaments interact, thus enabling microvilli to contract.

14–D. The zonula occludens around the entire apical perimeter of epithelial cells is formed by fusion of transmembrane proteins in the outer leaflets of the adjacent plasma membranes. It seals off the lumen from the intercellular space.

15–A. A cilium possesses an axoneme, a core of microtubules in a precise arrangement. There are nine outer doublet microtubules arranged in a ring around a pair of single microtubules (the 9 + 2 pattern).

16–D. Both basal bodies and centrioles have the 9 + 0 pattern of triplet microtubules and are similar in appearance, but their functions differ. Basal bodies give rise to cilia and flagella.

17–C. Bullous pemphigoid is an autoimmune disease. Affected individuals form antibodies against their own hemidesmosomes.

18–A. The immotile cilia syndrome results from a genetic defect that prevents synthesis of dynein ATPase, resulting in cilia that cannot actively move. Men are sterile because their sperm are not motile (the flagella in their tails lack this enzyme). Women may be infertile because cilia along their oviducts may fail to move oocytes towards the uterus.

19–D. Adenocarcinomas are epithelial tumors that originate in glandular epithelia. Carcinomas originate from surface epithelia.

6

Connective Tissue

I. Overview—Connective Tissue

—is formed primarily of **extracellular matrix** in which various types of **cells** are embedded.

—has many different **functions:** It provides structural **support** for organs and cells; serves as a **medium for exchange** of nutrients and wastes between the blood and tissues; aids in **protection** against microorganisms and in **repair** of damaged tissues; and provides a site for **storage of fat**.

II. Extracellular Matrix

—consists of ground substance, fibers, and tissue fluid.

—provides a medium for the transfer of nutrients and waste materials between connective tissue cells and the bloodstream.

A. Ground substance

—is a colorless, transparent, gel-like material that fills the space between the cells and fibers of connective tissues.

—is a complex mixture of **glycosaminoglycans, proteoglycans,** and **glycoproteins** (see Chapter 4 for details on the structure and function of these components).

—acts as a lubricant and helps prevent the penetration of tissues by foreign particles.

B. Fibers

—include collagen, reticular, and elastic fibers.

—are long, slender protein polymers.

—are present in different proportions in different types of connective tissue.

1. Collagen fibers

—are composed of **type I collagen,** which consists of many closely packed **tropocollagen** fibrils having an average diameter of 75 nm (see Figure 4.3).

—are produced in a two-stage process involving intracellular events (in fibroblasts) and extracellular events (see Figure 4.2).

—have great tensile strength and impart both flexibility and strength to tissues containing them.

—are present in bone, skin, cartilage, tendon, and other structures.

2. Reticular fibers

—are extremely **thin** (0.5–2.0 μm in diameter) and are formed of loosely packed 45-nm tropocollagen fibrils.

—are composed primarily of **type III collagen** and have a higher sugar content than collagen fibers.

—constitute the architectural framework of certain organs and glands.

—stain black with silver salts because of their high carbohydrate content.

3. Elastic fibers

—are coiled, branching fibers (0.2–1.0 μm in diameter), which sometimes form loose networks.

—are composed of microfibrils of **elastin** and **fibrillin** embedded in amorphous elastin (see Chapter 4 IV B).

—may be stretched up to 150% of their resting length.

—require special staining in order to be observed by light microscopy.

III. Connective Tissue Cells

—include many types with different functions.

—may be formed locally and remain in the connective tissue or be formed elsewhere and be located only transiently in connective tissue (Figure 6.1).

A. Fibroblasts

—arise from undifferentiated mesenchymal cells.

—are the **predominant cells** located in connective tissue.

—normally have an oval nucleus and often have two nucleoli.

1. Active fibroblasts

—are spindle-shaped with long tapering ends (fusiform).

—contain well-developed rough endoplasmic reticulum (RER) and Golgi complex.

—are **synthetically active,** producing procollagen and other precursors of the extracellular-matrix components.

2. Quiescent fibroblasts

—are smaller and contain less RER than active fibroblasts.

—may be flattened and possess several processes.

—are synthetically **inactive**.

—may revert to the active state if stimulated (e.g., during wound healing).

B. Pericytes

—are derived from undifferentiated mesenchymal cells and may retain the **pluripotential role** of the embryonic mesenchymal cell.

—are smaller than fibroblasts and are **located mostly along capillaries**.

—display few mitochondria and very little RER in electron micrographs.

C. Adipose cells

—arise from undifferentiated mesenchymal cells.

—are surrounded by a basal lamina.

—are responsible for the **synthesis and storage of fat**.

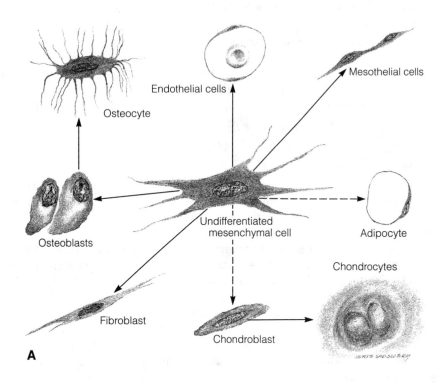

A

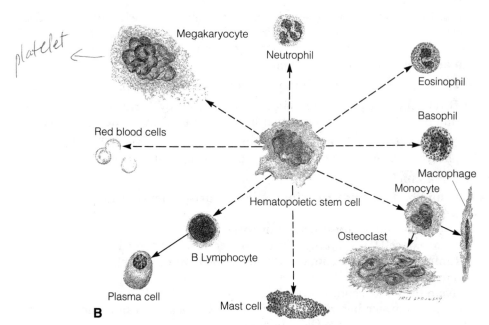

B

platelet ←

Figure 6.1. Diagram showing origin of connective tissue cells. (*A*) Cells arising from undifferentiated mesenchymal cells are formed in connective tissue and remain there. (*B*) Cells arising from hematopoietic stem cells are formed in the bone marrow and reside transiently in connective tissue. *Dotted lines* indicate that intermediate cell types occur between those shown.

1. Unilocular adipose cells

–have their cytoplasm and nucleus squeezed into a thin rim around the periphery to accommodate a single large fat droplet.

–have plasmalemma **receptors** for insulin, growth hormone, norepinephrine, and glucocorticoids **to control the uptake and release of free fatty acids and triglycerides**.

–generally do not increase in number after a limited postnatal period.

2. Multilocular adipose cells

–contain many small fat droplets.

D. Macrophages

–originate in the bone marrow as **monocytes** (see Chapter 10 VI E). Monocytes circulate in the bloodstream and migrate into the connective tissue, where they mature into functional macrophages.

–are generally spherical with a diameter of 10–30 μm.

–are the principle **phagocytosing cells** of connective tissue responsible for removing large particulate matter and assisting in the **immune response** (see Chapter 12 II D). They also secrete substances that function in **wound healing**.

–display filopodia, phagocytic vacuoles, lysosomes, and residual bodies when activated.

–may fuse, when stimulated, to form **foreign-body giant cells**. These **multinucleated cells** surround and phagocytose large, foreign bodies.

E. Lymphoid cells

–arise from lymphoid stem cells during hematopoiesis (see Chapter 10 VI G).

–are located throughout the body in the subepithelial connective tissue.

–accumulate in the respiratory system, gastrointestinal tract, and especially in areas of chronic inflammation.

1. T lymphocytes (T cells) initiate the **cell-mediated immune response**.

2. B lymphocytes (B cells), following activation by antigen, differentiate into plasma cells, which function in the **humoral immune response**.

3. Null cells lack the surface determinants characteristic of T and B lymphocytes but may have **cytotoxic activity** against tumor cells.

F. Plasma cells

–are **antibody-manufacturing cells** that arise from activated B lymphocytes.

–are responsible for **humoral immunity**.

–are ovoid cells containing an eccentrically placed nucleus possessing **clumps of heterochromatin,** which appear to be arranged in a spoke-wheel fashion.

–have a deeply basophilic cytoplasm, due to an abundance of RER, and a prominent area adjacent to the nucleus that appears pale and contains the Golgi complex (negative Golgi image).

G. Mast cells

–arise from myeloid stem cells during hematopoiesis.

–possess a small, ovoid, pale-staining nucleus and cytoplasm filled with coarse, deeply stained metachromatic granules.

−display a surface with folds, well-developed Golgi complex, scant RER, and many dense lamellated granules in electron micrographs.

−mediate **immediate (type I) hypersensitivity reactions** (anaphylactic reactions) as follows:

1. The first exposure to an allergen leads to production of IgE antibodies, which bind to receptors on the surface of mast cells (and basophils), causing them to become **sensitized**. Common antigens that may evoke this response include plant **pollens, insect venoms,** certain **drugs,** and **foreign serum**.

2. During a **second exposure** to the same allergen, the membrane-bound IgE binds the allergen, triggering **degranulation** of mast cells and the release of various mediators.

3. The released mediators include **heparin, eosinophil chemotactic factor** (ECF), **histamine,** and **leukotriene C**.

   a. **Histamine** increases the permeability of blood vessels.

   b. **Leukotriene C,** also known as slow-reacting substance of anaphylaxis (SRS-A), induces slow contraction of smooth muscle.

H. Granulocytes

−are white blood cells that possess cytoplasmic granules.
−arise from myeloid stem cells during hematopoiesis.

1. Neutrophils

−**phagocytose, kill, and digest bacteria** at sites of acute inflammation. **Pus** is an accumulation of dead neutrophils at an inflammatory site.

2. Eosinophils

−bind to antigen–antibody complexes on the surface of parasites (e.g., helminths) and then release cytotoxins that damage the parasites.
−are attracted by eosinophil chemotactic factor, which is secreted by mast cells and basophils, to sites of allergic inflammation and release enzymes that cleave histamine and leukotriene C, thus **moderating the allergic reaction**.

IV. Classification of Connective Tissue

−is based either on the proportion of cells to fibers (connective tissue proper) or on specialized properties.

A. Connective tissue proper

1. Loose connective tissue

−is also called **areolar** tissue.
−possesses relatively fewer fibers but more cells than dense connective tissue.
−**is well vascularized, flexible, and not very resistant to stress.**
−is **more abundant** than dense connective tissue.

2. Dense connective tissue

−contains more fibers but fewer cells than loose connective tissue.
−is classified by the orientation of its fiber bundles into two types:

a. Dense irregular connective tissue

–contains fiber bundles that have no definite orientation.

–constitutes most dense connective tissue.

–is characteristic of the **dermis** and **capsules of many organs**.

b. Dense regular connective tissue

–contains fiber bundles that are arranged in a uniform parallel fashion.

–possesses attentuated fibroblasts, which occupy narrow spaces within the fiber bundles.

–is present only in **tendons** and **ligaments**.

B. Specialized connective tissue

1. Mucous tissue

–is a type of loose connective tissue that is the main constituent of the umbilical cord, where it is known as **Wharton's jelly**.

–consists of a jelly-like matrix with some collagen fibers in which large stellate-shaped fibroblasts are embedded.

2. Mesenchymal tissue

–is the connective tissue present in **embryos**.

–consists of a gel-like amorphous matrix containing only a few scattered reticular fibers in which star-shaped, pale-staining mesenchymal cells are embedded. Mitotic figures are often observed in these cells.

3. Elastic tissue

–comprises coarse, branching elastic fibers with a sparse network of collagen fibers and some fibroblasts filling the interstitial spaces.

–is present in the dermis, lungs, elastic cartilage, ligaments of the back (ligamenta flava) and of the neck (ligamentum nuchae), as well as in large (conducting) blood vessels, where it forms fenestrated sheaths.

4. Reticular tissue

–contains a network of branched reticular fibers in which macrophages are dispersed.

–invests liver sinusoids and smooth muscle cells and forms the stroma of bone marrow and lymphatic organs.

5. Adipose tissue

–is the primary site for storage of energy (in the form of **triglycerides**).

–has a rich neurovascular supply.

a. White adipose tissue

–is composed of **unilocular** adipose cells.

–constitutes almost all the adipose tissue in adults and is found throughout the body.

–**stores and releases lipids** as follows:

(1) Adipose cells synthesize **lipoprotein lipase,** which is transferred to the luminal aspect of the capillary endothelium.

(2) Dietary fat is transported to adipose tissue as **very low density lipoproteins** (VLDLs) and **chylomicrons,** which are hydrolyzed into fatty acids by lipoprotein lipase.

(3) The free fatty acids enter the adipose cells and are stored as triglycerides (in fat droplets) by mechanisms that are not well understood. Adipose cells also synthesize fatty acids from glucose.

(4) Lipid storage is stimulated by **insulin,** which increases the synthesis of lipoprotein lipase and the uptake of glucose by adipose cells.

(5) Stored triglycerides are hydrolyzed by **hormone-sensitive lipase,** which is activated by cAMP. The free fatty acids are then transported to the capillary lumen.

b. Brown adipose tissue

—is composed of **multilocular** adipose cells, which contain numerous large mitochondria.

—is capable of **generating heat by uncoupling oxidative phosphorylation**.

—is found in infants (also in hibernating animals) and is much reduced in adults.

6. Cartilage and bone

—are discussed in Chapter 7.

V. Clinical Considerations

A. Edema

—is a pathologic process resulting in an **increased volume of tissue fluid**.

—may be caused by venous obstruction or decreased venous blood flow (as in congestive heart failure); increased capillary permeability (due to injury); starvation; excessive release of histamine; and obstruction of lymphatic vessels.

—that is responsive to localized pressure (i.e., depressions persist after release of pressure) is called **pitting edema**.

B. Keloids

—are swellings in the skin as a result of increased collagen formation in hyperplastic scar tissue.

—are most prevalent in African Americans.

C. Tumors of adipose tissue

1. Lipomas are common **benign** tumors of adipocytes.

2. Liposarcomas are **malignant** tumors of adipocytes. They may occur anywhere in the body but are seen most often in retroperitoneal tissues and in various areas of the leg.

D. Anaphylactic shock

—results from the effects of powerful mediators released during an **immediate hypersensitivity reaction** following a second exposure to an allergen.

—can occur within seconds or minutes after contact with an allergen.

—is marked by shortness of breath, falling blood pressure, and other symptoms typical of shock.

—may be life threatening if untreated.

E. Obesity

—occurs when energy input (in the form of food) exceeds energy output over a prolonged period.

1. **Hypertrophic obesity**

   —is characterized by an increase in the **size** of adipose cells due to an increase in the amount of fat stored in each cell.

   —usually begins later in life (**adult-onset**).

2. **Hyperplastic obesity**

   —is characterized by an increase in the **number** of adipose cells.

   —begins in childhood and usually is lifelong.

Review Test

Directions: Each of the numbered items or incomplete statements in this section is followed by answers or by completions of the statement. Select the **one** lettered answer or completion that is **best** in each case.

1. Which one of the following statements regarding collagen is FALSE?

(A) It is composed of tropocollagen
(B) Reticular fibers are composed of collagen
(C) It possesses a 64-nm to 67-nm periodicity
(D) Elastic fibers are composed of collagen

2. All of the following cells arise from undifferentiated mesenchymal cells within connective tissue and remain there permanently EXCEPT

(A) pericytes
(B) eosinophils
(C) fibroblasts
(D) osteoblasts
(E) adipocytes

3. Which of the following statements concerning mast cells is FALSE?

(A) They synthesize IgE
(B) They possess deeply stained basophilic granules
(C) They release heparin
(D) They arise from myeloid stem cells

4. Dense regular connective tissue is present in

(A) organ capsules
(B) basement membranes
(C) tendons
(D) skin

5. Which of the following statements about reticular fibers is FALSE?

(A) They are composed of type I collagen
(B) They have thinner fibrils and a higher sugar content than collagen fibers
(C) They are synthesized in a two-stage process involving intracellular and extracellular events
(D) They stain black with silver salts

6. Of the following cell types found in connective tissue, which is most often present along capillaries and resembles fibroblasts?

(A) Plasma cell
(B) Lymphocyte
(C) Macrophage
(D) Pericyte

7. Which one of the following statements about glycosaminoglycans is FALSE?

(A) They always contain hexosamine
(B) They include hyaluronic acid
(C) They are made up of disaccharides
(D) They are a primitive form of collagen
(E) They include chondroitin sulfate

8. Collagen precursors are secreted by

(A) plasma cells
(B) mast cells
(C) fibroblasts
(D) lymphocytes

9. All of the following are glycoproteins EXCEPT

(A) fibronectin
(B) osteonectin
(C) laminin
(D) collagen
(E) tenascin

10. The adipocytes of white adipose tissue contain all of the following components EXCEPT

(A) a flattened nucleus
(B) receptors for insulin
(C) many small fat droplets
(D) a thin rim of cytoplasm
(E) a basal lamina

11. Which one of the following cells arises from monocytes?

(A) Plasma cells
(B) Fibroblasts
(C) Lymphocytes
(D) Macrophages

12. Which one of the following statements regarding proteoglycans is FALSE?

(A) They consist of a core of fibrous protein covalently bound to glycosaminoglycans
(B) They are attached to ribonucleic acid
(C) They are binding sites for signaling molecules
(D) They impart selective permeability to certain regions of the extracellular matrix

13. Foreign-body giant cells are formed by the coalescence of

(A) macrophages
(B) lymphocytes
(C) fibroblasts
(D) adipose cells

14. Which one of the following statements concerning loose connective tissue is FALSE?

(A) It is more abundant than dense connective tissue
(B) It has a higher proportion of cells to fibers than does dense connective tissue
(C) It acts as a medium for exchange of nutrients and wastes between the blood and tissues
(D) It provides structural support for organs and cells
(E) It consists of fibers in which various types of cells are embedded

Directions: Each group of items in this section consists of lettered options followed by a set of numbered items. For each item, select the **one** lettered option that is most closely associated with it. Each lettered option may be selected once, more than once, or not at all.

Questions 15–18

Match each property below with the cell type to which it applies.

(A) Macrophages
(B) Eosinophils
(C) Plasma cells
(D) Mast cells
(E) Adipose cells

15. Arise from activated B lymphocytes

16. Release histamine and leukotriene C

17. Function in the immune response and wound healing

18. Increase in number in hyperplastic obesity

Questions 19 and 20

Match each cell component below with the corresponding letter in the photomicrograph.

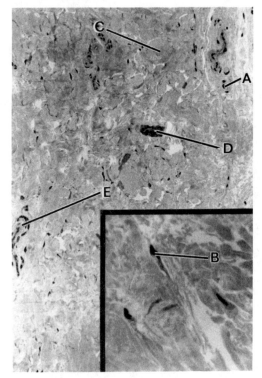

Reprinted from Gartner LP and Hiatt JL: *Atlas of Histology*. Baltimore, Williams & Wilkins, 1987, p 39.

19. Fibroblast

20. Collagen bundle

Answers and Explanations

1–D. Elastic fibers are composed of elastin microfibrils rather than collagen.

2–B. Eosinophils arise from myeloid stem cells during hematopoiesis and migrate to sites of inflammation within the connective tissue.

3–A. All immunoglobulins (antibodies), including IgE, are produced by plasma cells. Mast cells bind IgE antibodies via receptors in their plasma membrane.

4–C. Tendons are composed of dense regular connective tissue containing collagen fibers arranged in a uniform parallel fashion.

5–A. Reticular fibers are composed primarily of type III collagen. They contain thinner fibrils and have a higher sugar content than collagen fibers, which are formed of type I collagen.

6–D. Pericytes are pluripotential cells that resemble fibroblasts (although they are smaller) and are located adjacent to capillaries.

7–D. Glycosaminoglycans are long, unbranched polysaccharides composed of repeating disaccharide units. Although they often are linked to a core protein, they are unrelated to collagen, which is composed of tropocollagen fibrils.

8–C. Collagen precursors (procollagen) are secreted by fibroblasts, which also secrete other components of the connective tissue matrix.

9–D. Collagen, the most abundant protein of the extracellular matrix, is not a glycoprotein.

10–C. White adipose tissue is composed of unilocular adipocytes; these cells contain a single large fat droplet, which squeezes the cytoplasm and nucleus to the periphery. Brown adipose tissue is composed of multilocular adipocytes, which contain many small fat droplets.

11–D. Monocytes leave the bloodstream and migrate into the connective tissue where they mature into functional macrophages.

12–B. Proteoglycans are not attached to ribonucleic acid.

13–A. Foreign-body giant cells result when macrophages coalesce.

14–E. Both loose and dense connective tissue are composed of three elements: an amorphous ground substance, fibers, and various types of cells. The ground substance (a complex mixture of glycosaminoglycans, proteoglycans, and glycoproteins) plus fibers (most commonly collagen fibers) constitute the extracellular matrix.

15–C. Plasma cells arise from activated B lymphocytes.

16–D. Mast cells release histamine, leukotriene C, and other mediators.

17–A. Macrophages function as phagocytes and also secrete substances that assist in the immune response and in wound healing.

18–E. The number of adipose cells increases in hyperplastic obesity.

19–B. Fibroblast.

20–C. Collagen bundle. The photomicrograph is of dense irregular collagenous connective tissue.

7

Cartilage and Bone

I. Cartilage

—is a specialized type of **fibrous connective tissue**.

—has a **firm extracellular matrix** that is less pliable than that of connective tissue proper.

—contains characteristic cells (**chondrocytes**) embedded in the matrix.

—is **avascular**.

—**functions** primarily to support soft tissues and in the development and growth of long bones.

—is classified into three types—**hyaline cartilage, elastic cartilage,** and **fibrocartilage**—that vary in certain matrix components.

A. Hyaline cartilage

—is located in adults at the articular ends of long bones, in the walls of larger respiratory passages (nose, larynx, trachea, and bronchi), and at the ventral ends of ribs.

—serves as a temporary skeleton in the fetus until it is replaced with bone.

1. Matrix of hyaline cartilage

—is composed of an amorphous ground substance, containing **proteoglycan aggregates** and **chondronectin** (a glycoprotein), in which is embedded **type II collagen** (see Table 4.2).

—is about 40% collagen (dry weight).

—that is adjacent to chondrocytes—called the **territorial (capsular) matrix**—is poor in collagen but rich in glycosaminoglycans. This part of the matrix stains deeply basophilic and metachromatic; it also reacts more intensely with periodic acid–Schiff (PAS) stain than does the **interterritorial (intercapsular) matrix**.

2. Perichondrium

—is a layer of dense irregular connective tissue that surrounds hyaline cartilage except at articular surfaces.

—consists of an **outer fibrous layer,** containing **type I collagen,** fibroblasts, and blood vessels, and an **inner cellular layer,** containing chondrogenic cells.

–**provides the blood supply** for the avascular cartilagenous tissue.

3. **Chondrocytes**
   –are mature cartilage cells, which are embedded in **lacunae** in the matrix.
   –arise by differentiation of mesenchymal **chondrogenic cells** into **chondroblasts,** which are the earliest cells to produce cartilage matrix. Once these cells become totally enveloped by matrix, they are referred to as **chondrocytes** (see Figure 6.1A).
   –manufacture the cartilage matrix through which nutrients and waste materials pass to and from the cells.
   –contain an extensive Golgi complex, abundant rough endoplasmic reticulum (RER), lipid droplets, and glycogen.
   –located **superficially** are ovoid and positioned with their longitudinal axis parallel to the cartilage surface.
   –located **deeper** are more spherical in shape and may occur in groups of four to eight cells (**isogenous groups**).

4. **Histogenesis of hyaline cartilage**
   –is **stimulated** by thyroxine, testosterone, and growth hormone.
   –is **inhibited** by cortisone, hydrocortisone, and estradiol.
   –is similar to that of elastic cartilage and fibrocartilage.
   –occurs by the following two processes:

   a. **Interstitial growth**
      –results from cell **division of preexisting chondrocytes**.
      –occurs only during early stages of cartilage formation and in articular cartilage and the epiphyseal plates of long bones.

   b. **Appositional growth**
      –results from differentiation of chondrogenic cells in the perichondrium to form new chondrocytes, which lay down a new layer of cartilage matrix at the periphery.

5. **Degeneration of hyaline cartilage**
   –occurs when chondrocytes hypertrophy and die and the matrix becomes calcified, a process that increases with age.
   –is a normal part of endochondral bone formation.

6. **Regeneration of hyaline cartilage**
   –is very poor, except in young children.
   –results from activity of the perichondrium.

B. **Elastic cartilage**
   –possesses a perichondrium and is identical to hyaline cartilage except for its network of elastic fibers, which impart a **yellowish** color to it.
   –is less prone to degeneration than hyaline cartilage.
   –is located where **flexible support** is required (e.g., in the pinna of the ear, auditory tube, and epiglottis).

C. **Fibrocartilage**
   –**lacks** an identifiable perichondrium.
   –has properties between those of dense connective tissue and hyaline cartilage.

  −is characterized by alternating rows of fibroblast-derived chondrocytes and thick bundles of **type I collagen** fibers.

  −is always found in conjunction with hyaline cartilage, other fibrous tissues, or both.

  −is located where **support and tensile strength** are required (e.g., in the intervertebral disks, articular disks, and pubic symphysis, and at the insertions of some tendons and ligaments).

II. Bone

−is a specialized type of connective tissue with a **calcified** extracellular matrix.

−contains characteristic cells embedded in the matrix.

−is the primary constituent of the adult skeleton.

−**functions** to protect vital organs, support fleshy structures, and provide a calcium reserve, containing about 99% of the body's calcium.

−is a dynamic tissue that constantly undergoes changes in shape. **Applied pressure results in bone resorption,** whereas **applied tension results in bone formation**.

A. Bone matrix

1. Inorganic (calcified) portion of the bone matrix

−is composed of calcium, phosphate, bicarbonate, citrate, magnesium, potassium, and sodium.

−represents 50% of the dry weight of bone.

−consists primarily of **hydroxyapatite crystals,** which have the composition $Ca_{10}(PO_4)_6(OH)_2$.

2. Organic portion of the bone matrix

−consists primarily of **type I collagen** (95%).

−has a ground substance that contains **chondroitin sulfate** and **keratan sulfate**.

B. Periosteum

−is a layer of **noncalcified** connective tissue covering bone on its **external** surfaces, except at synovial articulations.

−is composed of an outer, fibrous dense collagenous layer and an inner, cellular (osteogenic) layer.

−contains **Sharpey's fibers** (type I collagen), which attach the periosteum to the bone.

−functions to **distribute blood vessels to bone**.

C. Endosteum

−is a thin specialized connective tissue that **lines the marrow cavities**.

−supplies **osteoprogenitor cells** and **osteoblasts** for bone growth and repair.

D. Bone cells

1. Osteoprogenitor cells

−are spindle-shaped cells, derived from embryonic mesenchyme, that are **located in the periosteum and endosteum**.

−are capable of differentiating into osteoblasts.

−may change into chondrogenic cells **at low oxygen tensions**.

2. Osteoblasts

–are derived from osteoprogenitor cells.

–synthesize and secrete **osteoid** (uncalcified bone matrix).

–on bony surfaces may resemble a layer of cuboidal, basophilic cells as they secrete organic matrix (see Figure 6.1A).

–possess cytoplasmic processes with which they contact the processes of other osteoblasts and osteocytes.

–have a well-developed RER and Golgi complex when **synthetically active**.

–become entrapped in **lacunae** but maintain contact with other cells via their cytoplasmic processes; once this happens, the cells are known as osteocytes.

3. Osteocytes

–are **mature bone cells** housed in their own lacunae.

–have narrow cytoplasmic processes extending through **canaliculi** in the calcified matrix (see Figure 6.1A).

–maintain communication with each other via **gap junctions** between their processes.

–are nourished and maintained by nutrients and metabolites within canaliculi.

–contain abundant heterochromatin, a paucity of RER, and a small Golgi complex.

4. Osteoclasts

–are large, motile, multinucleated cells (up to 50 nuclei) that are derived from **fusion of monocytes**.

–usually have an **acidophilic** cytoplasm.

–function in **osteolysis** (resorption of bone).

–are located in **Howship's lacunae** in areas of bone resorption.

a. Morphology—osteoclasts

–display four regions (ruffled border, clear zone, vesicular region, and basal region) in electron micrographs.

(1) Ruffled border

–is composed of irregular finger-like cytoplasmic projections extending into Howship's lacunae.

–is the site of active **bone resorption**.

(2) Clear zone

–is the region of cytoplasm that surrounds the ruffled border.

–contains **microfilaments,** which help osteoclasts to maintain contact with the bony surface and to isolate the region of osteolytic activity.

b. Bone resorption

–involves the following events:

(1) Osteoclasts secrete **acid,** thus creating an acidic environment that decalcifies the surface layer of bone.

(2) Acid hydrolases, collagenases, and other proteolytic enzymes secreted by osteoclasts then **degrade the organic portion** of the bone.

(3) Osteoclasts resorb the organic and inorganic residues of the bone matrix.

E. Classification of bone

—is based on both gross and microscopic properties.

1. Gross observation

—of cross-sections of bone reveals two types:

a. Spongy (cancellous) bone is composed of interconnected trabeculae, which surround cavities filled with bone marrow. The trabeculae contain osteocytes and are lined with a single layer of osteoblasts.

b. Compact bone has no trabeculae or bone marrow cavities.

2. Microscopic observation

—of bone reveals two types:

a. Primary bone

—is also known as **immature** or **woven** bone.

—contains many osteocytes and large, irregularly arranged collagen bundles.

—has a **low mineral content.**

—is the first compact bone produced during fetal development and bone repair.

—**is remodeled and replaced by secondary bone** except in a few places (e.g., tooth sockets, near suture lines in skull bones, and at insertion sites of tendons).

b. Secondary bone

—is also known as **mature** or **lamellar** bone.

—is the compact bone of adults.

—has a calcified matrix arranged in regular concentric layers, or **lamellae,** which are 3–7 μm thick and surround haversian canals. Each complex comprising lamellae and a haversian canal is called a **haversian system,** or **osteon.**

—often contains an amorphous **cementing substance** between adjacent haversian systems.

—contains osteocytes in lacunae, which are located between, and occasionally within, lamellae.

F. Organization of lamellae

—in compact bone (e.g., diaphysis of long bones) is characteristic and consists of four elements.

1. Haversian systems (osteons)

—are long cylinders that run more-or-less parallel to the long axis of the diaphysis.

—are composed of 4–20 lamellae surrounding a central haversian canal, which contains blood vessels, nerves, and loose connective tissue.

—are interconnected by **Volkmann's canals,** which also connect to the periosteum and endosteum and **carry the neurovascular supply.**

2. Interstitial lamellae

—are irregularly shaped lamellae located between haversian systems.

–are remnants of remodeled haversian systems.

3. Outer and inner circumferential lamellae

–are located at the external and internal surfaces of the diaphysis, respectively.

G. Histogenesis of bone

–occurs by two processes—**intramembranous** and **endochondral bone formation**. Both processes produce bone that appears histologically identical.

–is accompanied by bone resorption. The combination of bone formation and resorption—termed **remodeling**—occurs throughout life, although it is slower in secondary than in primary bone.

1. Intramembranous bone formation

–is the process by which most of the **flat bones** (e.g., skull bones, mandible, and maxilla) are formed.

–involves the following sequence of events:

a. Mesenchymal cells, condensed into **primary ossification centers,** differentiate into osteoblasts, which begin secreting **osteoid**.

b. As calcification occurs, osteoblasts are trapped in their own matrix, becoming osteocytes. These centers of developing bone are called **trabeculae** (fused spicules).

c. Fusion of the bony trabeculae produces **spongy bone** as blood vessels invade the area and other undifferentiated mesenchymal cells give rise to the bone marrow.

d. The periosteum and endosteum develop from portions of the mesenchymal layer that do not undergo ossification.

2. Endochondral bone formation

–is the process by which **long bones** are formed.

–begins in a piece of **hyaline cartilage** that serves as a small **model** for the bone.

–occurs in two stages involving development of primary and secondary centers of ossification.

a. Primary center of ossification

–develops at the **midriff of the diaphysis** of the cartilaginous model by the following sequence of events:

(1) Vascularization of the perichondrium at this site causes the transformation of chondrogenic cells to osteoprogenitor cells, which differentiate into osteoblasts. The perichondrium is now called the periosteum.

(2) Osteoblasts elaborate the **subperiosteal bone collar,** via intramembranous bone formation, deep to the periosteum.

(3) Chondrocytes within the core of the cartilaginous model **hypertrophy** and begin to degenerate, and the lacunae become confluent, forming spaces.

(4) Osteoclasts create perforations in the bone collar, permitting the **periosteal bud** (blood vessels, osteoprogenitor cells, and mesenchymal cells) to enter the newly formed spaces in the cartilaginous model. Cartilage in the walls of these spaces then becomes calcified.

(5) Bone matrix is elaborated and calcified on the surface of the calcified cartilage, forming a **calcified cartilage–calcified bone complex.** In histologic sections the calcified cartilage stains **basophilic,** whereas the calcified bone stains **acidophilic.**

(6) The subperiosteal bone collar becomes thicker and elongates toward the epiphyses.

(7) Osteoclasts begin to resorb the calcified cartilage–calcified bone complex, thus enlarging the primitive marrow cavity.

(8) Repetition of this sequence of events results in bone formation spreading toward the epiphyses.

b. Secondary centers of ossification

—develop at the **epiphyses** in a sequence of events similar to those described above for the primary center.

—begin when osteoprogenitor cells invade the epiphyses and differentiate into osteoblasts, which elaborate bone matrix to replace the disintegrating cartilage. When the epiphyses are filled with bone tissue, cartilage remains in two areas, the articular surfaces and the epiphyseal plates.

(1) Articular cartilage persists and does not contribute to bone formation.

(2) Epiphyseal plates continue to grow by adding new cartilage at the epiphyseal end while it is being replaced at the diaphyseal end.

(3) Diaphyseal and epiphyseal bone become continuous (thus connecting the two epiphyseal marrow cavities) at about 20 years.

3. Zones of epiphyseal plates

—are histologically distinctive and arranged in the following order:

a. Zone of reserve is located at the epiphyseal side of the plate and possesses small, randomly arranged, inactive chondrocytes.

b. Zone of proliferation is a region of rapid mitotic divisions, giving rise to rows of isogenous cell groups.

c. Zone of cell hypertrophy and maturation is the region where the chondrocytes are greatly enlarged.

d. Zone of calcification is the region where remnants of cartilage matrices become calcified and chondrocytes die.

e. Zone of ossification is where bone is elaborated upon the calcified cartilage, followed by the resorption of the calcified bone–calcified cartilage complex.

4. Calcification

—begins with the deposition of calcium phosphate on collagen fibrils.

—is stimulated by certain proteoglycans and **osteonectin,** a Ca^{2+}-binding glycoprotein.

—occurs by a mechanism that is not well understood.

100 / *Cell Biology and Histology*

H. Bone repair—a bone fracture

—damages the matrix, bone cells, and blood vessels in the region.

—is accompanied by localized hemorrhaging and blood-clot formation.

1. Proliferation of osteoprogenitor cells

—occurs in the periosteum and endosteum surrounding the fracture.

—results in a cellular tissue that surrounds the fracture and penetrates between the ends of the damaged bone.

2. Formation of a bony callus

—occurs both internally and externally at a fracture site.

a. Fibrous connective tissue and hyaline cartilage are formed in the fracture zone.

b. Endochondral bone formation replaces the cartilage with primary bone.

c. Intramembranous bone formation also produces primary bone in the area.

d. The irregularly arranged trabeculae of primary bone join the ends of the fractured bone, forming a **bony callus**.

e. The primary bone is resorbed and replaced with secondary bone as the fracture heals.

I. Role of vitamins in bone formation

1. Vitamin D

—is necessary for **absorption of calcium** from the small intestine. Vitamin D deficiency results in poorly calcified (soft) bone, a condition known as **rickets** in children and **osteomalacia** in adults.

—is also necessary for **bone formation** (ossification).

—excess causes bone resorption.

2. Vitamin A

—deficiency inhibits proper bone formation and growth, while an excess accelerates ossification of the epiphyseal plates.

—deficiency or excess results in **small stature**.

3. Vitamin C

—is necessary for **collagen formation**.

—deficiency results in **scurvy,** characterized by poor bone growth and inadequate fracture repair.

J. Role of hormones in bone formation

1. Parathyroid hormone

—activates osteoclasts to **resorb bone,** thus **elevating blood calcium levels**.

—excess (hyperparathyroidism) renders bone **more susceptible to fracture** and subsequent deposition of calcium in arterial walls and certain organs such as the kidney.

2. Calcitonin

—is produced by parafollicular cells of the thyroid gland.

—eliminates the ruffled border of osteoclasts.

—**inhibits bone–matrix resorption,** thus preventing the release of calcium.

3. Pituitary growth hormone

—is produced in the anterior lobe of the pituitary gland.

—stimulates overall growth, especially that of epiphyseal plates.

—excess during growing years causes **pituitary giantism** and in adults causes **acromegaly**.

—deficiency during growing years causes **dwarfism**.

III. Joints

A. Synarthroses

—are **immoveable joints** composed of connective tissue, cartilage, or bone.

—unite the first rib to the sternum and connect the skull bones.

B. Diarthroses

—are also called **synovial joints**.

—permit **maximum movement**.

—generally unite long bones.

—are surrounded by a two-layered **capsule,** enclosing and sealing the articular cavity, which contains **synovial fluid**—a colorless, viscous fluid rich in hyaluronic acid and proteins.

1. External (fibrous) capsular layer

—is a tough, fibrous layer of dense connective tissue.

2. Internal (synovial) capsular layer

—is also called the **synovial membrane**.

—is lined by a layer of squamous to cuboidal epithelial cells on its internal surface.

    a. Type A cells are intensely phagocytic and have a well-developed Golgi complex, many lysosomes, and sparse RER.

    b. Type B cells resemble fibroblasts and have a well-developed RER; these cells probably secrete **synovial fluid**.

IV. Clinical Considerations

A. Osteoporosis

—is a **decrease in bone mass** associated with a normal ratio of mineral to matrix.

—results from decreased bone formation, increased bone resorption, or both.

—commonly occurs in old age because of diminished secretion of growth hormone; in immobile patients because of lack of physical stress on the bone; and in postmenopausal women because of diminished estrogen secretion.

B. Osteomalacia

—results from calcium deficiency in **adults**.

—is characterized by **deficient calcification** of newly formed bone and **decalcification** of already calcified bone.

—may be severe during pregnancy since the calcium requirements of the fetus may cause calcium losses from the mother.

C. Rickets

—results from calcium deficiency in **children**.

—is characterized by **deficient calcification** in newly formed bone.

—generally is accompanied by deformation of the bone spicules in epiphyseal plates so that bones grow more slowly than normal and are deformed.

D. Acromegaly

—results from an **excess of pituitary growth hormone** in adults.

—is characterized by **very thick bones** in the extremities and in portions of the facial skeleton.

Review Test

Directions: Each of the numbered items or incomplete statements in this section is followed by answers or by completions of the statement. Select the **one** lettered answer or completion that is **best** in each case.

1. Which of the following statements regarding osteoclasts is FALSE?

(A) They are multinucleated cells
(B) They produce proteolytic enzymes
(C) They are found occupying Howship's lacunae
(D) They are derived from osteoprogenitor cells

2. Which one of the following statements regarding the periosteum is FALSE?

(A) It facilitates the nutrition of new bone
(B) It produces osteoblasts
(C) It is responsible for appositional bone growth
(D) Its outer layer contains osteoprogenitor cells

3. Which one of the following statements regarding osteocytes is FALSE?

(A) They communicate with each other via haversian canals
(B) They live for a long time
(C) They possess long cytoplasmic processes
(D) They are housed in their own lacunae

4. Which one of the following statements regarding hyaline cartilage is TRUE?

(A) It is a vascular structure
(B) It contains type IV collagen
(C) It grows by appositional growth only
(D) It is located at the articular ends of long bones

5. A 7-year-old boy with fragile long bones may have a diet deficient in

(A) potassium
(B) calcium
(C) iron
(D) carbohydrates

6. Which of the following statements regarding bone is FALSE?

(A) Bone matrix contains type I collagen
(B) Sharpey's fibers attach the periosteum to bone
(C) Haversian canals are interconnected via Volkmann's canals
(D) Bone growth occurs via interstitial growth only

7. Which of the following statements regarding periosteal buds is FALSE?

(A) They invade the cartilage model in endochondral bone formation
(B) They carry blood vessels and osteoprogenitor cells
(C) They are involved in elaborating primary bone
(D) They are involved in elaborating secondary bone

8. Which of the following statements concerning synovial membranes is FALSE?

(A) They secrete hyaluronic acid
(B) They contain two cell types on the internal surface
(C) They possess an outer fibrous layer
(D) They are found in synarthroses
(E) They are found in diarthroses

9. Acromegaly, a disease characterized by very thick bones in the extremities and face, results from

(A) hypervitaminosis A
(B) excessive growth hormone
(C) hypovitaminosis A
(D) hypervitaminosis D

10. Which of the following statements regarding cartilage is FALSE?

(A) Its growth rate is increased by testosterone and thyroxine
(B) Its growth rate is decreased by cortisone and estradiol
(C) Fibrocartilage is the cartilage model in endochondral bone formation
(D) The matrix of elastic cartilage is identical to that of hyaline cartilage except that it contains elastic fibers

11. Endochondral bone formation involves all of the following events EXCEPT

(A) transformation of chondrogenic cells into osteoprogenitor cells
(B) condensation of mesenchymal cells
(C) resorption of a calcified cartilage–calcified bone complex
(D) formation of cartilage

Directions: Each group of items in this section consists of lettered options followed by a set of numbered items. For each item, select the **one** lettered option that is most closely associated with it. Each lettered option may be selected once, more than once, or not at all.

Questions 12–15

Match each physiologic effect with the vitamin or hormone causing it.

(A) Vitamin A
(B) Vitamin C
(C) Vitamin D
(D) Calcitonin
(E) Parathyroid hormone

12. Stimulates the mobilization of calcium ions from bone

13. Promotes absorption of calcium from the small intestine

14. Inhibits bone resorption

15. Acts as an enzyme cofactor in collagen synthesis

Questions 16–20

Match each area below with the corresponding letter in the photomicrographs.

16. Subperiosteal bone collar

17. Zone of proliferation

18. Calcified cartilage–calcified bone complex

19. Zone of calcifying cartilage

20. Zone of maturation and hypertrophy

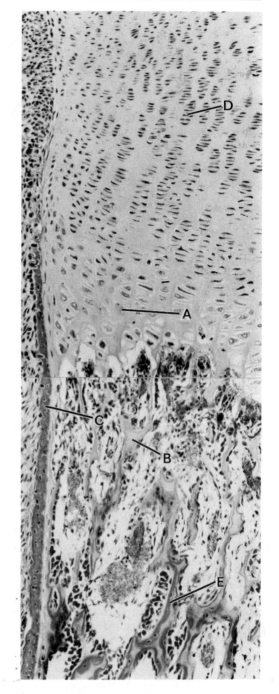

Reprinted from Gartner LP and Hiatt JL: *Atlas of Histology.* Baltimore, Williams & Wilkins, 1987, p 61.

Answers and Explanations

1–D. Osteoclasts are multinucleated cells that produce proteolytic enzymes and occupy Howship's lacunae. They are derived from monocytes.

2–D. Only the inner layer of the periosteum possesses osteoprogenitor cells.

3–A. Osteocytes communicate with each other via narrow cytoplasmic processes that extend through canaliculi.

4–D. Hyaline cartilage is avascular, contains type II collagen, and grows both interstitially and appositionally.

5–B. Because calcium is maintained at a constant level in the blood and the tissues, a diet deficient in calcium leads to calcium loss from the bones. As a result, the bones become fragile.

6–D. Both intramembranous and endochondral bone formation involve appositional growth only.

7–D. Periosteal buds carry blood vessels and osteogenic cells into the cartilage model in endochondral bone formation. These buds are then responsible for giving rise to cells that produce primary bone, which is later remodeled and replaced by secondary bone.

8–D. Synovial membranes have a squamous to cuboidal epithelium on their inner surface composed of type A and type B cells; the latter secrete synovial fluid, which is rich in hyaluronic acid. Synovial membranes surround the articular surface of diarthroses (moveable joints).

9–B. Excessive growth hormone causes acromegaly. Excessive vitamin D causes bone resorption. Both an excess and deficiency of vitamin A results in short stature.

10–C. Hyaline cartilage is the cartilage model associated with endochondral bone formation.

11–B. Intramembranous bone formation begins with condensation of mesenchymal cells, which differentiate into osteoblasts. During endochondral bone formation, new cartilage is formed at the epiphyseal end and replaced by bone at the diaphyseal end of the epiphyseal plates.

12–E. Parathyroid hormone stimulates osteoclasts to resorb bone, thus releasing calcium ions.

13–C. Vitamin D is necessary for absorption of calcium from the small intestine.

14–D. Calcitonin inhibits bone resorption and thus has the opposite effect of parathyroid hormone.

15–B. Vitamin C is a necessary cofactor for enzymes that catalyze hydroxylation of specific amino acid residues during formation of collagen.

16–C. Subperiosteal bone collar.

17–D. Zone of proliferation.

18–E. Calcified cartilage–calcified bone complex.

19–B. Zone of calcifying cartilage.

20–A. Zone of maturation and hypertrophy.

8

Muscle

I. Overview—Muscle

−is classified by its morphology and function into three types: **striated, cardiac,** and **smooth** muscle.

−is formed of differentiated cells that possess **contractile filaments** containing **actin** and **myosin**.

−contraction may be **voluntary** (skeletal muscles) or **involuntary** (cardiac and smooth muscles).

II. Structure of Skeletal Muscle

A. Connective tissue investments

−connect skeletal muscle to bone and other tissues.

−include **aponeuroses** and **tendons**.

−convey neural and vascular elements to individual muscle fibers.

1. Epimysium surrounds an entire muscle.

2. Perimysium surrounds **fascicles** (small bundles) of muscle cells.

3. Endomysium surrounds individual muscle cells and is composed of **reticular fibers** and an **external lamina**.*

B. Types of skeletal muscle fibers

−are referred to as **red** (slow), **white** (fast), and **intermediate**. All three types often are present in a given muscle.

−vary in their content of **myoglobin** (which, like hemoglobin, can bind oxygen); in the **number of mitochondria** they contain; in their **concentration of various enzymes;** and in their **rate of contraction** (Table 8.1)

−are modulated by **innervation**. If a red fiber is denervated and its innervation is replaced with a nerve from a white fiber, the red fiber will become a white fiber.

---

*The term **external lamina** is synonymous with the term **basal lamina** but is used in reference to nonepithelial cells.

Table 8.1. Characteristics of Red and White Muscle Fibers*

Type	Myoglobin Content	Number of Mitochondria	Enzyme Content	Contraction
Red (slow)	High	Many	High in oxidative enzymes; low in ATPase	Slow but repetitive; not easily fatigued
White (fast)	Low	Few	Low in oxidative enzymes; high in ATPase and phosphorylases	Fast and easily fatigued

*Intermediate fibers have characteristics between those of red and white fibers.

C. Skeletal muscle cells (Figure 8.1)

- are long, cylindrical, **multinucleated** cells (also called muscle fibers), which are enveloped by an external lamina and reticular fibers. The cytoplasm is referred to as **sarcoplasm**.
- are enveloped by the **sarcolemma** (plasma membrane), which forms deep tubular invaginations, called **T (transverse) tubules,** that extend into the cells.
- consist, in large part, of collections of **myofibrils,** cylindrical units 1–2 μm in diameter that extend the entire length of the cell. The precise alignment of myofibrils results in a characteristic banding pattern, which is visible by light microscopy as alternating dark **A bands** and light **I bands;** the latter are bisected by dark **Z disks** (see Figure 8.1*D*).

1. Sarcomere

- is the regular repeating region between successive Z disks.
- constitutes the **functional unit of contraction** in skeletal muscle.

2. Myofibrils

- consist of longitudinally arranged, cylindrical bundles of **thick** and **thin myofilaments,** which are observable by electron microscopy (see Figure 8.1*E*).
- are held in alignment by the intermediate filaments **desmin** and **vimentin,** which tether the periphery of Z disks in adjacent myofibrils to one another.

3. Sarcoplasmic reticulum (SR)

- is a modified smooth endoplasmic reticulum (SER) that surrounds myofilaments and forms a meshwork around each myofibril.
- forms a pair of dilated **terminal cisternae,** which encircle the myofibrils at the junction of each A and I band.
- functions in the **regulation of muscle contraction** by sequestering calcium ions (leading to relaxation) or releasing calcium ions (leading to contraction).

4. Triads

- are specialized complexes consisting of a narrow central T tubule flanked on each side by terminal cisternae of the sarcoplasmic reticulum.
- are located at the A–I junction in mammalian skeletal muscle cells.
- function to **help provide uniform contraction** throughout muscle fibers.

SKELETAL MUSCLE

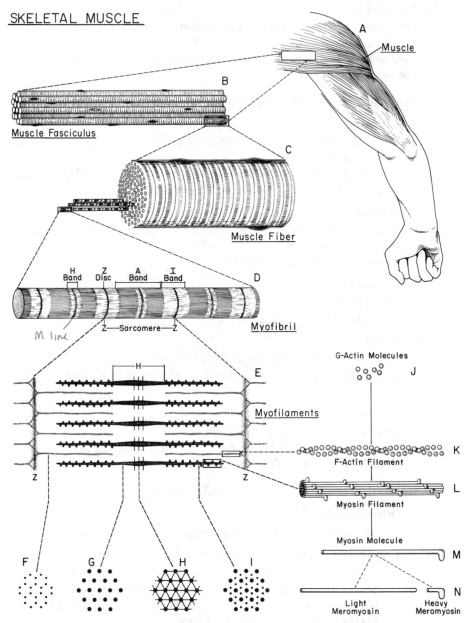

Figure 8.1. Diagram of skeletal muscle and its components as observed by light and electron microscopy. (Adapted from Fawcett DW: *Bloom and Fawcett's Textbook of Histology,* 11th ed. Philadelphia, WB Saunders Company, 1986, p 282.)

5. Myofilaments

- —are the basic structural components of myofibrils and are visible in the electron microscope.
- —include **thick** filaments (15 nm in diameter and 1.5 μm long) and **thin** filaments (7 nm in diameter and 1.0 μm long).
- —lie parallel to the long axis of the myofibril in a precise arrangement that is responsible for the sarcomere banding pattern.

D. Satellite cells

 −are **regenerative cells** that lie within the external lamina surrounding skeletal muscle cells.
 −are able to differentiate, fuse with one another, and form skeletal muscle cells when the need arises.

E. Skeletal muscle cross-striations (see Figure 8.1*D* and *E*)

1. A bands

 −are anisotropic with polarized light but stain darkly with ordinary histologic stains.
 −contain **both thin and thick filaments,** which overlap and interdigitate. Six thin filaments surround each thick filament (see Figure 8.1*I*).

2. I bands

 −are isotropic with polarized light and appear lightly stained in routine histologic preparations.
 −contain only **thin filaments**.

3. H bands

 −are light regions that transect A bands.
 −consist of **thick filaments** only.

4. M lines

 −are narrow, dark regions at the center of H bands.
 −are formed by cross-connections at the centers of adjacent thick filaments.

5. Z disks (lines)

 −are dense regions bisecting each I band.
 −contain **α-actinin,** which binds to thin filaments and anchors them to the Z disks with the assistance of **nebulin**.

F. Molecular organization of myofilaments

1. Thin filaments

 −are composed of F actin, tropomyosin, troponin, and associated proteins.

a. F actin (see Figure 8.1*J* and *K*)

 −is a double-helical polymer of G-actin monomers. Each monomer possesses an **active site** that can interact with myosin.
 −is present as filaments (with a diameter of 5 μm), which exhibit **polarity,** having a (+) and (−) end.

b. Tropomyosin

 −is a linear molecule 41 nm in length.
 −molecules bind head-to-tail, forming filaments that are located in the grooves of the F-actin helix.

c. Troponin

 −is associated with each tropomyosin molecule and is composed of three polypeptides:
 (1) Troponin T (TnT) forms the tail of the molecule and functions in binding the complex to tropomyosin.

(2) Troponin C (TnC) forms part of the globular head and possesses four **binding sites for calcium**. It appears to be related to a specialized form of **calmodulin**.

(3) Troponin I (TnI) binds to actin, inhibiting interaction of myosin and actin.

2. Thick filaments

−each contain 200–300 myosin molecules arranged in an antiparallel fashion, and three associated proteins—myomesin, titin, and C protein.

a. Myosin (see Figure 8.1*L* and *M*)

−is composed of two identical heavy chains and two pairs of light chains.

−molecule resembles a "double-headed golf club."

(1) Myosin heavy chains consist of a long rod-like **"tail"** and a globular **"head."** The tails of the heavy chains are wound around each other in an α-helical configuration.

(a) The tails function in self-assembly of myosin molecules into bipolar thick filaments.

(b) The heads contain actin-binding sites and function in contraction.

(2) Myosin light chains are of two types; one molecule of each type is associated with the globular head of each heavy chain.

(3) Digestion of myosin

(a) The enzyme **trypsin** cleaves myosin into **light meromyosin,** which contains part of the tail portion, and **heavy meromyosin,** which contains the two heads and the remainder of the tail (**S2 segment**).

(b) Treatment of heavy meromyosin with **papain** digests the S2 segment, releasing the two globular heads (**S1 fragments**). Isolated S1 fragments have **ATPase** activity, but interaction with actin is required for release of the noncovalently bound ADP and P_i (see below).

b. Myomesin

−is a protein located at the M line that is believed to function in **cross-linking adjacent thick filaments** to one another so as to maintain their spacing and arrangements.

c. Titin

−is a large linear protein that displays axial periodicity.

−forms an elastic lattice that parallels the thick and thin filaments and is believed to **anchor thick filaments to Z disks**.

d. C protein

−binds to thick filaments at the M lines.

III. Contraction of Skeletal Muscle

A. Sliding-filament model (Table 8.2)

−describes the movement of thick (myosin) and thin (actin) filaments during contraction of skeletal muscle.

Table 8.2. Effect of Contraction on Skeletal Muscle Cross-bands

Bands	Myofilament Component	Change in Bands During Contraction
I	Thin only	Shorten
H	Thick only	Shorten
A	Thick and thin	No change in length
Z disks	Thin only (attached by α-actinin)	Move closer to each other

1. During contraction, thick and thin **filaments do not shorten but increase their overlap**.

2. Thin filaments slide past thick filaments, penetrating more deeply into the A band, which remains constant in length.

3. I bands and H bands decrease in length as Z disks are drawn closer together.

B. Initiation and regulation of contraction

1. **Depolarization and Ca^{2+} release**

–triggers binding of actin and myosin and thus **stimulates muscle contraction**.

a. The sarcolemma is **depolarized** at the **myoneural junction**.

b. **T tubules** convey the wave of depolarization to the myofibrils.

c. **Ca^{2+} ions** are released into the cytosol at the A–I junctions via **voltage-gated Ca^{2+}-release channels (junctional feet)** in the SR terminal cisternae. As long as the Ca^{2+} level is sufficiently high, the contraction cycle will continue.

2. **Activation of actin by Ca^{2+}**

a. In the **resting state,** the myosin-binding sites on thin (actin) filaments are partially covered by tropomyosin. Troponin I also is bound to actin and hinders myosin–actin interaction.

b. Binding of Ca^{2+} by troponin C results in a **conformational change** that breaks the TnI–actin bond and causes tropomyosin to shift its position slightly, thus unmasking the myosin-binding sites (the **active state**).

C. Contraction cycle

–involves the following steps:

1. **ATP binds to myosin head** (S1 fragment), which causes dissociation of the myosin head from actin.

2. The **bound ATP is hydrolyzed** to ADP and P_i, which remain attached to the myosin head. The head pivots and becomes aligned with another actin monomer.

3. Bound P_i is released, and a **cross-bridge forms between the myosin head and actin**. This step is dependent on Ca^{2+} ions.

4. Bound ADP is released, and the **myosin head undergoes a conformational change** that results in a pulling of the actin (thin) filament toward the center of the sarcomere.

5. This sequence of steps is repeated many times, leading to **extensive overlay of the thick and thin filaments** and shortening of the whole fiber.

D. Relaxation

—occurs when the Ca^{2+} concentration in the cytosol is reduced enough that TnC loses its bound Ca^{2+}. As a result, tropomyosin returns to its resting position, masking actin's binding sites and restoring the resting state.

—depends on a Ca^{2+} **pump** in the sarcoplasmic reticulum, which transports Ca^{2+} ions from the cytosol to the inner surface of the SR membrane where they are bound by **calsequestrin.**

E. Energy source for skeletal muscle contraction

—is **ATP,** which is formed from ADP by several processes.

1. In **white fibers,** ATP is generated by anaerobic **glycolysis,** resulting in a build-up of lactic acid.

2. In **red fibers,** ATP is generated by **oxidative phosphorylation.**

3. ATP also is formed by transfer of the high-energy phosphate in creatine phosphate to ADP in a reaction catalyzed by **phosphocreatine kinase.**

F. Motor unit

—consists of a neuron and all of the muscle cells it innervates. A **muscle** may contract with varying degrees of strength, but a **muscle cell obeys the "all or none law"** (i.e., it either contracts or does not contract).

—has the following characteristics:

1. All muscle cells of a single motor unit **contract in unison.**

2. The force generated by the muscle as a whole is determined by the **number** of motor units undergoing simultaneous contraction.

3. Muscles requiring **fine coordination** (e.g., external muscles of the eye) are composed of many small motor units.

4. Muscles responsible for **coarse movement** (e.g., levator scapulae muscle) are composed of a small number of very large motor units.

IV. Innervation of Skeletal Muscle

—consists of **motor** nerve endings (myoneural junctions) and two types of **sensory** nerve endings (muscle spindles and Golgi tendon organs). Both types of sensory nerve endings function in **proprioception.**

A. Myoneural junction

—is also called the **motor end-plate** or **neuromuscular junction.**

—is a specialized region (**synapse**) where a motor nerve terminates on a skeletal muscle cell.

1. Structural components

a. Axon terminal

—lacks myelin but is covered by a **Schwann cell** on its nonsynaptic surface.

—membrane on the synaptic surface is known as the **presynaptic membrane.**

—houses mitochondria, synaptic vesicles (containing the neurotransmitter **acetylcholine**), and SER elements.

b. Synaptic cleft
   – is a narrow space between the presynaptic membrane of the axon terminal and **postsynaptic membrane** (sarcolemma) of the muscle cell.
   – contains an amorphous external lamina matrix derived from the muscle cell.

c. Muscle cell near myoneural junction
   (1) Sarcolemma contains invaginations, called **junctional folds,** which are lined by an external lamina and extend inward from the synaptic cleft.
   (2) Acetylcholine receptors are located on the sarcolemma.
   (3) Sarcoplasm is richly endowed with mitochondria, ribosomes, and rough endoplasmic reticulum.

2. Conduction of nerve impulses
   – **across the myoneural junction** involves the following sequence of events:

   a. Depolarization of the presynaptic membrane causes opening of **voltage-gated Ca^{2+} channels** through which extracellular Ca^{2+} ions enter the axon terminal.

   b. The rise in cytosolic Ca^{2+} ions triggers the synaptic vesicles to release acetylcholine in multimolecular quantities (**quanta**).

   c. The released acetylcholine binds to receptors in the postsynaptic membrane, resulting in **depolarization** of the sarcolemma and generation of an **action potential**.

   d. Acetylcholinesterase located in the external lamina lining the junctional folds of the motor end-plate degrades acetylcholine, thus ending the depolarizing signal to the muscle cell.

   e. Recycling of acetylcholine occurs as **choline** is returned to the axon terminal to be recombined with acetyl CoA (from mitochondria) and stored in synaptic vesicles.

   f. Membranes of the emptied synaptic vesicles are recycled via **clathrin-coated endocytic vesicles** (see Figure 3.4).

B. Muscle (neuromuscular) spindle
   – is an elongated, fusiform sensory organ within skeletal muscle that functions primarily as a **stretch receptor**.
   – is bounded by a connective tissue capsule, which encloses the fluid-filled **periaxial space** and 8–10 modified skeletal muscle fibers (**intrafusal fibers**).
   – is surrounded by normal skeletal muscle fibers (**extrafusal fibers**).
   – is anchored via the capsule to the perimysium and endomysium of the extrafusal fibers.

1. Intrafusal fibers
   – are **modified** skeletal muscle fibers that have centrally located nuclei and myofibrils located at both ends of the cell.
   – possess nuclear regions (incapable of contraction) that act as **sensory regions**.

    a. Nuclear bag fibers have nuclei that are grouped in the center of the cell.

    b. Nuclear chain fibers have nuclei that are arranged like the links of a chain. These are thin cells and outnumber the nuclear bag fibers.

2. Innervation of muscle spindles

    a. Afferent (sensory) nerves

    —with two types of nerve endings innervate intrafusal fibers.

      (1) Annulospiral endings (type Ia nerve fibers) wind around the **center** of both types of intrafusal fibers.

      (2) Flower spray endings (type IIa nerve fibers) terminate on either side of the annulospiral endings.

    b. γ-Efferent (motor) nerves

    —terminate as motor end-plates on the myofibrillar regions of the intrafusal fibers.

3. Stimulation of muscle spindles

    a. Stretching of a muscle

    —also stretches the muscle spindle and thus stimulates the afferent nerve endings to send impulses to the central nervous system.

      (1) Annulospiral nerve endings respond to the **rate** of stretching (**phasic response**).

      (2) Flower spray nerve endings respond to the **duration** of stretching (**tonic response**).

    b. Depolarization of γ-efferent neurons

    —also stimulates the intrafusal nerve endings; the rate and duration of the stimulation are monitored as described for stretching.

    c. Muscle overstimulation

    —results from stretching at too great a frequency or for too long.

    —causes stimulation of **α-efferent neurons** to the muscle, initiating contraction and thus counteracting the stretching.

C. Golgi tendon organ

—functions as a **tension receptor** in tendons.

—is composed of encapsulated **collagen fibers** that are surrounded by terminal branches of **type Ib sensory nerves**.

—is stimulated when the muscle contracts too strenuously and the tension on the tendon increases. Impulses sent from the type Ib neurons then **inhibit** the α-efferent (motor) neurons to the muscle, thus preventing further contraction.

V. Cardiac Muscle

—exhibits some properties and structures that are identical to those of skeletal muscle, whereas others differ (Table 8.3).

A. General features—cardiac muscle cells

—**contract spontaneously and display a rhythmic beat,** which is modified by hormonal and neural (sympathetic and parasympathetic) stimuli.

—may branch at their ends to form connections with adjacent cells.

Table 8.3. Comparison of Skeletal, Cardiac, and Smooth Muscle

Property	Skeletal Muscle	Cardiac Muscle	Smooth Muscle
Shape and size of cells	Long Cylindrical	Blunt-ended Branched	Short Spindle-shaped
Number and location of nucleus	Many Peripheral	One or two Central	One Central
Striations present	Yes	Yes	No
T tubules and sarcoplasmic reticulum	Has triads at A–I junctions	Has diads at Z disks	Has caveolae (but no T tubules) and some SER
Gap junctions present	No	Yes (in intercalated disks)	Yes (in sarcolemma); are called nexus
Sarcomeres present	Yes	Yes	No
Regeneration	Is restricted	None	Is extensive
Voluntary contraction	Yes	No	No
Distinctive characteristic	Peripheral nuclei	Intercalated disks	Lack of striations

  —contain **one** (occasionally **two**) **centrally located nuclei.**
  —commonly contain **glycogen granules,** especially at either pole of the nucleus.
  —possess thick and thin filaments arranged in **poorly defined myofibrils**.
  —exhibit a cross-banding pattern identical to that of skeletal muscle.
  —contract and relax by the same mechanisms that operate in skeletal muscle.
  —**do not regenerate;** injuries to cardiac muscle are repaired by the formation of fibrous connective (scar) tissue by fibroblasts.

B. Structural components of cardiac muscle cells

  —differ from those of skeletal muscle as follows:

  1. T tubules

  —are larger than those of skeletal muscle and are lined by an external lamina.
  —are located at Z disks in mammalian cardiac muscle rather than at A–I junctions as they are in skeletal muscle.

  2. Sarcoplasmic reticulum

  —is poorly defined.
  —contributes to the formation of **diads,** which consist of one T tubule and one profile of sarcoplasmic reticulum.
  —**leaks calcium ions** into the sarcoplasm at a slow rate during relaxation, resulting in automatic rhythms. Calcium ions also enter cardiac muscle cells from the extracellular environment via T tubules and the sarcolemma. Hence, calcium ions from both the SR and the extracellular space are necessary for contraction of cardiac muscle cells.

  3. Mitochondria

  —lie parallel to the I bands and often are adjacent to lipids.

  4. Atrial granules

  —are present in atrial cardiac muscle cells, which also exhibit the structural features previously described.

−contain the precursor of **atrial natriuretic peptide,** which acts to **decrease resorption of sodium and water in the kidneys**.

5. Intercalated disks

−are complex step-like junctions forming end-to-end attachments between adjacent cardiac muscle cells.

a. Transverse portion of intercalated disks

−runs across muscle fibers at right angles.

−possesses three specializations: **fasciae adherentes,** which are analogous to zonulae adherentes and to which actin filaments attach; **desmosomes** (maculae adherentes); and **gap junctions** (see Chapter 5 II).

b. Lateral portion of intercalated disks

−has numerous large gap junctions, which facilitate **ionic coupling** between cells and aid in coordinating contraction; as a result cardiac muscle behaves as a **functional syncytium**.

−also contains desmosomes.

6. Connective tissue elements

−are located between cardiac muscle fibers.

−support a rich capillary bed that supplies sufficient nutrients and oxygen to maintain the high metabolic rate of cardiac muscle.

7. Purkinje fibers

−are **modified** cardiac muscle cells located in the **atrioventricular bundle of His** of the heart.

−are specialized for **conduction** and contain only a few peripherally located myofibrils.

−are large, pale cells filled with glycogen and containing many mitochondria.

−form gap junctions, fasciae adherentes, and desmosomes with cardiac muscle cells (but not through typical intercalated disks).

VI. Smooth Muscle

−differs in many ways from skeletal and cardiac muscle (see Table 8.3).

A. Structure—smooth muscle cells

−are **nonstriated, fusiform** cells containing a single nucleus.

−range in length from 20 μm in small blood vessels to 500 μm in the pregnant uterus.

−are surrounded by a reticular fiber network and external lamina.

−may be arranged in layers, small bundles, or helically (in arteries).

−actively divide and **regenerate**.

1. Nucleus

−is centrally located.

−may not be visible in each cell in cross-sections of smooth muscle because some nuclei lie outside the plane of section.

−exhibits a **cork-screw shape** and is **deeply indented** in longitudinal sections of contracted smooth muscle.

2. Cytoplasmic organelles

a. Mitochondria, RER, and the Golgi complex are concentrated near the nucleus and are involved in synthesis of type III collagen, elastin, glycosaminoglycans, external lamina, and growth factors.

b. Sarcolemmal vesicles (caveolae) are present along the periphery of smooth muscle cells and may function in the uptake and release of calcium ions.

c. Smooth endoplasmic reticulum is sparse but may be associated with sarcolemmal vesicles.

3. Filaments in smooth muscle

a. Contractile filaments (actin and myosin)
–are **not** organized into myofibrils.
–are attached to peripheral and cytoplasmic densities.
–are aligned obliquely to the longitudinal axis of smooth muscle cells.

b. Intermediate filaments
–are attached to cytoplasmic densities.
–are composed of **desmin** in **nonvascular** smooth muscle cells and of **vimentin** in **vascular** smooth muscle cells.

4. Cytoplasmic densities
–are believed to be **analogous to Z disks.**
–contain α-actinin and function as **filament-attachment sites**.

5. Gap junctions
–in the **sarcolemma** facilitate the spread of excitation between smooth muscle cells. These gap junctions are referred to as **nexus.**

B. Contraction of smooth muscle
–**occurs more slowly and lasts longer** than contraction of skeletal muscle because the rate of ATP hydrolysis is slower.
–results from interaction of myosin and actin as in skeletal muscle but is regulated by different mechanisms.

1. Contraction cycle
–is stimulated by a transient increase in cytosolic Ca^{2+} levels.
–involves the following steps:

a. Ca^{2+} ions bind to **calmodulin,** altering its conformation.

b. The Ca^{2+}–calmodulin complex activates myosin light-chain kinase, which catalyzes phosphorylation of one of the light chains of myosin.

c. The globular head of phosphorylated myosin then interacts with actin, which stimulates the myosin ATPase, resulting in contraction (as in skeletal muscle). As long as myosin is in its active phosphorylated form, the contraction cycle will continue.

d. Dephosphorylation of myosin prevents myosin–actin interaction and leads to relaxation.

2. Initiation of contraction
–in **vascular smooth muscle** usually is triggered by a **nerve impulse,** with little spread of the impulse from cell to cell.

−in **visceral smooth muscle** is triggered by stretching of the muscle it-self (**myogenic**); the signal spreads from cell to cell.

−in the **uterus** during the terminal stages of pregnancy is triggered by **oxytocin**.

−in smooth muscle elsewhere in the body is triggered by **epinephrine**.

C. Innervation of smooth muscle

−generally tends to modify the activity of smooth muscle.

−is by **sympathetic** (noradrenergic) nerves and **parasympathetic** (cholinergic) nerves of the autonomic nervous system, which act in an antagonistic fashion to stimulate or depress activity of the muscle.

VII. Contractile Nonmuscle Cells

−include myoepithelial cells and myofibroblasts.

A. Myoepithelial cells

−are located between the epithelium and basal lamina of certain glands and in gland ducts.

−can **contract,** thus expressing secretory material from the glandular epithelium into the ducts and out of the gland.

−are generally similar in morphology to smooth muscle cells, but they have a **basket-like shape** and several radiating **processes**.

−arise from ectoderm, whereas most smooth muscle cells arise from mesoderm.

−are attached to the underlying basal lamina via hemidesmosomes.

−contain **actin, myosin,** and intermediate (cytokeratin) filaments, as well as cytoplasmic and peripheral densities to which these filaments attach.

−contract by a **calmodulin-mediated process** similar to that in smooth muscle cells.

−in lactating **mammary glands** contract in response to **oxytocin**.

−in **lacrimal glands** contract in response to **acetylcholine**.

B. Myofibroblasts

−resemble fibroblasts but possess higher amounts of **actin** and **myosin** and are capable of **contraction**.

−may contract during wound healing to decrease defect size (**wound contraction**).

VIII. Clinical Considerations

A. Duchenne muscular dystrophy (DMD)

−is caused by a **sex-linked,** recessive genetic defect that results in the inability to synthesize **dystrophin,** a protein that normally is present in small amounts in the sarcolemma. The function of dystrophin is not yet known.

−is the most common and serious degenerative disorder occurring in young men.

−is characterized by the replacement of degenerating skeletal muscle cells by fatty and fibrous connective tissue.

−may also affect cardiac muscle.

−results in death usually before the age of 20.

B. Amyotrophic lateral sclerosis (ALS)

   —is also called **Lou Gehrig's disease**.

   —is marked by degeneration of motor neurons of the spinal cord, resulting in muscle atrophy.

   —results in death, usually due to respiratory muscle failure.

C. Myasthenia gravis

   —is an **autoimmune** disease, in which **antibodies block acetylcholine receptors** of myoneural junctions, reducing the number of sites available for initiation of sarcolemma depolarization.

   —is characterized by gradual weakening of skeletal muscles, especially the most active ones (e.g., extraocular muscles and muscles of the tongue, face, and extremities).

   —exhibits clinical signs that include thymic hyperplasia (thymoma) and the presence of circulating antibodies to acetylcholine receptors.

   —may lead to death, most commonly because of respiratory compromise and pulmonary infections.

D. Myocardial infarct

   —is an irreversible necrosis of cardiac muscle cells due to prolonged ischemia.

   —may result in death if the cardiac muscle damage is extensive.

E. Rigor mortis

   —is a postmortem rigidity appearing as hardening of skeletal muscles.

   —is caused by the inability of muscle cells to synthesize ATP; as a result myosin cannot dissociate from actin, and the muscles remain in a contracted state.

F. Botulism

   —is a form of **food poisoning** caused by ingestion of *Clostridium botulinum* toxin, which inhibits acetylcholine release at myoneural junctions.

   —is marked by muscle paralysis, vomiting, nausea, visual disorders, and (if untreated) death.

G. Metabolic disturbances

   —such as shock, major trauma, disuse of muscle, or starvation lead to rapid degradation of myofibrillar proteins into their component amino acids, resulting in diminished muscle mass.

Review Test

Directions: Each of the numbered items or incomplete statements in this section is followed by answers or by completions of the statement. Select the **one** lettered answer or completion that is **best** in each case.

1. Which of the following cells is under voluntary control?

(A) Skeletal muscle cells
(B) Smooth muscle cells
(C) Visceral smooth muscle cells
(D) Cardiac muscle cells
(E) Myoepithelial cells

2. Which of the following statements concerning mature skeletal muscle is FALSE?

(A) The cells are multinucleated
(B) A sarcomere is the region between two successive Z disks
(C) The nuclei are centrally located
(D) Myofibrils display a characteristic cross-banding pattern
(E) Actin, myosin, troponin, and tropomyosin are present

3. Which one of the following statements concerning Z disks is FALSE?

(A) They are absent in cardiac muscle
(B) They are absent in smooth muscle
(C) They are present in skeletal muscle
(D) They bisect I bands
(E) They function to anchor thin filaments

4. Contraction in all types of muscle requires calcium ions. Which of the following muscle components can bind or sequester calcium ions?

(A) Rough endoplasmic reticulum
(B) Tropomyosin
(C) Troponin
(D) Active sites on actin

5. Which of the following statements concerning smooth muscle is TRUE?

(A) It has triads associated with its contraction
(B) It has diads associated with its contraction
(C) It has cells with a single centrally located nucleus
(D) It has cells that contain no sarcolemmal vesicles

6. Thick filaments are anchored to Z disks by

(A) C protein
(B) nebulin
(C) titin
(D) myomesin
(E) α-actinin

7. The endomysium is a connective tissue investment that surrounds

(A) individual muscle fibers
(B) muscle fascicles
(C) individual myofibrils
(D) an entire muscle
(E) small bundles of muscle cells

8. Which of the following statements concerning triads in mammalian skeletal muscle is TRUE?

(A) They are located in the Z disk
(B) They consist of two terminal cisternae of the sarcoplasmic reticulum separated by a T tubule
(C) They can be observed with the light microscope
(D) They are characterized by a T tubule that sequesters calcium ions
(E) They consist of two T tubules separated by a central terminal cisterna

9. Which one of the following statements concerning cardiac muscle cells is TRUE?

(A) They are spindle shaped
(B) They require an external stimulus to undergo contraction
(C) They are multinuclear cells
(D) They are joined together end-to-end by intercalated disks
(E) They possess numerous caveolae

10. Which of the following statements concerning the sarcoplasmic reticulum is FALSE?

(A) It is associated with each myofibril in a skeletal muscle cell
(B) It binds calcium ions
(C) It forms part of the diad in cardiac muscle
(D) It communicates with the extracellular space at the surface of the sarcolemma
(E) It releases calcium ions after receiving a signal for contraction

11. A 22-year-old woman complains of fatigue of the muscles in her face, eyes, and tongue. Clinical tests indicate thymoma and the presence of circulating antibodies to acetylcholine receptors. A possible diagnosis is

(A) Duchenne muscular dystrophy
(B) Tay-Sachs disease
(C) myasthenia gravis
(D) amyotrophic lateral sclerosis
(E) defective LDL receptors

12. Which of the following statements about the specialized receptors that monitor the stretching of skeletal muscles is TRUE?

(A) Phasic responses refer to the duration of the stretching
(B) Tonic responses refer to the rate of stretching
(C) These receptors are known as Golgi tendon organs
(D) These receptors are known as muscle spindles
(E) These receptors contain extrafusal fibers surrounded by a capsule

13. Contraction of skeletal muscle requires ATP. Which of the following statements concerning the role of ATP in contraction is TRUE?

(A) ATP binds to the tail portion of myosin
(B) ATP binds to the globular head portion of myosin
(C) ATP binds to the TnT subunit of troponin
(D) ATP must be hydrolyzed before binding to myosin
(E) ATP is hydrolyzed only subsequent to the contractile process

14. Which of the following statements concerning transverse tubules is TRUE?

(A) They are present in smooth muscle
(B) They are extensions of the sarcolemma
(C) They are extensions of the sarcoplasmic reticulum
(D) They form part of the diads present in skeletal muscle
(E) They serve to sequester calcium ions

15. Acetylcholine is necessary for the depolarization of skeletal muscle cells. Which of the following statements concerning acetylcholine action is TRUE?

(A) Acetylcholine is released in quanta from the postsynaptic membrane by exocytosis
(B) Acetylcholine is released in quanta from the sarcolemma membrane by exocytosis
(C) Acetylcholine enters the caveoli to depolarize the sarcolemma
(D) Acetylcholine attaches to transmitter-gated ion channels of the sarcolemma
(E) Acetylcholine enters clathrin-coated vesicles prior to its degradation by acetylcholinesterase

Directions: Each group of items in this section consists of lettered options followed by a set of numbered items. For each item, select the **one** lettered option that is most closely associated with it. Each lettered option may be selected once, more than once, or not at all.

Questions 16 and 17

Match each numbered structure below with the corresponding letter in the micrograph.

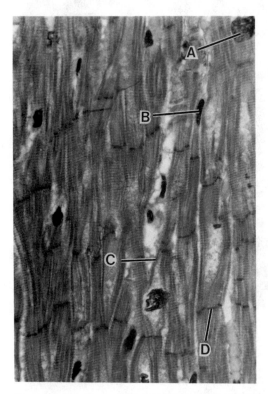

Reprinted from Gartner LP and Hiatt JL: *Atlas of Histology*. Baltimore, Williams & Wilkins, 1987, p 93.

16. Intercalated disk

17. Myofibrils

Questions 18–20

Match each numbered structure below with the corresponding letter in the electron micrograph.

Reprinted from Gartner LP and Hiatt JL: *Atlas of Histology*. Baltimore, Williams & Wilkins, 1987, p 83.

18. Z disk

19. I band (thin filaments only)

20. H band (thick filaments only)

Answers and Explanations

1–A. Skeletal muscle cells are under voluntary control; other cells listed are all regulated by the autonomic nervous system.

2–C. In mature skeletal muscle the nuclei are located peripherally, just deep to the sarcolemma.

3–A. Z disks are a characteristic feature of cardiac and skeletal muscle, but they are absent from smooth muscle. The latter contains cytoplasmic densities that are functionally analogous to Z disks and act as filament-attachment sites.

4–C. Binding of calcium ions to the TnC subunit of troponin leads to the unmasking of myosin-binding sites on actin (thin filaments).

5–C. Smooth muscle cells contain one centrally located nucleus.

6–C. Titin forms an elastic lattice that anchors thick filaments to Z disks.

7–A. The endomysium is a thin connective tissue layer, composed of reticular fibers and an external lamina, that invests individual muscle fibers (cells). The epimysium surrounds the entire muscle, and the perimysium surrounds bundles (fascicles) of muscle fibers.

8–B. A triad in skeletal muscle is composed of three components, a T tubule and two terminal cisternae of the SR that flank it. The SR sequesters calcium ions, not T tubules.

9–D. Cardiac muscle cells are joined together end-to-end by a unique junctional specialization called the intercalated disk.

10–D. The sarcoplasmic reticulum does not communicate with the extracellular space at the surface of the sarcolemma; the T tubule does.

11–C. Myasthenia gravis is an autoimmune disease in which antibodies to acetylcholine receptors are produced. As a result, neural stimulation of contraction at the myoneural junction is impaired.

12–D. Stretch receptors of muscle are called muscle spindles. The intrafusal of fibers are surrounded by a capsule.

13–B. ATP binds to the globular head of myosin and then is hydrolyzed by the (actin-activated) myosin ATPase. Hydrolysis of ATP provides the energy for muscle contraction.

14–B. Transverse tubules are present within skeletal and cardiac muscle and are formed by deep invaginations of the sarcolemma. They help to spread depolarization throughout muscle fibers, leading to release of calcium ions and contraction.

15–D. Acetylcholine is released in quanta from the presynaptic membrane by exocytosis; it enters the synaptic cleft and attaches to transmitter-gated ion channels of the sarcolemma.

16–D. Intercalated disk.

17–C. Myofibrils.

18–B. Z disk.

19–C. I band.

20–D. H band.

9

Nervous Tissue

I. Overview—Nervous Tissue

—is divided **anatomically** into the **central nervous system** (CNS), which includes the brain and spinal cord, and the **peripheral nervous system** (PNS), which includes the nerves outside the CNS and their associated ganglia.

—is divided **functionally** into a **sensory** component, which transmits electrical impulses (signals) to the CNS, and a **motor** component, which transmits impulses from the CNS to various structures of the body. The motor component is further divided into the **somatic** and **autonomic** systems.

—contains two types of cells: **nerve cells** (**neurons**), which can conduct electrical impulses, and **glial (neuroglial) cells,** which support, nurture, and protect the neurons.

II. Histogenesis of the Nervous System

—involves the following embryonic structures:

A. Neuroepithelium

—thickens and differentiates to form the neural plate.

B. Neural plate

—invaginates and thickens to form the **neural groove**.

C. Neural tube

—is the cylindrical structure that results from fusion of the edges of the neural groove.

—enlarges at its cranial end to form the **brain;** the remaining portion gives rise to the **spinal cord**.

D. Neural crest cells

—stream off the edges of the neural groove before formation of the neural tube.

—migrate throughout the body and give rise to cranial and spinal ganglia, Schwann cells of peripheral axons, and several other structures (e.g., chromaffin cells of the adrenal medulla, melanocytes).

III. Cells of Nervous Tissue

A. Neurons

—consist of a **cell body** and its processes, which usually include multiple **dendrites** and a single **axon**.

—comprise the smallest and largest cells of the body ranging from 5 μm to 150 μm in diameter.

1. Morphologic classification of neurons

a. Unipolar neurons

—possess a single process.

—are rare in vertebrates.

b. Bipolar neurons

—possess a single axon and a single dendrite.

—are present in several sense organs.

c. Multipolar neurons

—possess a single axon and more than one dendrite.

—are the **most common type** of neuron in vertebrates.

d. Pseudounipolar neurons

—possess a single process, extending from the cell body, that subsequently branches into an axon and dendrite.

—originate embryologically as bipolar cells whose axon and dendrite later fuse into a single process.

—are present in spinal and cranial ganglia.

2. Functional classification of neurons

a. Sensory neurons

—receive stimuli from the internal and external environment.

—conduct impulses **to the CNS** for processing and analysis.

b. Interneurons

—connect other neurons in a chain or sequence.

—most commonly connect sensory and motor neurons.

—also regulate signals transmitted to neurons.

c. Motor neurons

—conduct impulses **from the CNS** to other neurons, muscles, and glands.

3. Neuronal cell body

—is also called the **soma** (pl., somata) or **perikaryon.** (*Neuroayton*)

—is the region of a neuron containing the nucleus, various cytoplasmic organelles and inclusions, and cytoskeletal components.

a. Nucleus

—is large, spherical, and pale-staining.

—is **centrally located** in the soma of most neurons.

—contains abundant euchromatin and a large nucleolus ("owl-eye" nucleus).

b. Cytoplasmic organelles and inclusions

(1) **Nissl bodies** are composed of polysomes and rough endoplasmic reticulum (RER). They are observed as clumps by light microscopy and are most abundant in large motor neurons.

(2) **Golgi complex** is located near the nucleus, and **mitochondria** are scattered throughout the soma cytoplasm.

(3) **Melanin-containing granules** are present in some neurons in the CNS and in dorsal root and sympathetic ganglia.

(4) **Lipofuscin-containing granules** are present in some neurons and increase in number with age.

(5) **Lipid droplets** occasionally are present.

c. Cytoskeletal components

(1) **Neurofilaments** (10 nm in diameter) are abundant and run throughout the soma cytoplasm. They are intermediate filaments composed of three distinct polypeptides.

(2) **Microtubules** (24 nm in diameter) are also present in the soma cytoplasm.

4. Dendrites

—**receive stimuli** (signals) from sensory cells or axons or other neurons and convert these into small electrical impulses, which are **transmitted toward** the soma.

—have **arborized terminals** (except in bipolar neurons), which permit a neuron to receive stimuli simultaneously from many other neurons.

—possess a cytoplasm similar to that in the soma except that it lacks a Golgi complex.

5. Axons

—conduct impulses **away from** the soma to the axon terminals without any diminution in their strength.

—vary in diameter and length in different types of neurons and may be as long as 100 cm.

—originate from the **axon hillock,** a specialized region of the soma that lacks RER and ribosomes but contains many microtubules and neurofilaments.

—may have **collaterals,** branching at right angles from the main trunk.

—possess a cytoplasm (**axoplasm**) that is devoid of rough endoplasmic reticulum and a Golgi complex.

—are surrounded by a plasma membrane called the **axolemma**.

—terminate in many small branches (**axon terminals**) from which impulses are passed to another neuron or other type of cell.

B. Neuroglial cells

—comprise several cell types **located only in the CNS**.

—outnumber neurons by approximately 10 to 1.

—are embedded in a web of tissue composed of modified ectodermal elements; the entire supporting structure is termed the **neuroglia**.

—function in various ways **to support and protect neurons;** they do not conduct impulses or form synapses with other cells.

—are revealed in histologic sections only with special gold and silver stains.

1. Astrocytes

—are the largest of the neuroglial cells.

—contain many processes, some of which possess expanded pedicles (**vascular feet**) that surround blood vessels.

—form, along with other components of the neuroglia, a protective **sealed barrier** between the pia mater and the nervous tissue of the brain and spinal cord.

—provide **structural support** for nervous tissue.

—proliferate to form **scar tissue** after injury to the CNS.

a. Protoplasmic astrocytes

—reside mostly in **gray matter**.

—have branched processes that envelop blood vessels, neurons, and synaptic areas.

—contain some intermediate filaments composed of **glial fibrillar acidic protein** (GFAP).

—help to establish the **blood–brain barrier** and may contribute to its maintenance.

b. Fibrous astrocytes

—reside mostly in **white matter**.

—have long, slender processes with few branches.

—contain many intermediate filaments composed of GFAP.

2. Oligodendrocytes

—are neuroglial cells that live **symbiotically** with neurons (i.e., each cell type is affected by the metabolic activities of the other).

—are necessary for the survival of neurons.

—are located in both **gray matter and white matter**.

—possess a small, round, condensed nucleus and only a few short processes.

—have an electron-dense cytoplasm containing ribosomes, numerous microtubules, many mitochondria, rough endoplasmic reticulum, and a large Golgi complex.

—produce **myelin** for axons in the CNS. Each cell can produce myelin for several axons. (Schwann cells, derived from the neural crest, myelinate axons of the PNS, but each Schwann cell produces myelin for only one axon.)

—form a protective barrier around neurons, serving to insulate them.

3. Microglia

—are small, **phagocytic** neuroglial cells, which are derived from monocytes.

—have a very condensed, elongated nucleus and many short, branching processes.

4. Ependymal cells

—form the **epithelium** that lines the neural tube and ventricles of the brain.

—often have **cilia,** which aid in moving the cerebrospinal fluid.

IV. Synapses

—are sites of **functional apposition** at which signals are transmitted from one neuron to another, or from a neuron to another type of cell (e.g., muscle cell).

—are classified according to the site of synaptic contact and the method of signal transmission.

A. Site of synaptic contact

1. **Axodendritic synapse** is between an axon and a dendrite.

2. **Axosomatic synapse** is between an axon and a soma. The CNS primarily contains axodendritic and axosomatic synapses.

3. **Axoaxonic synapse** is between axons.

4. **Dendrodendritic synapse** is between dendrites.

B. Method of signal transmission

1. **Chemical synapses**

   —involve the release of a chemical **neurotransmitter** by the presynaptic cell. This chemical acts on the postsynaptic cell to generate an action potential.

   —are the most common neuron–neuron synapse and the only neuron–muscle synapse.

   —exhibit a **delay** in signal transmission of about 0.5 milliseconds—the time required for secretion and diffusion of neurotransmitter.

2. **Electrical synapses**

   —involve movement of ions from one neuron to another via **gap junctions,** which transmit the action potential of the presynaptic cell directly to the postsynaptic cell.

   —are much less numerous than chemical synapses.

   —permit **nearly instantaneous** signal transmission.

C. Synaptic morphology

1. **Axon terminals**

   —may vary morphologically depending of the site of synaptic contact.

   a. **Boutons terminaux** are bulbous expansions that occur singly at the end of axon terminals.

   b. **Boutons en passage** are swellings along the axon terminal; at each swelling a synapsis can occur.

2. **Presynaptic membrane**

   —is the thickened axolemma of the neuron that is transmitting the impulse.

   —contains **voltage-gated Ca^{2+} channels,** which regulate the entry of calcium ions into the axon terminal.

3. **Postsynaptic membrane**

   —is the thickened plasma membrane of the neuron or other target cell that is receiving the impulse.

4. **Synaptic cleft**

   —is the narrow space (20–30 nm in width) between the pre- and postsynaptic membranes across which neurotransmitters diffuse.

5. Synaptic vesicles
—are small, membrane-bounded structures (40–60 nm in diameter).
—are located in the axoplasm of the transmitting neuron.
—**discharge neurotransmitters** into the synaptic cleft by exocytosis.

D. Neurotransmitters
—are produced, stored, and released by presynaptic neurons.
—diffuse across the synaptic cleft and bind to receptors in the postsynaptic membrane, leading to generation of an **action potential**.

1. **Acetylcholine** is the neurotransmitter at myoneural junctions, all parasympathetic synapses, and preganglionic sympathetic processes.

2. **Norepinephrine** is the neurotransmitter at postganglionic sympathetic synapses.

3. **Glutamic acid, γ-aminobutyric acid (GABA), dopamine, serotonin, glycine,** and other compounds also can act as neurotransmitters.

4. **Endorphins** and **enkephalins,** which are produced in the brain, exhibit analgesic properties, probably by inhibiting neurons that transmit pain signals. They may also act as neurotransmitters in selected neurons.

V. Nerve Fibers
—are individual axons enveloped by a myelin sheath or by Schwann cells in the PNS (oligodendrocytes in the CNS).

A. Myelin sheath
—is produced by **oligodendrocytes** in the CNS and by **Schwann cells** in the PNS.
—consists of a **lipoprotein material** organized into a sheath that is formed by several layers of the plasma membrane of an oligodendrocyte or Schwann cell wrapping around the axon.
—is not continuous along the length of the axon but is interrupted by gaps called **nodes of Ranvier.**
—is of constant thickness along the length of an axon, but its thickness usually increases as axonal diameter increases.
—is removed by standard histologic techniques. Methods using **osmium tetroxide** preserve the myelin sheath and stain it black.
—displays the following microscopic features:

1. Major dense lines
—are visible in electron micrographs of myelinated fibers.
—represent **fusions** between the cytoplasmic surfaces of the plasma membranes of Schwann cells (or oligodendrocytes).

2. Intraperiod lines
—also are displayed in electron micrographs.
—represent **close contact,** but not fusion, of the extracellular surfaces of adjacent Schwann-cell (or oligodendrocyte) plasma membranes.

3. Clefts (incisures) of Schmidt–Lantermann
—are visible in light micrographs of **peripheral** myelinated nerves.
—are cone-shaped, oblique **discontinuities** in the myelin sheath.

B. Nodes of Ranvier

–are regions along the axon that **lack myelin** and represent discontinuities between adjacent Schwann cells (or oligodendrocytes).

1. The axon at the nodes is covered by Schwann cells in the PNS, although it may not be covered by oligodendrocytes in the CNS.

2. The axolemma at the nodes contains **many Na$^+$ pumps** and displays a characteristic electron density.

C. Internodes

–are the segments of a nerve fiber **between adjacent nodes of Ranvier**.
–vary in length from 0.08 to 1 mm, depending on the size of the Schwann cells (or oligodendrocytes) associated with the fiber.

D. Types of nerve fibers

–vary in their **rates** of impulse transmission and other characteristics.

1. Type A fibers

–are **thick, myelinated,** and have long internodes.
–conduct impulses at a **high velocity** (15–100 m/sec).

2. Type B fibers

–have a smaller diameter and a thinner myelin sheath than type A fibers.
–conduct impulses at a **moderate velocity** (3–14 m/sec).

3. Type C fibers

–are **thin, unmyelinated,** and ensheathed by Schwann cells or oligodendrocytes.
–conduct impulses at a relatively **slow velocity** (0.5–2 m/sec).

VI. Nerves (Figure 9.1)

–are cordlike bundles of nerve fibers surrounded by connective tissue sheaths.
–are **visible to the naked eye** and usually appear **whitish** due to the presence of myelin.

A. Connective tissue investments

1. Epineurium is the layer of fibrous dense connective tissue (**fascia**) that forms the external coat of nerves.

2. Perineurium is the layer of dense connective tissue that surrounds each bundle of nerve fibers (**fascicle**). Its inner surface consists of layers of flattened epithelioid cells joined by **tight junctions** (zonulae occludentes), which prohibit passage of most macromolecules.

3. Endoneurium is the thin layer of reticular fibers, produced mainly by Schwann cells, and surrounds individual nerve fibers.

B. Functional classification of nerves

1. Sensory nerves contain **afferent** fibers and only carry sensory signals from the internal and external environment to the CNS.

2. Motor nerves contain **efferent** fibers and only carry signals from the CNS to effector organs.

3. Mixed nerves, which are the most common, contain both afferent and efferent fibers and thus carry both sensory and motor signals.

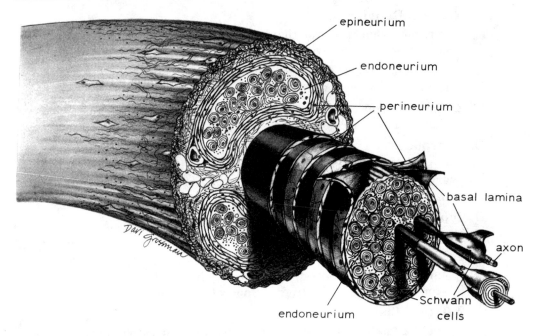

Figure 9.1. Drawing of a peripheral nerve in cross-section showing the various connective tissue sheaths. Each bundle of nerve fibers, or fascicle (one is extended in drawing), is covered by perineurium. (Reprinted from Kelly DE et al: *Bailey's Textbook of Microscopic Anatomy,* 18th ed. Baltimore, Williams & Wilkins, 1984, p 353.)

VII. Ganglia

—are encapsulated aggregations of **neuronal cell bodies** located outside the CNS.

A. Autonomic ganglia

—are **motor** ganglia where pre- and postganglionic neurons of the autonomic nervous system synapse.

B. Craniospinal ganglia

—are **sensory** ganglia associated with most cranial nerves and the dorsal roots of spinal nerves (**dorsal root ganglia**).

—contain the cell bodies of sensory neurons, which are **pseudounipolar** and transmit sensory signals from receptors to the CNS without synapsing.

VIII. Histophysiology of Nervous Tissue

A. Resting membrane potential

—exists across the plasma membrane of all cells.

—does not vary with time in most cells, which are said to be electrically inactive.

—undergoes controlled changes in neurons and muscle cells—the most important **electrically active cells**.

—is established and maintained as follows:

1. The concentration of K^+ **ions** is about 20 times greater inside neurons than in the extracellular fluid, whereas the concentration of Na^+ **ions** is about 10 times greater extracellularly than intracellularly.

2. Because the plasma membrane is highly permeable to K^+ ions, they diffuse outward, setting up a positive charge on the outer surface of the membrane, which is offset by the negative charge within the cell resulting from impermeable, negatively charged (anionic) molecules. This is the major mechanism for establishing the membrane potential.

3. The **resting membrane potential** exists when there is no net movement of K^+ ions (i.e., when outward diffusion of K^+ ions is just balanced by the external positive charge acting against further diffusion).

4. The interior of the cell is about 40–100 mV **negative** compared with the outside.

5. The ionic concentration gradients across the membrane are maintained in part by the **Na^+–K^+ pump,** which actively transports Na^+ ions to the outside in exchange for K^+ ions (see Chapter 1 III B 1).

B. Action potential

—is the electrical activity that occurs in a neuron as an impulse is propagated along the axon.

—is observed as a **movement of negative charges along the outside of an axon**.

—is an **all-or-nothing event with a constant amplitude and duration**.

—may be generated in axons at rates up to 1000 per second.

1. Generation of the action potential

a. An excitatory stimulus on a postsynaptic neuron partially **depolarizes** a portion of the plasma membrane (the potential is **less negative**).

b. Once the membrane potential reaches a critical **threshold, voltage-gated Na^+ channels** in the membrane open, permitting Na^+ ions to enter the cell (Figure 9.2).

c. The influx of Na^+ ions leads to **reversal of the resting potential** in the immediate area (i.e., the external side becomes negative).

d. The Na^+ channels close spontaneously and are inactivated for 1–2 milliseconds (**refractory period**).

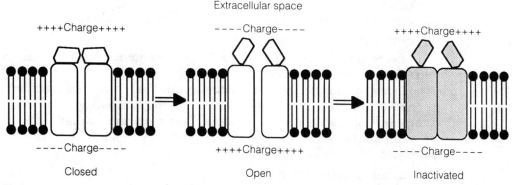

Figure 9.2. Model of the voltage-gated Na^+ channel showing the transition between its closed, open, and inactivated states. In the resting state, the channel-blocking segment and gating keep the channel closed to entry of extracellular Na^+ ions. Depolarization of the membrane causes a conformational change that opens the channel to influx of Na^+ ions. The channel closes spontaneously and becomes inactive within a millisecond after opening.

    e. Opening of **voltage-gated K$^+$ channels** also is triggered by depolarization. Because these remain open longer than the Na$^+$ channels, exit of K$^+$ ions during the refractory period **repolarizes** the membrane back to its resting potential (there is typically a brief period of **hyperpolarization**).

    f. The ion channels then return to their normal states, so the cell can respond to another stimulus.

  2. Propagation of the action potential

    —results from longitudinal diffusion of Na$^+$ ions (which enter the cell at the initial site of excitation) toward the axon terminals (**orthodromic spread**). This movement depolarizes the adjacent region of membrane, giving rise to a new action potential at this site.

    —does **not** result from diffusion of Na$^+$ ions toward the soma (**antidromic spread**) because the Na$^+$ channels are inactivated in this region.

    —occurs most rapidly in myelinated fibers, which exhibit **saltatory conduction**—a process whereby the action potential "jumps" from one node of Ranvier to the adjacent one.

C. Axonal transport

  —of proteins, organelles, and vesicles occurs at high, intermediate, and low velocities depending on the nature of the transported materials.

  1. Anterograde transport carries material away from the soma.

  2. Retrograde transport carries material toward the soma for reutilization, recycling, or degradation.

D. Trophic function of nervous tissue

  1. Denervation of a muscle or gland leads to its atrophy.

  2. Reinnervation restores its structure and function.

IX. Somatic and Autonomic Nervous Systems

  —are **functional** concepts relating to all the neural elements involved in transmission of impulses from the CNS to the somatic and visceral components of the body, respectively.

A. Somatic nervous system

  —contains motor fibers that innervate voluntary **skeletal muscle**.

B. Autonomic nervous system (ANS)

  —contains motor fibers that control and regulate **smooth muscle, cardiac muscle,** and some **glands**.

  —establishes and maintains **homeostasis** of the body's visceral functions.

  —is divided anatomically and functionally into the **sympathetic** and **parasympathetic systems,** which generally **function antagonistically** in a given organ (i.e., when the sympathetic system stimulates an organ, the parasympathetic inhibits it, and vice versa).

  1. Autonomic nerve chains

    a. Cell bodies of preganglionic neurons are located in the CNS and extend their **preganglionic fibers** (axons) to an **autonomic ganglion** located outside of the CNS.

b. In the ganglion, the preganglionic fibers synapse with postganglionic neurons, which typically are multipolar and surrounded by satellite cells.

c. Postganglionic fibers leave the ganglion and terminate in the **effector organ**.

2. **Sympathetic system**

   −is also called the **thoracolumbar outflow** of the ANS because the cell bodies of preganglionic fibers are located in the thoracic and first two lumbar segments of the spinal cord.

   −contains ganglia that are arranged in a chain-like fashion on either side of the spinal cord.

   −usually stimulates contraction of smooth muscle in blood vessels, resulting in **vasoconstriction**.

3. **Parasympathetic system**

   −is also called the **craniosacral outflow** of the ANS because the cell bodies of preganglionic fibers are located in certain cranial nerve nuclei within the brain and in some sacral segments of the spinal cord.

   −contains **intramural ganglia** located in the walls of viscera and in **Meissner's** and **Auerbach's plexuses**.

   −usually stimulates secretion (**secretomotor** function).

X. Central Nervous System

A. White matter and gray matter

−are both present in the CNS.

1. **White matter** contains mostly myelinated nerve fibers but also some unmyelinated fibers and neuroglial cells.

2. **Gray matter** contains neuronal cell bodies, many unmyelinated fibers, some myelinated fibers, and neuroglial cells.

   a. Spinal cord gray matter

   −appears in the shape of an **H** in cross-sections of the spinal cord.

   (1) A small **central canal,** lined by ependymal cells, is in the center of the crossbar in the H. This is a remnant of the embryonic neural tube.

   (2) The dorsal vertical bars of the H (**dorsal horns**) consist of **sensory** fibers extending from the dorsal root ganglia and cell bodies of interneurons.

   (3) The ventral vertical bars of the H (**ventral horns**) consist of the fibers of large multipolar **motor** neurons.

   b. Brain gray matter

   −is located at the periphery (cortex) of the cerebrum and cerebellum, with white matter deep to it in these structures.

   −also forms the **basal ganglia,** which are located deep within the cerebrum surrounded by white matter.

B. Meninges

−are membranes formed of connective tissue that **surround the brain and spinal cord**.

1. Dura mater

–is the outermost meninx.

–is composed of dense fibrous connective tissue.

–is continuous with the periosteum of the skull but is separated from the periosteum of the vertebrae by the **epidural space**.

–is lined by simple squamous epithelium on its internal surface in the skull and on both surfaces in the spinal column.

2. Arachnoid

–is the middle meninx lying between the dura mater and pia mater.

–is covered on both surfaces by simple squamous epithelium.

–has **trabeculae,** which form a loose network between it and the pia mater.

a. Subarachnoid space

–lies between the arachnoid and pia mater and is bridged by trabeculae.

–**is filled with cerebrospinal fluid.**

b. Arachnoid villi

–are specialized regions of the arachnoid that protrude into the dura mater and terminate in venous spaces.

–function to return cerebrospinal fluid into the venous sinuses.

3. Pia mater

–is the innermost meninx and contains many blood vessels.

–is covered with squamous cells of mesenchymal origin.

–is separated from the underlying nervous tissue by neuroglial elements.

C. Choroid plexus

–consists of folds of the pia mater that extend into the third, fourth, and lateral ventricles of the brain.

–is composed of a core of connective tissue covered by simple cuboidal (ependymal) epithelium having many microvilli. The connective tissue is vascular and contains many **fenestrated capillaries**.

–**secretes cerebrospinal fluid.**

D. Cerebrospinal fluid

–bathes and nourishes the brain and spinal cord and protects them by acting as a cushion to absorb shock.

–fills the ventricles of the brain, central canal of the spinal cord, subarachnoid space, and perivascular space.

–contains little protein but is rich in sodium, potassium, and chloride ions.

–is continuously **produced by the choroid plexus** and is **reabsorbed by the arachnoid villi,** which transport it into the superior sagittal sinus. Blockage in reabsorption leads to **hydrocephalus**.

E. Cerebral cortex

–is the thin layer of **gray matter on the surface of the cerebral hemispheres**.

–is folded into many **gyri** with about two-thirds of its area buried in fissures.

1. Functions—the cerebral cortex

−is responsible for diverse functions including the following:

a. Memory and learning

b. Association and analysis of information

c. Initiation of motor responses

d. Integration of incoming nerve signals from the external and internal environment

2. Layers of the neocortex

a. Molecular layer (I) is the superficial layer located just below the pia mater. It contains only a few somata but has abundant neuroglia.

b. External granular layer (II) is composed of granule cells and neuroglia.

c. External pyramidal layer (III) consists of pyramidal cells, granule cells, and neuroglia.

d. Internal granular layer (IV) is a narrow band of small and large granule cells and neuroglia.

e. Internal pyramidal layer (V) houses medium and large pyramidal cells and neuroglia.

f. Multiform layer (VI) contains cells of various shapes including fusiform (**Martinotti cells**) and neuroglia.

F. Cerebellar cortex

−is the thin layer of **gray matter on the surface of the cerebellum**.

1. Functions—the cerebellar cortex

−has three main functions:

a. Maintenance of balance and equilibrium

b. Maintenance of muscle tone

c. Coordination of the activity of skeletal muscles

2. Layers of the cerebellar cortex

a. Molecular layer

−is the outermost layer underlying the pia mater.

−contains unmyelinated fibers from the granular layer; dendritic arborizations of Purkinje cells; stellate cells (located superficially); and basket cells.

b. Purkinje cell layer

−is composed of large Purkinje cells, which are present **only** in the cerebellar cortex.

(1) Purkinje cells have a flask-shaped soma, centrally located nucleus, highly branched (arborized) dendrites, and a single axon.

(2) Myelinated axons of Purkinje cells enter the granular layer, and their arborized dendrites form a single plane in the molecular layer.

(3) Each Purkinje cell may receive several hundred thousand excitatory and inhibitory impulses to sort and integrate.

c. Granular layer

—is composed of closely packed small granule cells interspersed with regions devoid of cells, known as **cerebellar islands** (glomeruli). These regions house synapses between axons entering the cerebellum and dendrites of granule cells.

XI. Degeneration and Regeneration of Nervous Tissue

A. Death of neurons

—occurs as the result of injury to or disease affecting the somata.

—results in degeneration and permanent loss of nervous tissue because **neurons cannot divide**.

—in the CNS may be followed by proliferation of the neuroglia, which fills in areas left by dead neurons.

B. Transection of axons

—induces changes in the soma including **chromatolysis** (dissolution of Nissl bodies with a concomitant loss of cytoplasmic basophilia); increase in soma volume; and movement of the nucleus to a peripheral position.

1. Degeneration of distal axonal segment

a. The axon and its myelin sheath, which are separated from the soma, degenerate completely, and the remnants are removed by macrophages.

b. Schwann cells proliferate, forming a **solid cellular column** distal to the injury, which remains attached to the effector organ.

2. Regeneration of proximal axonal segment

a. The distal end, closest to the wound, initially degenerates, and the remnants are removed by macrophages.

b. Growth at the distal end then begins (0.5–3 mm/day) and progresses toward the columns of Schwann cells.

c. Regeneration is successful if the sprouting axon penetrates a Schwann-cell column and reestablishes contact with the effector organ.

XII. Clinical Considerations

A. Congenital malformations in the CNS

—resulting from abnormal organogenesis include the following:

1. Spina bifida and spina bifida anterior

—result from failure of the neural groove to close completely (thus, the neural tube is deformed) and to separate from the ectoderm.

a. **Spina bifida** is characterized by defective closure of the spinal column through which the spinal cord and meninges may protrude in severe cases.

b. **Spina bifida anterior** is characterized by defective closure of the vertebral arch(es), which may be associated with defective development of the abdominal and thoracic viscera.

2. Anencephaly

—results from lack of closure of the anterior neuropore, the opening in the anterior portion of the neural tube in the early embryo.

–is characterized by a poorly formed brain and absent cranial vault.

3. Hirschsprung's disease

–is also called **congenital megacolon**.

–results from failure of neural crest cells to migrate into the wall of the gut.

–is marked by the **absence of Auerbach's plexus,** a part of the parasympathetic system innervating a distal segment of the colon. This loss of motor function leads to dilation and hypertrophy of the colon.

B. Neuroglial tumors

–constitute 50% of the intracranial tumors.

–are derived from astrocytes, oligodendrocytes, and ependymocytes.

–range in severity from slow-growing **benign oligodendrogliomas** to rapid-growing fatal **malignant astrocytomas**.

Review Test

Directions: Each of the numbered items or incomplete statements in this section is followed by answers or by completions of the statement. Select the **one** lettered answer or completion that is **best** in each case.

1. Neural crest cells give rise to all of the following EXCEPT

(A) dorsal root ganglia
(B) adrenal medulla
(C) sympathetic ganglia
(D) preganglionic autonomic nerves

2. All of the following can act as neurotransmitters EXCEPT

(A) γ-aminobutyric acid (GABA)
(B) endorphins
(C) epinephrine
(D) tyrosine
(E) serotonin

3. Somata possess which one of the following structures?

(A) Microtubules
(B) Neurofilaments
(C) Nissl bodies
(D) Mitochondria
(E) All of the above

4. Which of the following statements regarding nerve action potentials is FALSE?

(A) Membrane potentials are reversed by Na^+ ions entering the cell
(B) Exit of K^+ ions causes the membrane to return to its normal state
(C) Axons generate action potentials at the rate of 1–2 per second
(D) Na^+ channels become inactivated during the refractory period

5. Which of the following statements about the perineurium is TRUE?

(A) It is a fascia surrounding many bundles of nerve fibers
(B) It is a layer of connective tissue, lined by epithelioid cells that surrounds a bundle (fascicle) of nerve fibers
(C) It is a thin layer of reticular fibers covering individual nerve fibers
(D) It is a fascia that excludes macromolecules and forms the external coat of nerves

6. Acetylcholine is the neurotransmitter in all of the following EXCEPT

(A) preganglionic parasympathetic synapses
(B) preganglionic sympathetic synapses
(C) postganglionic parasympathetic synapses
(D) postganglionic sympathetic synapses
(E) myoneural junctions

7. Nissl bodies are composed of

(A) synaptic vesicles and acetylcholine
(B) polyribosomes and rough endoplasmic reticulum
(C) lipoprotein and melanin
(D) neurofilaments and microtubules
(E) smooth endoplasmic reticulum and mitochondria

8. The axon hillock contains

(A) rough endoplasmic reticulum
(B) ribosomes
(C) microtubules
(D) Golgi complex
(E) synaptic vesicles

9. Which of the following statements regarding the cerebrospinal fluid is FALSE?

(A) It is secreted by the choroid plexus
(B) It bathes and nourishes the brain and spinal cord
(C) It is composed mostly of protein with little sodium and potassium
(D) It is resorbed by arachnoid villi
(E) Blockage of its pathway leads to hydrocephalus

10. Synaptic vesicles possess which of the following characteristics?

(A) Manufacture neurotransmitter
(B) Enter the synaptic cleft
(C) Become incorporated into the presynaptic membrane
(D) Become incorporated into the postsynaptic membrane

11. Following transection of an axon, the soma undergoes which of the following changes?

(A) The nucleus becomes smaller
(B) The soma shrinks
(C) Nissl bodies are increased
(D) The nucleus becomes eccentric

12. Myelination of peripheral nerves is accomplished by

(A) astrocytes
(B) oligodendrocytes
(C) Schwann cells
(D) neural crest cells
(E) basket cells

13. A patient with Hirschsprung's disease will present with which of the following symptoms?

(A) Absent cranial vault
(B) Exposed spinal cord
(C) Headache
(D) Large, dilated colon

14. Cell bodies of preganglionic sympathetic neurons are located

(A) in the brain
(B) in the sacral spinal cord
(C) in the thoracic spinal cord
(D) throughout the CNS
(E) in autonomic ganglia only

15. All of the following cell types are found in the brain EXCEPT

(A) Purkinje cells
(B) stellate cells
(C) Schwann cells
(D) small granule cells
(E) pyramidal cells

16. Which of the following statements related to the meninges is FALSE?

(A) The arachnoid possesses trabeculae
(B) The pia mater is the vascular layer of the meninges
(C) The dura mater is continuous with the periosteum of the skull
(D) The pia mater is covered with squamous cells
(E) The arachnoid produces cerebrospinal fluid

17. The autonomic nervous system innervates all of the following EXCEPT

(A) muscles in the wall of the stomach
(B) biceps muscle
(C) cardiac muscle
(D) pituitary gland
(E) muscles responsible for vasoconstriction

Directions: The group of items in this section consists of lettered options followed by a set of numbered items. For each item, select the **one** lettered option that is most closely associated with it. Each lettered option may be selected once, more than once, or not at all.

Questions 18–20

Match each of the numbered structures below with the corresponding letter in the photomicrograph.

18. Perineurim

19. Axon

20. Myelin space

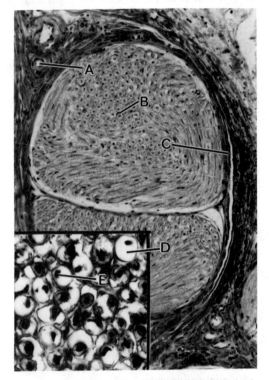

Reprinted from Gartner LP and Hiatt JL: *Atlas of Histology*. Baltimore, Williams & Wilkins, 1987, p 107.

Answers and Explanations

1–D. Neural crest cells migrate throughout the body and give rise to ganglia and other structures, including portions of the adrenal medulla, but they do not contribute to the development of preganglionic autonomic nerves.

2–D. Although tyrosine is a precursor of several neurotransmitters (dopamine, norepinephrine, and epinephrine), it does not function as a neurotransmitter itself. Epinephrine (also called adrenalin) is synthesized and released into the blood by the adrenal medulla; like norepinephrine, it acts as a neurotransmitter at adrenergic synapses. GABA and glycine are neurotransmitters at many inhibitory synapses in the CNS. Besides acting as neurotransmitters, endorphins and enkephalins function as natural analgesics, probably inhibiting synapses in sensory pathways carrying pain impulses.

3–E. Somata possess microtubules, neurofilaments, Nissl bodies, and mitochondria.

4–C. Axons generate as many as 1000 action potentials per second.

5–B. Each bundle of nerve fibers is surrounded by the perineurium, which consists of dense connective tissue underlain by a layer of epithelioid cells. Tight junctions between these cells exclude most macromolecules. The external coat of nerves, the epineurium, surrounds many fascicles but does not exclude macromolecules. The layer of reticular fibers that covers individual nerve fibers is the endoneurium; it also does not exclude macromolecules.

6–D. Postganglionic sympathetic synapses are adrenergic (norepinephrine is the neurotransmitter). Pre- and postganglionic parasympathetic synapses, preganglionic sympathetic synapses, and myoneural junctions are all cholinergic (acetylcholine is the neurotransmitter).

7–B. Nissl bodies are large granular basophilic bodies composed of polysomes and rough endoplasmic reticulum. They are found only in neurons (in the soma cytoplasm).

8–C. The axon hillock contains microtubules, which are arranged in bundles.

9–C. Cerebrospinal fluid is poor in protein but rich in sodium, potassium, and chloride.

10–C. Synaptic vesicles release neurotransmitter into the synaptic cleft by exocytosis. In this process, the vesicle membrane is incorporated into the presynaptic membrane. Although these vesicles contain neurotransmitter, they do not manufacture it.

11–D. The reaction of the soma to axonal transection includes movement of the nucleus to the periphery, an increase in soma volume, and dissolution of Nissl bodies (called chromatolysis).

12–C. Schwann cells produce myelin in the PNS, whereas oligodendrocytes produce myelin in the CNS. Astrocytes, neural crest cells, and basket cells do not produce myelin.

13–D. Hirschsprung's disease is characterized by a dilated and hypertrophied colon caused by the absence of the parasympathetic myenteric ganglia located in Auerbach's plexus.

14–C. Cell bodies of preganglionic sympathetic neurons are located in the thoracic spinal cord and first two segments of the lumbar spinal cord. Postganglionic sympathetic cell bodies are located in autonomic ganglia located on both sides of the spinal cord; they are not found in the brain or anywhere in the CNS.

15–C. Schwann cells are located only in the PNS. The cerebral cortex contains pyramidal cells, and the cerebellar cortex contains Purkinje, stellate, and small granule cells.

16–E. The cerebrospinal fluid circulates in the subarachnoid space between the arachnoid and the pia mater, but it is produced by the choroid plexus.

17–B. The biceps muscle, a skeletal muscle under voluntary control, is innervated by the somatic nervous system. The autonomic nervous system innervates smooth muscle in the viscera and blood vessels, cardiac muscle, and certain glands, including the pituitary gland.

18–C. Perineurium.

19–E. Axon.

20–D. Myelin space (myelin has been dissolved away).

10

Blood and Hematopoiesis

I. Overview—Blood

—is a specialized type of connective tissue composed of formed elements (**erythrocytes, leukocytes,** and **platelets**) and a fluid intercellular material (**plasma**).

—occupies a volume of approximately **5 liters** in an average human adult.

—circulates in a closed system of vessels, **transporting** nutrients, waste products, hormones, proteins, ions, and so on.

—**regulates body temperature** and assists in regulation of the **osmotic balance** and **acid–base balance**.

—possesses cells that have a short life span and are replaced continuously by a process known as **hematopoiesis**.

II. Blood Constituents

A. Plasma

—consists of 90% **water,** 9% **organic compounds** (proteins, amino acids, hormones, and so on), and 1% **inorganic salts**.

1. Main plasma proteins

a. Albumin is a small protein (60,000 MW) that preserves osmotic pressure within the vascular system and assists in transporting certain metabolites.

b. γ-Globulins are antibodies (immunoglobulins) [see Chapter 12].

c. α-Globulins and β-globulins transport metal ions (e.g., iron and copper) and lipids (in the form of lipoproteins).

d. Fibrinogen is converted into **fibrin** during blood clotting.

e. Complement proteins (C1 through C9) function in nonspecific host defense and in initiating the inflammatory process.

2. Serum

—is the **yellowish fluid** remaining after blood has clotted.

—is similar in composition to plasma but lacks fibrinogen and other clotting factors.

B. Formed elements of blood (Table 10.1)

1. Erythrocytes

– are also known as **red blood cells** (RBCs).

– are round, **anucleated,** biconcave cells.

– **stain light salmon pink** with either Wright's or Giemsa stain.

– have a life span of about 120 days in the circulation. As they age, erythrocytes become fragile and present membrane surface factors recognized by splenic macrophages, which destroy the old red blood cells.

– have determinants for the **A, B, and O blood groups** located on the external surface of the plasma membrane. These determinants are composed of carbohydrates.

– contain several **cytoskeletal proteins** (ankyrin, band 4.1 protein, spectrin, and actin), which maintain cell shape (see Chapter 1 V A).

– possess no organelles but are filled with **hemoglobin.**

– contain **soluble enzymes** responsible for glycolysis and the hexose monophosphate shunt, which produce ATP needed for the cells' energy requirements.

a. Hematocrit

– is an estimation of the **volume of packed erythrocytes per unit volume of blood,** expressed as a percentage.

– has the following normal values: 40%–50% in adult men; 35%–45% in adult women; 35% in children up to 10 years; and 45%–60% in newborns.

b. Hemoglobin (Hb)

– is a protein composed of four polypeptide chains, each of which is covalently linked to a heme group. Four normal chains, which differ in their amino acid sequence, occur in humans and are designated α, β, γ, and δ.

– occurs in several normal forms differing in their **chain composition**. The predominant form of adult hemoglobin is **HbA$_1$** ($\alpha_2\beta_2$); a minor form is **HbA$_2$** ($\alpha_2\delta_2$). Fetal hemoglobin is designated **HbF** ($\alpha_2\gamma_2$).

Table 10.1. Size and Number of Formed Elements in Human Blood

	Diameter (μm)			
Cell Type	**Smear**	**Section**	**No./mm^3**	**% of Leukocytes**
Erythrocyte	7–8	6–7	5×10^6 (men)	—
			4.5×10^6 (women)	
Agranulocytes				
Lymphocyte	8–10	7–8	1500–2500	20–25
Monocyte	12–15	10–12	200–800	3–8
Granulocytes				
Neutrophil	9–12	8–9	3500–7000	60–70
Eosinophil	10–14	9–11	150–400	2–4
Basophil	8–10	7–8	50–100	0.5–1
Platelet	2–4	1–3	250,000–400,000	—

Reprinted from Gartner LP and Hiatt JL: *Atlas of Histology.* Baltimore, Williams & Wilkins, 1987, p 70.

—exists in several abnormal forms such as **HbS,** which results from a point mutation in the β chain (substitution of valine for glutamic acid). Erythrocytes containing HbS are sickle-shaped and fragile, resulting in **sickle cell anemia**.

c. Transport of O_2 and CO_2

—to and from the tissues of the body is carried out by erythrocytes.

(1) In the lungs, where the partial pressure of O_2 is high, hemoglobin preferentially binds O_2, forming **oxyhemoglobin**.

(2) In the tissues, where the partial pressure of O_2 is low and that of CO_2 is high, oxyhemoglobin releases O_2 and binds CO_2, forming **carboxyhemoglobin**.

(3) Because carbon monoxide (CO) binds very strongly to hemoglobin, it can prevent binding of O_2 and lead to **CO asphyxiation** if inhaled in sufficient amounts.

2. Leukocytes (see Table 10.1)

—are also known as **white blood cells** (WBCs).

—possess varying numbers of **azurophilic granules,** which are considered to be lysosomes and contain various hydrolytic enzymes.

a. Granulocytes (Table 10.2)

—include **neutrophils, eosinophils,** and **basophils**.

Table 10.2. Selected Characteristics of Granulocytes

Characteristic	Neutrophils	Eosinophils	Basophils
Nuclear shape	Lobulated (3 or 4 lobes)	Bilobed	S-shaped
Number of azurophilic granules	Many	Few	Few
Specific granules			
Size	Small	Large	Large
Color*	Light pink	Dark pink	Dark blue to black
Contents	Alkaline phosphatase Collagenase Lactoferrin Lysozyme Phagocytin	Acid phosphatase Arylsulfatase β-Glucuronidase Cathepsin Major basic protein Peroxidase Phospholipase RNase	Eosinophil chemotactic factor (ECF) Heparin Histamine Peroxidase
Life span	1 week	Few hours in blood; 2 weeks in connective tissue	Very long (1–2 years in mice)
Main functions	Phagocytose, kill, and digest bacteria	Moderate inflammatory reactions by inactivating histamine and leukotriene C	Mediate inflammatory responses in a manner similar to mast cells
Special properties	Form H_2O_2 during phagocytosis	Are decreased in number by corticosteroids	Have receptors for IgE on their plasma membrane

*Cells stained with Giemsa or Wright's stain.

—possess **specific granules,** which stain with certain dyes and contain type-specific contents.

—generate ATP via the glycolytic pathway, Krebs cycle (basophils), and anaerobic hexose monophosphate shunt (neutrophils).

b. Agranulocytes (Table 10.3)

—comprise **lymphocytes** and **monocytes** primarily.

—lack specific granules.

—include **null cells,** which constitute about 5% of the circulating lymphocytes. These cells resemble lymphocytes but lack their characteristic surface determinants. Pluripotential hematopoietic stem cells and natural killer (NK) cells are two types of null cell.

3. Platelets (see Table 10.1)

—are also known as **thrombocytes.**

—are **anucleated,** disk-shaped fragments that arise from megakaryocytes in the bone marrow.

—display a clear, peripheral region (**hyalomere**) and a region containing purple granules (**granulomere**) in stained blood smears.

—are surrounded by a **glycocalyx,** which coats the plasma membrane (see Chapter 1 II C). **Calcium ions** and **adenosine diphosphate** (ADP) increase the "stickiness" of the glycocalyx, enhancing adherence of platelets to each other and to vessel walls.

—function in **blood coagulation** by aggregating at lesions in vessel walls and producing various factors that aid in **clot formation.**

—are responsible for **clot retraction** and contribute to **clot removal.**

a. Hyalomere

—contains the following components:

(1) Actin and **myosin,** which permit platelets to contract

(2) Bundles of microtubules, which encircle platelets just deep to the plasma membrane and help maintain their oval shape

(3) Surface-opening (connecting) tubule system, which facilitates the delivery of platelet contents to the outside

(4) Dense tubular system, which consists of irregularly shaped tubules and probably sequesters calcium ions

Table 10.3. Selected Characteristics of Agranulocytes

Characteristic	Monocytes	T Lymphocytes	B Lymphocytes
Plasma membrane	Form filopodia and pinocytic vesicles	Have T-cell receptors	Have Fc receptors and antibodies
Number of azurophilic granules	Many	Few	Few
Life span	Less than 3 days in blood	Several years	Few months
Main functions	Become macrophages in connective tissue	Generate cell-mediated immune response; secrete numerous growth factors	Generate humoral immune response

b. Granulomere

–contains the following granules:

(1) α Granules (300–500 nm in diameter), which contain fibrinogen, platelet thromboplastin, coagulation factors V and VIII and platelet-derived growth factor.

(2) δ Granules (250–300 nm in diameter), also called **dense bodies,** which contain pyrophosphate, ADP, ATP, histamine, serotonin, and calcium ions.

(3) λ Granules (200–250 nm in diameter), which contain lysosomal enzymes; these may assist in clot removal.

III. Blood Coagulation

–contributes to hemostasis.

–is so controlled that normally it occurs **only in regions where the endothelium is damaged**.

–is dependent on the activation of at least 13 plasma proteins, the **coagulation factors** (Figure 10.1), which take part in a cascade of reactions.

–requires the presence of **platelet membranes** and **calcium ions** (factor IV).

–occurs via two interrelated pathways, the **intrinsic** and **extrinsic pathways**. The final steps in both pathways involve the transformation of prothrombin to **thrombin,** an enzyme that catalyzes the conversion of **fibrinogen** (factor I) to **fibrin monomers,** which coalesce to form a **reticulum of clot** (see Figure 10.1).

A. Extrinsic pathway

–is initiated within seconds (**rapid onset**) after trauma that releases **tissue thromboplastin** (factor III).

–occurs in response to damage to the blood vessels.

B. Intrinsic pathway

–is initiated within several minutes (**slow onset**) after trauma.

–occurs in response to trauma to blood vessels or when platelets or factor XII are exposed to collagen in the vessel wall.

–depends on **von Willebrand's factor** (vWF) and **factor VIII,** which form a complex that binds to subendothelial collagen and to receptors on platelet membranes, thus promoting **platelet aggregation** and **adherence to collagen** in the vessel wall.

IV. Bone Marrow

A. Yellow marrow

–is located in the long bones of adults.

–is highly infiltrated with fat.

–is **not** hematopoietic, although it has the potential to become so if necessary.

B. Red marrow

–is located in the epiphyses of long bones, as well as in flat, irregular, and short bones.

–is a highly vascular tissue composed of a **stroma,** large venous **sinusoids,** and many islands of **hematopoietic cells**.

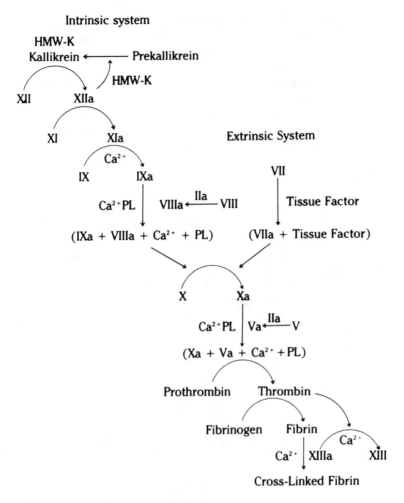

Figure 10.1. Diagram of the cascade of reactions that occurs during blood coagulation. HMWK = high molecular-weight kinogen. Coagulation factors are factor I = fibrinogen; factor II = prothrombin; factor III = tissue thromboplastin; factor IV = calcium; factor V = proaccelerin; factor VII = stable factor; factor VIII = antihemophilic factor; factor IX = plasma thromboplastin (Christmas factor); factor X = Stuart factor; factor XI = plasma thromboplastin antecedent; factor XII = Hageman factor; and factor XIII = fibrin stabilizing factor. (Reprinted with permission from Kjeldsberg C et al: *Practical Diagnosis of Hematologic Disorders.* Chicago, ASCP Press, 1989, p 527.)

—is the site where the various blood cells **differentiate** and **mature** postnatally.

1. **Sinusoids**
   —are large (45–80 μm in diameter) vascular channels with a highly attenuated endothelial lining.
   —are associated on their extravascular surfaces with reticular fibers and **adventitial reticular cells,** which manufacture these fibers.

2. **Stromal cells**
   —include macrophages, adventitial reticular cells, fibroblasts, and endothelial cells.
   —produce and release various **hematopoietic growth factors**.

a. Macrophages

—are located in the extravascular areas near the sinusoids and extend their processes between the endothelial cells into the sinusoidal lumina.

b. Adventitial reticular cells

—are believed to subdivide the bone-marrow cavity into smaller compartments, which are occupied by islands of hematopoietic cells.

—may accumulate fat (instead of fat cells), thus transforming red marrow into yellow marrow.

V. Prenatal Hematopoiesis

—occurs successively in the yolk sac, liver, spleen, and bone marrow.

A. Mesoblastic (yolk-sac) stage

—begins at 2 weeks postconception via the formation of blood islands in the **yolk-sac mesoderm.**

—is characterized by development of the mesoderm cells into large primitive **erythroblasts,** which give rise to **nucleated** erythrocytes. Leukocytes are not present during this stage.

B. Hepatic and splenic stages

—begin about 6 weeks postconception.

1. Erythroblasts are similar to those of the postnatal individual, but **erythrocytes are nucleated**.

2. Leukocytes begin to develop at approximately 2 months' gestation in the liver and somewhat later in the spleen.

C. Bone-marrow stage

1. The bone marrow first participates in hematopoiesis at about 6 months' gestation and assumes an increasingly larger role thereafter.

2. The liver and spleen cease to participate in hematopoiesis about the time of birth.

VI. Postnatal Hematopoiesis

—involves three classes of cells: **stem, progenitor,** and **precursor**.

A. Comparison of stem, progenitor, and precursor cells

1. Stem cells

—are capable of **self-renewal** and when necessary can undergo enormous proliferation.

—can differentiate into **multiple** cell lineages.

—are present in the circulation (as members of the null cell population) and in the bone marrow.

2. Progenitor cells

—have reduced potentiality and generally are **committed to a single cell lineage**.

—**proliferate and differentiate** into precursor cells in the presence of appropriate growth factors.

—are morphologically indistinguishable from stem cells; both classes of cells appear similar to small lymphocytes.

3. Precursor cells

—are all the cells in each lineage that display **distinct morphologic characteristics**.

—are formed in succession in each cell lineage.

B. Initial steps in blood formation

1. Pluripotential hematopoietic stem cells (PHSCs)

—give rise to multipotential hematopoietic stem cells in the bone marrow.

2. Multipotential hematopoietic stem cells

—are of two types: **CFU-S** (colony-forming unit–spleen) and **CFU-Ly** (colony-forming unit–lymphocyte).

—undergo cell division and differentiation in the bone marrow to form progenitor cells.

a. CFU-S, the myeloid stem cell, is the multipotential stem cell that gives rise to erythrocytes, granulocytes, monocytes, and platelets.

b. CFU-Ly, the lymphoid stem cell, is the multipotential stem cell that gives rise to T and B lymphocytes and natural killer cells.

C. Erythrocyte formation (erythropoiesis)

—begins with formation of two types of progenitor cells: **BFU-E,** derived from CFU-S, and **CFU-E,** which arises from BFU-E.

—occurs at the rate of about one trillion erythrocytes daily in a normal human adult.

1. Erythroid progenitor cells

a. BFU-E (burst-forming unit–erythroid) has a high rate of mitotic activity and responds to high concentrations of **erythropoietin,** a hormone that stimulates erythropoiesis.

b. CFU-E (colony-forming unit–erythroid) responds to low concentrations of erythropoietin and gives rise to the first histologically recognizable erythrocyte precursor, the **proerythroblast**.

2. Erythrocyte precursor cells

—include a series of cell types (the **erythroid series**) that differentiate sequentially to form mature erythrocytes (Table 10.4).

D. Granulocyte formation

—begins with production of three uni- or bipotential progenitor cells, all of which are descendants of CFU-S.

—occurs at the rate of about one million granulocytes daily in a normal human adult.

1. Granulocyte progenitor cells

—give rise to histologically identical **myeloblasts** and **promyelocytes** in all three cell lineages.

a. CFU-Eo is the progenitor of the **eosinophil** lineage.

b. CFU-Ba is the progenitor of the **basophil** lineage.

Table 10.4. Selected Characteristics of Erythrocyte Precursor Cells

Characteristic	Proerythro-blast	Basophilic Erythroblast	Polychro-matophilic Erythroblast	Orthochro-matophilic Erythroblast	Reticulo-cyte
Nucleus					
Shape	Round	Round	Round and small	Round	None
Color*	Burgundy red	Burgundy red	Dense blue	Dark, may be extruding	None
Chromatin network	Very fine	Fine	Coarse	Pyknotic	None
Number of nucleoli	3–5 (pale gray)	1–2	None	None	None
Mitosis	Yes	Yes	Yes	No	No
Cytoplasmic color*	Pale gray with blue clumps	Grayish pink with in-tensely blue clumps	Yellowish pink with bluish back-ground	Pink with trace of blue	Pink[†]
Hemoglobin	None (ferritin is present)	Some	Abundant	Abundant	Abundant

*Cells stained with Giemsa or Wright's stain.
[†]Cells stained supravitally with brilliant cresyl blue display a reticulum. Ubiquitin destroys the remaining organelles.

 c. CFU-NM is the common progenitor of the **neutrophil** and **monocyte** lineages. CFU-NM gives rise to two unipotential progenitors: **CFU-N** (neutrophil) and **CFU-M** (monocyte).

2. Granulocyte precursor cells

 —are histologically similar in the early stages of all three lineages.
 —develop characteristic types of granules during the myelocyte stage and a distinctive nuclear shape during the stab (band) stage (Table 10.5).

E. Monocyte formation

 —diverges from neutrophil pathway with formation of **CFU-M,** which arises from the common progenitor CFU-NM.
 —involves only two precursor cells: **monoblasts** and **promonocytes**.
 —occurs at the rate of about ten trillion monocytes daily in a normal human adult.

1. Promonocytes

 —are reported to be large cells (16–18 μm in diameter).
 —contain a somewhat kidney-shaped, acentrically located nucleus, numer-ous azurophilic granules (lysosomes), an extensive Golgi complex, many mitochondria, and considerable rough endoplasmic reticulum (RER).
 —undergo cell division and subsequently develop into **monocytes**.

2. Monocytes

 —leave the bone marrow to enter the circulation.
 —**differentiate into macrophages** after they enter connective tissue.

F. Platelet formation

 —begins with formation of the unipotential progenitor **CFU-Meg,** which arises from CFU-S.

Table 10.5. Selected Characteristics of Neutrophil Precursor Cells

Characteristic	Myeloblast	Promyelocyte	Neutrophilic Myelocyte	Neutrophilic Metamyelocyte	Neutrophilic Stab Cell
Cell diameter (μm)	12–14	16–24	10–12	10–12	11–12
Nucleus					
Shape	Large, round	Large, round	Flat (acentric)	Kidney-shaped (acentric)	Horseshoe-shaped
Color*	Reddish blue	Reddish blue	Blue to dark blue	Dark blue	Dark blue
Chromatin network	Very fine	Fine	Coarse	Very coarse	Very coarse
Number of nucleoli	2 or 3 (pale gray)	1 or 2 (pale gray)	1 (?)	None	None
Mitosis	Yes	Yes	Yes	No	No
Cytoplasmic appearance*	Blue clumps in pale blue background; cytoplasmic blebs at cell periphery	Bluish hue; no cytoplasmic blebs	Pale blue	Blue	Similar to mature neutrophils
Granules	None	Azurophilic	Azurophilic and specific	Azurophilic and specific	Azurophilic and specific

*Cells stained with Giemsa or Wright's stain.

—involves a single precursor cell, the **megakaryoblast,** and the mature **megakaryocyte,** which remains in the bone marrow and sheds platelets.

1. Megakaryoblasts

—are large cells (25–40 μm in diameter), whose single, large nucleus may be indented or lobed, but displays a fine chromatin network.

—divide endomitotically (i.e., no daughter cells are formed), becoming huge with the ploidy of the nucleus increasing to as much as 64*N*.

—have a basophilic, nongranular cytoplasm containing large mitochondria, numerous polysomes, some rough endoplasmic reticulum, and a fairly well-developed Golgi complex.

—give rise (endomitotically) to megakaryocytes.

2. Megakaryocytes

—are extremely large cells (40–100 μm in diameter), whose single, large **polypoid** nucleus is highly indented.

—possess a well-developed Golgi complex associated with the formation of α granules, lysosomes, and dense bodies (δ granules).

—contain numerous mitochondria and a sizable rough endoplasmic reticulum network.

—lie just outside the sinusoids in the bone marrow.

—form **platelet demarcation channels,** which fragment to proplatelets (clusters of adhering platelets) or single platelets; these are released into the sinusoidal lumen.

G. Lymphocyte formation (lymphopoiesis)

—begins with differentiation of CFU-Ly, the lymphoid stem cell, into two uni-potential, **immunoincompetent** progenitor cells, **CFU-LyB** and **CFU-LyT**. These **prelymphocytes** undergo processing to become mature, immunocompetent cells.

1. B-lymphocyte maturation

—involves acquisition by pre-B lymphocytes of cell-surface markers, including membrane-bound **antibodies,** which confer **immunocompetence**.

—occurs in birds in an organ called the bursa of Fabricius (hence, **B** lymphocytes).

—occurs in mammals in a bursal equivalent site, most probably the bone marrow.

2. T-lymphocyte maturation

—involves migration of progenitor T lymphocytes to the **thymus,** where they acquire cell-surface markers, including **T-cell receptors,** which confer **immunocompetence** (see Chapter 12 II A 1). Most of the newly formed T lymphocytes are destroyed in the thymus and never enter the circulation.

3. Mature B and T lymphocytes

—migrate from the bone marrow and thymus, respectively, to various peripheral organs (e.g., the lymph nodes and spleen) where they establish **clones** of immunocompetent (antigen-specific) lymphocytes. (The immunologic functions and interactions of the various classes of B and T lymphocytes are discussed in Chapter 12).

VII. Hematopoietic Growth Factors

—include **erythropoietin,** several acidic glycoproteins called **colony-stimulating factors** (CSFs), and various **interleukins**.

—facilitate and stimulate formation of blood cells.

—may act as hormones, circulating in the bloodstream, or as local factors produced in the bone marrow.

—act at low concentrations and bind to specific membrane receptors on their target cells.

—may act on a single type of target cell or on several types.

—have various effects on target cells including the following: control their mitotic rate; enhance their survival; control the number of times they divide before they differentiate; and promote their differentiation.

A. Erythropoietin

—is produced in the kidney (perhaps by mesangial cells).

—**acts on CFU-E and BFU-E** to induce formation of erythrocytes.

B. Interleukin 3 (IL-3)

—is also known as **multi-CSF**.

—is produced by T lymphocytes and bone-marrow stromal cells.

—**acts on PHSC, CFU-S, and most myeloid progenitor cells** to trigger formation of erythrocytes, granulocytes, and monocytes.

C. Granulocyte–macrophage colony-stimulating factor (GM-CSF)

   —is produced by T lymphocytes, macrophages, fibroblasts, and bone-marrow stromal cells.

   —**acts on granulocyte and monocyte progenitor cells** to trigger formation of granulocytes and monocytes.

D. Granulocyte colony-stimulating factor (G-CSF)

   —is produced by macrophages, fibroblasts, and bone-marrow stromal cells.

   —**acts on CFU-NM and CFU-N** to stimulate differentiation into neutrophil precursor cells.

E. Macrophage colony-stimulating factor (M-CSF)

   —is produced by macrophages and bone-marrow stromal cells.

   —**acts on CFU-NM and CFU-M** to stimulate differentiation into monocytes, which become macrophages.

   —also facilitates antitumor activity.

F. Interleukin 7 (IL-7)

   —is produced by bone-marrow stromal cells.

   —**acts on lymphocyte progenitor cells** stimulating formation of B and T lymphocytes.

VIII. Clinical Considerations

A. Erythrocytic disorders

   1. Hereditary spherocytosis

   —is discussed in Chapter 1 VI E 2.

   2. Sickle cell anemia

   —is caused by a point mutation in the DNA encoding the hemoglobin molecule, leading to production of an **abnormal hemoglobin** (HbS). Affected individuals are resistant to malaria.

   —occurs almost exclusively among people of African descent (1 in 600 are affected in the United States).

   —is marked by crystallization of hemoglobin under low oxygen tension, giving erythrocytes the characteristic sickle shape. **Sickled erythrocytes are fragile and undergo a higher rate of destruction** than normal cells in the spleen.

   —is associated with hypoxia, increased bilirubin levels, a low red blood count, and capillary stasis.

   3. Iron-deficiency anemia

   —is one of the most common of all nutritional deficiencies.

   —is usually caused by malabsorption, chronic blood loss, or increased demand for iron (e.g., during pregnancy).

   —is **usually asymptomatic,** although in severe cases immunocompetence and brain function may be impaired.

   4. Pernicious anemia

   —is caused by a severe **deficiency of vitamin B_{12},** resulting from impaired production of **gastric intrinsic factor** by the parietal cells of the stomach. This factor is required for proper absorption of vitamin B_{12}.

a. Symptoms

- **(1)** Generalized weakness and loss of consciousness (presenting symptoms)
- **(2)** Neurologic impairments such as bilateral tingling and numbness in the hands and feet, as well as loss of proprioception in the toes

b. Signs

- **(1)** Demyelination of peripheral and spinal cord axons
- **(2)** Presence of numerous large, abnormal erythroblasts in the bone marrow and reduced numbers of precursor cells later in the erythroid series

B. Leukocytic disorders

1. Infectious mononucleosis

- —mostly affects young individuals of high school and college age.
- —is caused by **Epstein-Barr virus,** which is related to herpes virus.
- —is characterized by fatigue, swollen and tender lymph nodes, fever, sore throat, and an increase in blood lymphocytes (many of them atypical lymphocytes).
- —may be transmitted by saliva (as in kissing).
- —may be serious in immunosuppressed or immunodeficient individuals whose B lymphocytes can undergo intense proliferation, leading to death.

2. Hodgkin's disease

- —is a malignancy involving neoplastic transformation of **lymphocytes** (most probably). (See Chapter 12 VII E for discussion of this disease.)

3. Leukemias

- —are characterized by the replacement of normal hematopoietic cells of the bone marrow by neoplastic cells.
- —are classified according to the **type** and **maturity** of the cells involved.

a. Acute leukemias

- —occur primarily in children and involve relatively **immature cells** (i.e., blasts).
- —show **rapid onset** of the following signs: anemia; high white blood cell count and/or many immature leukocytes in the circulation; depressed platelet count; tenderness of bones; enlarged lymph nodes, spleen, and liver; vomiting; and headache.

b. Chronic leukemias

- —occur primarily in adults and elderly individuals.
- —initially involve relatively **mature cells**.
- —show **slow onset** of a mild leukocytosis and enlarged, asymptomatic lymph nodes early; later signs include anemia, weakness, enlarged spleen and liver, and reduced platelet count.

C. Bleeding disorders

1. Thrombocytopenia

- —is a **reduced platelet count** (less than 100,000/mm^3).
- —can be caused by a decrease in the rate of platelet formation or an increase in their rate of destruction.

−is marked by **spontaneous bleeding** (when platelet count is less than 20,000/mm^3) with a protracted bleeding time but a normal coagulation time.

−involves bleeding of small veins, resulting in **petechiae** (small hemorrhagic spots on the skin and mucosa).

−is **common in AIDS patients** due to platelet injury by the immune system or to virus-mediated depression of megakaryocytes.

2. Coagulation disorders

−result from inherited or acquired defects of the coagulation factors. Factor VIII deficiency and von Willebrand's disease are the two most common inherited coagulation disorders.

a. Factor VIII deficiency (hemophilia A)

−is an X chromosome-linked disorder and affects men more frequently than women.

−occurs with variable degrees of severity, depending on the extent of the reduction in the level of factor VIII, which is produced by hepatocytes.

−results in excessive bleeding, especially bleeding into the joints, in severe cases.

−is associated with a normal platelet count, normal bleeding time, and absence of petechiae, but thromboplastin time is increased.

b. Von Willebrand's disease

−is caused by a genetic defect that results in a **decrease in the amount of von Willebrand's factor,** which is required in the intrinsic pathway of coagulation.

−exhibits an autosomal dominant mode of transfer.

−in most cases is mild and does not lead to bleeding into the joints.

−in severe cases is characterized by excessive and/or spontaneous bleeding from mucous membranes and from wounds.

Review Test

Directions: Each of the numbered items or incomplete statements in this section is followed by answers or by completions of the statement. Select the **one** lettered answer or completion that is **best** in each case.

1. Which one of the following substances is NOT contained in serum?

(A) α Globulins
(B) γ Globulins
(C) Albumin
(D) Fibrinogen

2. Which of the following proteins associated with the erythrocyte plasma membrane is responsible for maintaining the cell's biconcave disk shape?

(A) HbA_1
(B) HbA_2
(C) Porphyrin
(D) Spectrin
(E) α-Actinin

3. Energy production in mature erythrocytes is accomplished by which one of the following?

(A) Glycolytic pathway
(B) Mitochondria
(C) Electron transport system
(D) Citric acid cycle (Krebs cycle)

4. All of the following cells possess azurophilic granules EXCEPT

(A) monocytes
(B) lymphocytes
(C) myeloblasts
(D) promyelocytes
(E) basophils

5. BFU-E gives rise to which of the following blood cells?

(A) Neutrophils
(B) Basophils
(C) Monocytes
(D) Erythrocytes
(E) Lymphocytes

6. Leukotriene is produced by which of the following cells?

(A) Neutrophils
(B) Basophils
(C) Eosinophils
(D) Erythrocytes
(E) Lymphocytes

Directions: Each group of items in this section consists of lettered options followed by a set of numbered items. For each item, select the **one** lettered option that is most closely associated with it. Each lettered option may be selected once, more than once, or not at all.

Questions 7–15

Match each description below with the corresponding lettered structure shown in the photomicrograph.

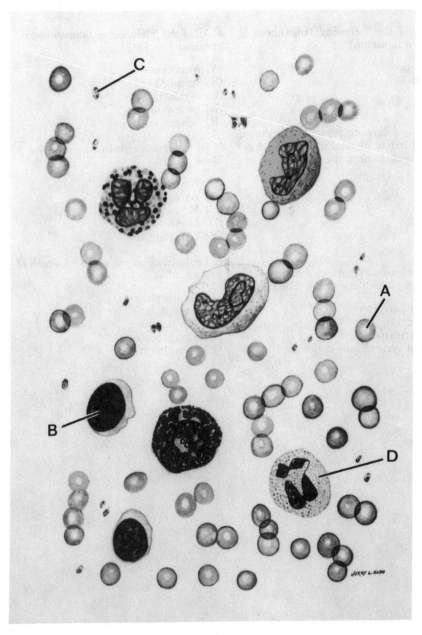

Reprinted from Gartner LP and Hiatt JL: *Atlas of Histology*. Baltimore, Williams & Wilkins, 1987, p 74.

7. Is immunocompetent

8. Is derived from CFU-Meg

9. Is derived from CFU-E

10. Is derived from CFU-NM

11. Is associated with demarcation channels

12. Is derived from myeloblasts

13. Is associated with antibody production

14. Possesses specific granules

15. Is derived from reticulocytes

Questions 16–19

Match each sign or symptom below with the disorder that is most closely associated with it.

(A) Hereditary spherocytosis
(B) Pernicious anemia
(C) Infectious mononucleosis
(D) Chronic leukemia
(E) Von Willebrand's disease

16. Excessive and/or spontaneous bleeding from mucous membranes

17. Defective spectrin

18. Swollen and tender lymph nodes, increase in serum lymphocytes, sore throat, and fever

19. Vitamin B_{12} deficiency

Answers and Explanations

1–D. Serum is the fluid that is expressed from plasma subsequent to clotting. Therefore, it contains globulins and albumin, but lacks fibrinogen, which is converted to fibrin during the clotting process.

2–D. Spectrin is associated with the erythrocyte cell membrane and assists in maintaining its biconcave disk shape.

3–A. Erythrocytes fulfill their energy requirement via the glycolytic pathway and the hexose monophosphate shunt.

4–C. Myeloblasts do not possess azurophilic granules (lysosomes).

5–D. BFU-E (burst-forming unit–erythroid) gives rise to CFU-E (colony-forming unit–erythroid) and thus is an erythrocyte progenitor cell.

6–B. Basophils produce and release several pharmacologic agents including leukotriene (a slow-reacting substance of anaphylaxis, SRS-A).

7–B. Lymphocyte.

8–C. Platelet.

9–A. Erythrocyte.

10–D. Neutrophil.

11–C. Platelet.

12–D. Neutrophil.

13–B. Lymphocyte.

14–D. Neutrophil.

15–A. Erythrocyte.

16–E. Von Willebrand's disease results from a decrease in von Willebrand's factor, which along with factor VIII promotes platelet aggregation and adherence to subendothelial collagen. Severe cases of this coagulation disorder are marked by excessive, sometimes spontaneous, bleeding from mucous membranes.

17–A. A genetic defect in spectrin that decreases its ability to bind to band 4.1 protein causes hereditary spherocytosis.

18–C. Infectious mononucleosis, caused by the Epstein-Barr virus, is characterized by swollen and tender lymph nodes, sore throat, fever, fatigue, and an increase in blood lymphocytes, many of which are atypical.

19–B. A severe deficiency of vitamin B_{12}, resulting from impaired production of gastric intrinsic factor, causes pernicious anemia.

11

Circulatory System

I. Blood Vascular System

—consists of the heart, arteries, veins, and capillaries.
—transports **oxygen and nutrients to the tissues.**
—carries away **carbon dioxide and waste products from the tissues.**
—circulates **hormones** from their site of synthesis to their target cells.

A. Heart

—is a four-chambered muscular organ composed of two **atria,** which receive the blood, and two **ventricles,** which pump the blood to the lungs or tissues.
—is surrounded by a fibroserous sac called the **pericardium.**
—receives **sympathetic** and **parasympathetic** nerve fibers, which **modulate the rate of the heartbeat** but do not initiate it.
—produces **atrial natriuretic peptide,** a hormone that increases secretion of sodium and water by the kidneys; inhibits renin release; and decreases blood pressure.

1. Cardiac layers (tunicae)

a. Endocardium

—lines the lumen of the heart and is composed of simple squamous epithelium (**endothelium**) and a thin layer of loose connective tissue.
—is continuous with the intima of the blood vessels leaving and entering the heart.
—is underlain by **subendocardium,** a connective tissue layer that contains veins, nerves, and Purkinje fibers.

b. Myocardium

—consists of layers of **cardiac muscle cells** arranged in a spiral fashion about the heart's chambers and inserted into the fibrous skeleton.
—contracts to propel blood into the arteries for distribution to the body tissues.

c. Epicardium

—is the outermost layer of the heart and constitutes the **visceral layer of the pericardium.**

–is composed of simple squamous epithelium (**mesothelium**) on the external surface; fibroelastic connective tissue containing nerves and the coronary vessels; and adipose tissue.

2. Fibrous skeleton of the heart

–consists of thick bundles of **collagen fibers** oriented in various directions.

–also contains occasional foci of fibrocartilage.

3. Heart valves

–are composed of a skeleton of fibrous connective tissue, arranged like an aponeurosis, and lined on both sides by endothelium.

–are attached to the **annuli fibrosi** of the fibrous skeleton.

–include the atrioventricular valves, aortic valve, and pulmonary trunk valve.

4. Impulse-generating and impulse-conducting system

–comprises several specialized structures whose coordinated functions act to **initiate and regulate the heartbeat**.

a. Sinoatrial (SA) node

–is the **"pacemaker"** of the heart, located in the wall of the right atrium near the entrance of the superior vena cava.

–**generates impulses** that initiate contraction of atrial muscle cells; the impulses are then conducted to the atrioventricular node.

b. Atrioventricular (AV) node

–is located in the wall of the right atrium, adjacent to the tricuspid valve.

–is similar in structure to the SA node and can become the pacemaker in case of SA-node pathology.

c. Atrioventricular bundle of His

–is a band of conducting tissue radiating from the AV node into the interventricular septum where it divides into two branches, enters the subendocardium, and continues as Purkinje fibers.

d. Purkinje fibers

–are large, modified cardiac muscle cells (see Chapter 8 V B 7).

–make contact with cardiac muscle cells at the apex of the heart via gap junctions, desmosomes, and fasciae adherentes.

B. Arteries

–conduct blood **away from the heart** to the organs and tissues.

–possess walls composed of three layers: the **tunica intima** (inner), **tunica media** (middle), and **tunica adventitia** (outer). The components of these layers and the variations among the different types of arteries are summarized in Table 11.1.

1. Types of arteries

a. Elastic arteries

–are also called **conducting arteries** because they conduct the blood directly from the heart.

Table 11.1. Comparison of Tunicae in Different Types of Arteries*

Tunica Components	Elastic Arteries	Muscular Arteries	Arterioles	Metarterioles
Intima				
Endothelium	+	+	+	+
Factor VIII in endothelium	+	+	+	–
Basal lamina	+	+	+	+
Subendothelial layer[†]	+	+	±	–
Internal elastic lamina	Incomplete	Thick, complete	Some elastic fibers	–
Media				
Fenestrated elastic membranes	40–70	–	–	–
Smooth muscle cells	Interspersed between elastic membranes	Up to 40 layers	1 or 2 layers	Discontinuous layer
External elastic lamina	Thin	Thick	–	–
Vasa vasorum	±	–	–	–
Adventitia				
Fibroelastic connective tissue	Thin layer	Thin layer	–	–
Loose connective tissue	–	–	+	±
Vasa vasorum	+	±	–	–
Lymphatic vessels	+	+	–	–
Nerve fibers	+	+	+	–

*+ indicates component is present and prominent; ± indicates component is present but is not prominent; – indicates component is absent.
[†]In elastic arteries, the subendothelial layer is composed of loose connective tissue containing fibroblasts, collagen, and elastic fibers. In arterioles, this layer is less prominent; the connective tissue is sparse and contains a few reticular fibers.

  –constitute the largest arteries and include the **aorta** and its **major branches**.
  –help to **reduce changes in blood pressure** associated with the heartbeat.
  –possess small vessels (**vasa vasorum**) and nerves in their tunicae adventitia and media.

b. Muscular arteries

  –are also called **distributing arteries** because they distribute the blood to various organs.
  –include most of the **named** arteries of the human body.
  –are medium sized, smaller than elastic arteries but larger than arterioles.
  –possess vasa vasorum and nerves in their tunica adventitia.

c. Arterioles

  –are unnamed.
  –are the **smallest** arteries with diameters less than 0.1 mm and a narrow lumen, whose diameter usually equals the wall thickness.
  –**regulate blood pressure.**

d. Metarterioles
—are narrow vessels that arise from arterioles.

—give rise to **capillaries,** which are surrounded by rings of smooth muscle cells (**precapillary sphincters**) at their origin. Constriction of precapillary sphincters prevents blood from entering the capillary bed.

2. Vasoconstriction
—primarily involves **arterioles** and reduces blood flow to a specific region.

—is stimulated by unmyelinated **sympathetic nerve fibers** (see Chapter 9 IX B).

3. Vasodilation
—is effected by **parasympathetic nerve fibers** as follows:

a. Acetylcholine released from the nerve terminals causes release of **endothelial-derived relaxing factor** (**EDRF**) from the vessel endothelium.

b. **EDRF** (now known to be **nitric oxide**) diffuses to smooth muscle cells in the vessel wall and activates their cGMP system, resulting in relaxation.

4. Sensory nerve endings
—are located in the walls of arteries.

a. Baroreceptors
—respond to changes in blood pressure.

—are primarily represented by the **carotid sinus**.

b. Chemoreceptors
—respond to changes in the blood levels of oxygen and carbon dioxide and to changes in blood pH.

—help to regulate respiration.

—are primarily represented by the **carotid** and **aortic bodies**.

C. Capillaries
—are small vessels (about 8–10 μm in diameter and less than 1 mm long).

—form a rich network (**capillary bed**) interposed between arterioles and venules.

—consist of a **single layer of endothelial cells** rolled into a cylinder, which is surrounded by a basal lamina and occasional **pericytes** (see Chapter 6 III B).

—exhibit **selective permeability,** permitting the exchange of oxygen, carbon dioxide, metabolites, and other substances between the blood and tissues.

—are classified into three types, depending on the structure of their endothelial cells and the continuity of the basal lamina.

1. Capillary endothelial cells
—are polygonal in shape with an attenuated cytoplasm and a nucleus that bulges into the capillary lumen.

—possess a Golgi complex near the nucleus, free ribosomes, mitochondria, and some rough endoplasmic reticulum (RER).

—contain intermediate filaments (**vimentin**) in the perinuclear zone; these probably have a supportive function.

−generally are joined by **fasciae occludentes** (tight junctions); some desmosomes and gap junctions also are present (see Chapter 5 III).

2. Types of capillaries

a. Continuous (somatic) capillaries (Figure 11.1)

−lack fenestrae in their walls.

Figure 11.1. Electron micrograph of a continuous capillary in heart muscle. The nucleus (*N*) of an endothelial cell bulges into the capillary lumen occupied by a red blood cell. *LD* = lamina densa; *LL* = lamina lucida. × 29,330. (Reprinted from Gartner LP and Hiatt JL: *Atlas of Histology*. Baltimore, Williams & Wilkins, 1987, p 124.)

–contain numerous **pinocytic vesicles** except in the central nervous system (CNS), where these capillaries contain only a limited number of pinocytic vesicles (a property that partly is responsible for the blood–brain barrier).

–have a **continuous** basal lamina.

–are located in nervous tissue, muscle, connective tissue, exocrine glands, and the lungs.

b. Fenestrated (visceral) capillaries

–are formed of endothelial cells whose walls are perforated with **fenestrae**. These openings are 60–80 nm in diameter and **bridged by a diaphragm** thinner than a cell membrane; in the renal glomerulus, the fenestrae are larger and lack a diaphragm.

–contain only a few pinocytic vesicles.

–have a **continuous** basal lamina.

–are located in endocrine glands, the intestine, the pancreas, and the glomeruli of the kidneys.

c. Sinusoidal capillaries

–possess many large **fenestrae that lack diaphragms**.

–have a diameter of 30–40 μm, which is much larger than that of continuous and fenestrated capillaries.

–lack pinocytic vesicles.

–have a **discontinuous** basal lamina.

–are located in the liver, spleen, bone marrow, lymph nodes, and adrenal cortex.

3. Permeability of capillaries

–is dependent on the morphology of their endothelial cells as well as the size, charge, and shape of the transported molecules.

–is altered during the inflammatory response by **histamine** and **bradykinin**.

a. Some substances **diffuse** and some are **actively transported** across the plasma membrane of capillary endothelial cells.

b. Other substances move across capillary walls via **small pores** (intercellular junctions) or **large pores** (fenestrae and pinocytic vesicles).

c. Leukocytes leave the bloodstream and enter the tissue spaces by squeezing through the intercellular junctions, a process called **diapedesis**.

4. Metabolic functions of capillaries

–are carried out by the endothelial cells and include the following:

a. Conversion of inactive angiotensin I to active angiotensin II (especially in the lung), which is a powerful vasoconstrictor and stimulates secretion of aldosterone (In the kidney, aldosterone promotes the retention of water [see Chapter 18 III A 4 b]).

b. Deactivation of various pharmacologically active substances (e.g., bradykinin, serotonin, thrombin, norepinephrine, prostaglandins)

c. Breakdown of lipoproteins to yield triglycerides and cholesterol

d. Release of prostacyclin, a potent vasodilator and inhibitor of platelet aggregation

5. Blood flow into capillary beds

—occurs either from **metarterioles** (with precapillary sphincters) or from **terminal arterioles**.

a. Metarterioles constitute the proximal portion of **central channels,** and **thoroughfare channels** (with no precapillary sphincter) form their distal portion.

b. Thoroughfare channels receive blood from capillary beds and drain into **small venules**.

6. Bypassing a capillary bed

a. Contraction of precapillary sphincters

—forces the blood flow through **central channels,** thus bypassing the capillary bed. In this case, blood flows directly from the metarteriole into a thoroughfare channel, and from there into a venule.

b. Arteriovenous anastomoses

—are small vessels or occasionally **glomera** (e.g., in the nail bed and fingertip) that directly connect arterioles to venules, thus bypassing the capillary bed.

—function in **thermoregulation** and in the control of blood flow and blood pressure.

D. Veins

—conduct blood away from the organs and tissues **to the heart**.

—contain about 70% of the total blood volume at any given time.

—possess walls composed of three layers: the **tunica intima** (inner), **tunica media** (middle), and **tunica adventitia** (outer), which is the thickest and most prominent. The components of these layers and the variations among different types of veins are summarized in Table 11.2.

—have thinner walls and larger, more irregular lumina than the corresponding arteries.

—may have valves in their lumina that prevent retrograde flow of the blood.

1. Large veins

—include the vena cava, pulmonary veins, and mesenteric vein.

—possess **cardiac muscle** in the tunica adventitia for a short distance as they enter the heart. This layer also contains vasa vasorum and nerves in large veins.

2. Small and medium-sized veins

—include veins the size of the external jugular vein.

—have a diameter of 1–9 mm.

3. Venules

—are unnamed.

—have a diameter of 0.2–1 mm.

—are involved in **exchange of metabolites** with tissues and in **diapedesis**.

II. Lymphatic Vascular System

—consists of peripheral lymphatic capillaries, lymphatic vessels of gradually increasing size, and lymphatic ducts.

Table 11.2. Comparison of Tunicae in Different Types of Veins*

Tunica Components	Large Veins	Medium and Small Veins	Venules
Intima			
Endothelium	+	+	+
Basal lamina	+	+	+
Valves	In some	In some	−
Subendothelial layer	+	+	−
Media			
Connective tissue	+	Reticular and elastic fibers	±
Smooth muscle cells	+	+	±
Adventitia			
Smooth muscle cells	Longitudinally oriented bundles	−	−
Collagen layers with fibroblasts	+	+	+

*+ indicates component is present and prominent; ± indicates component is present but is not prominent; − indicates component is absent.

—**collects excess tissue fluid (lymph) and returns it to the venous system.**
—drains most tissues with the exception of the nervous system and bone marrow.

A. Lymphatic capillaries

—are thin-walled vessels that begin as **blind-ended channels**.
—are composed of a single layer of **attenuated endothelial cells** that lack fenestrae and fasciae occludentes.
—are held open by small **elastic filaments,** which also anchor them to the surrounding connective tissue.
—possess a sparse basal lamina.

B. Large lymphatic vessels

—possess valves and are similar in structure to small veins, except that they have larger lumina and thinner walls.
—have lymph nodes interposed along their routes that filter the lymph.
—converge and become two large trunks, called the **thoracic duct** and **right lymphatic duct**.

C. Lymphatic ducts

—are similar in structure to large veins.
—contain longitudinal and circular layers of smooth muscle cells in their tunica media.
—have a poorly developed tunica adventitia, which contains vasa vasorum and nerves.
—empty into the junction of the internal jugular and subclavian veins.

III. Clinical Considerations

A. Aneurysm

—is a **ballooning out of an artery**.

–occurs because of a weakness in the arterial wall, which may result from an age-related displacement of elastic fibers by collagen.

–may be associated with **atherosclerosis, syphilis,** or **connective tissue disorders** such as Marfan's syndrome or Ehlers-Danlos syndrome.

–can be life-threatening since the weakness of the wall may cause the artery to burst.

B. Atherosclerosis

–is characterized by deposits of yellowish plaques (**atheromas**) in the intima of large and medium-sized arteries.

–may result in blockage of blood flow to the region supplied by an affected artery.

C. Rheumatic heart-valve disease

–is a **sequel to childhood rheumatic fever** (subsequent to streptococcal infection), which causes scarring of the heart valves.

–is characterized by reduced elasticity of the heart valves, making them unable to close (**incompetence**) or open (**stenosis**) properly.

–most commonly affects the **mitral valve,** followed by the aortic valve.

D. Ischemic (coronary) heart disease

–is usually caused by **coronary atherosclerosis,** which results in decreased blood flow to the myocardium.

–may result (depending on its severity) in angina pectoris, myocardial infarct, chronic ischemic cardiopathy, or sudden cardiac death.

E. Tetralogy of Fallot

–is a **congenital malformation** of the cardiovascular system consisting of a defective interventricular septum, hypertrophy of the right ventricle (due to a narrow pulmonary artery or valve), and transposed (dextroposed) aorta.

–should be **repaired surgically** early in life before the pulmonary constriction becomes exacerbated.

F. Varicose veins

–are abnormally tortuous, **dilated veins,** usually of the **leg**.

–are caused by a decline in muscle tone, degenerative alteration of the vessel wall, and valvular incompetence.

–generally occur in older persons.

–when they occur in the region of the anorectal junction, they are known as **hemorrhoids**.

G. Metastasis of malignant tumors

–is the transfer of tumor cells from an affected to an unaffected region of the body.

–occurs via **lymph vessels** (typical of carcinomas) or **blood vessels,** especially veins (more typical of sarcomas than carcinomas).

Review Test

Directions: Each of the numbered items or incomplete statements in this section is followed by answers or by completions of the statement. Select the **one** lettered answer or completion that is **best** in each case.

1. Which of the following statements concerning the epicardium is TRUE?

(A) It is continuous with the endocardium
(B) It is the visceral pericardium
(C) It possesses modified cardiac muscle cells
(D) It functions to increase intraventricular pressure

2. The atrial muscle of the heart produces a hormone that

(A) decreases blood pressure
(B) increases blood pressure
(C) causes vasoconstriction
(D) facilitates the release of renin
(E) facilitates sodium resorption in the kidneys

3. The generation of impulses in the normal heart is the responsibility of which of the following structures?

(A) Atrioventricular (AV) node
(B) Atrioventricular bundle of His
(C) Sympathetic nerves
(D) Sinoatrial (SA) node
(E) Purkinje fibers

4. Which of the following are present in elastic arteries but are not seen in muscular arteries?

(A) Elastic fibers
(B) Factor VIII
(C) Fenestrated membranes
(D) External elastic lamina
(E) Basement membrane

5. Which of the following statements concerning metarterioles is FALSE?

(A) They control blood flow into capillary beds
(B) They possess an incomplete layer of smooth muscle cells
(C) They receive blood from thoroughfare channels
(D) They possess precapillary sphincters

6. Which of the following statements concerning innervation of blood vessels is FALSE?

(A) The carotid body monitors blood oxygen levels
(B) The carotid sinus is a pressure receptor
(C) Acetylcholine acts directly on smooth muscles
(D) Vasoconstriction is controlled by sympathetic nerve fibers

7. Which of the following characteristics distinguishes somatic capillaries from visceral capillaries?

(A) Presence or absence of fenestrae
(B) Size of the lumen
(C) Thickness of the vessel wall
(D) Presence or absence of pericytes
(E) Thickness of the basal lamina

8. The blood–brain barrier is thought to occur because capillaries in the CNS have which of the following characteristics?

(A) Discontinuous basal lamina
(B) Fenestrae with diaphragms
(C) Fenestrae without diaphragms
(D) A few pinocytic vesicles
(E) No basement membrane

9. Capillaries function in all of the following ways EXCEPT

(A) exchange of metabolites
(B) exchange of gases
(C) control of blood pressure
(D) inhibition of clot formation

10. The lymphatic system drains all of the following tissues and organs EXCEPT

(A) muscles
(B) kidneys
(C) bone marrow
(D) liver
(E) intestines

Directions: Each group of items in this section consists of lettered options followed by a set of numbered items. For each item, select the **one** lettered option that is most closely associated with it. Each lettered option may be selected once, more than once, or not at all.

Questions 11–15

Match each structure below with the corresponding letter shown in the photomicrograph.

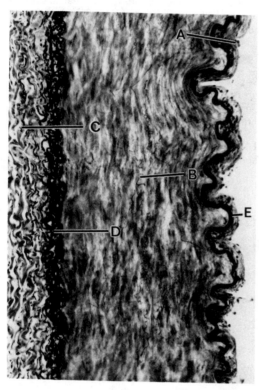

Reprinted from Gartner LP and Hiatt JL: *Atlas of Histology*. Baltimore, Williams & Wilkins, 1987, p 119.

11. Tunica media

12. External elastic lamina

13. Endothelium

14. Tunica adventitia

15. Internal elastic lamina

Questions 16–20

Match each sign below with the disorder that is most closely associated with it.

(A) Rheumatic fever
(B) Atherosclerosis
(C) Tetralogy of Fallot
(D) Varicose veins
(E) Ischemic heart disease
(F) Aneurysm

16. Dextroposed aorta

17. Yellowish plaques in the intima of arteries

18. Cardiac valve incompetence

19. Replacement of elastic fibers by collagen in vessel walls

20. Alterations in vessel walls and reduced muscle tone near the anorectal junction

Answers and Explanations

1–B. The epicardium, composed of fibrous connective tissue and fat and lined by mesothelium, forms the outer layer of the heart and the visceral layer of the pericardium.

2–A. Atrial natriuretic peptide, which decreases blood pressure, is produced mainly by cardiac muscle cells of the right atrium. It inhibits the release of renin and causes the kidneys to decrease the resorption of sodium and water.

3–D. Impulses are generated in the SA node is the "pacemaker" of the heart, from which they are conducted to the AV node The bundle of His and Purkinje fibers conduct impulses from the AV node to the cardiac muscle cells of the ventricles. Sympathetic nerves can increase the rate of the heartbeat but do not originate it.

4–D. A major difference in the walls of elastic and muscular arteries is the presence of fenestrated (perforated) membranes in the tunica media of elastic arteries; these are absent in muscular arteries. The media in both types of arteries contains smooth muscle cells and an external elastic lamina (thin in elastic arteries; thick in muscular arteries).

5–C. Metarterioles are drained by thoroughfare channels.

6–C. Acetylcholine stimulates the endothelial cells of a vessel to release nitric oxide (endothelial-derived relaxing factor), which causes relaxation of smooth muscle cells. Thus, acetylcholine does not act directly on smooth muscle cells.

7–A. Somatic capillaries lack fenestrae, whereas visceral capillaries are characterized by their presence. Both types of capillary possess a continuous basal lamina and are surrounded by occasional pericytes.

8–D. Capillaries in the CNS are of the continuous type and thus lack fenestrae but have a continuous basal lamina. In contrast to continuous capillaries in other parts of the body, they contain only a limited number of pinocytic vesicles; this characteristic is thought to be partly responsible for the blood–brain barrier.

9–C. Capillaries function as exchange vessels for metabolites and gases and in inhibiting clot formation. Capillaries do not control blood pressure.

10–C. Lymphatic vessels are not present in bone marrow. However, they do drain the muscles, kidneys, liver, and intestines.

11–B. Tunica media.

12–D. External elastic lamina.

13–A. Endothelium.

14–C. Tunica adventitia.

15–E. Internal elastic lamina.

16–C. Tetralogy of Fallot comprises multiple malformations including a dextroposed aorta, defective interventricular septum, hypertrophy of the right ventricle, and pulmonary stenosis.

17–B. Atherosclerosis is characterized by yellowish plaques in the intima of large and medium-sized arteries.

18–A. Rheumatic heart-valve disease, which is a sequel to childhood rheumatic fever, is marked by incompetence of the heart valves.

19–F. An aneurysm, the ballooning out of an artery, may result from an age-related displacement of elastic fibers in the vessel by collagen.

20–D. Hemorrhoids, a form of varicose veins in the region of the anorectal junction, result from a decline in muscle tone, degenerative changes in vein walls, and valvular incompetence.

12

Lymphoid Tissue

I. Overview—The Lymphoid System

–consists of several discrete organs (**thymus, spleen, tonsils,** and **lymph nodes**), **diffuse lymphoid tissue,** and white blood cells, most importantly **T lymphocytes** (T cells), **B lymphocytes** (B cells), and **macrophages**.

–functions primarily to defend the organism by mounting **humoral immune responses** against foreign substances (**antigens**) and **cell-mediated immune responses** against microorganisms, tumor and transplanted cells, and virus-infected cells.

II. Cells of the Immune System

–include **clones of T and B lymphocytes,** each comprising a small number of **identical** cells that have the ability to recognize and respond to a single antigenic determinant (**epitope**), or a small group of closely related epitopes. Exposure to antigen and one or more **cytokines** induces **activation** of resting T and B cells, leading to their proliferation and differentiation into **effector cells** (Figure 12.1).

–also include **antigen-presenting cells** (e.g., macrophages, lymphoid dendritic cells, Langerhans' cells, follicular dendritic cells, M cells, and B cells*), as well as **mast cells** and **granulocytes** (see Chapter 6 III G and H).

A. T lymphocytes — Thymus derived – have "clone" cells by time thymus shuts down

- local killer cells ic involved in tissue graft rejection

–include several functionally distinct subtypes.

–are responsible for **cell-mediated immune responses** and assist B cells in developing a humoral response to most antigens (called **thymic-dependent antigens**).

–only recognize antigen that has been processed by and is displayed (presented) on the surface of antigen-presenting cells (APCs).

---

*Although B cells can present epitopes to T cells, and thus are antigen-presenting cells, their role in this function probably is limited to the secondary (anamnestic) immune response rather than the primary immune response.

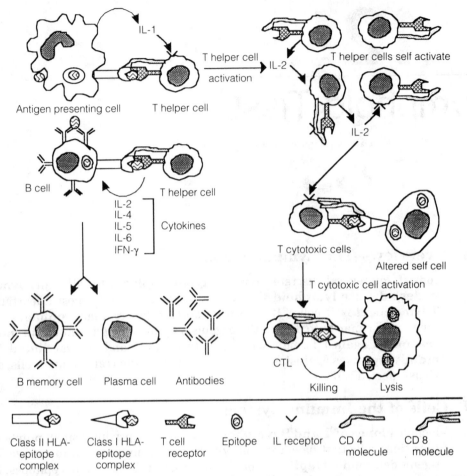

Class II HLA-epitope complex   Class I HLA-epitope complex   T cell receptor   Epitope   IL receptor   CD 4 molecule   CD 8 molecule

Figure 12.1. Schematic overview of cell-mediated and humoral immune response. Note the central role of T helper (T_H) cells and of various cytokines in stimulating formation of effector cells: plasma cells in the case of a humoral response and cytotoxic T lymphocytes in the case of a cell-mediated response.

1. Maturation of T lymphocytes

—occurs in the thymus and involves the following events:

  a. Immunoincompetent progenitor T lymphocytes migrate from the bone marrow to the thymus, where they are termed **thymocytes**.

  b. Within the thymic **cortex,** thymocytes undergo **gene rearrangements** and begin to express antigen-specific T-cell receptors, which are integral membrane proteins; that is, the cells become **immunocompetent**.

  c. The cortical thymocytes also begin to express thymus-induced **CD (cluster of differentiation) markers,** designated CD2, CD3, CD4, and CD8, on their surface.

  d. In the thymic **medulla,** some thymocytes lose CD4 and develop into CD8$^+$ cells; others lose CD8 and develop into CD4$^+$ cells.

  e. Medullary thymocytes also develop class I or class II HLA (human leukocyte antigen) molecules, the expression of the major histocompatibility genes.

2. T-lymphocyte subtypes

a. T helper (T_H) cells

—are **CD4$^+$** cells.

—synthesize and release numerous growth factors, known as **cytokines (lymphokines)**, following their activation.

(1) The cytokines **interleukin 2** (IL-2) and **interleukin 4** (IL-4) induce B cells to respond to an antigenic stimulus.

(2) Other cytokines produced by T_H cells and other cells **modulate the immune response** in diverse ways (Table 12.1).

b. T cytotoxic (T_C) cells

—are **CD8$^+$** cells.

—after priming by an antigenic stimulus are induced by **interleukin 2** (IL-2) to proliferate and differentiate into **cytotoxic T lymphocytes (CTLs)**, which mediate (via **perforin**) lysis of foreign cells and virus-infected self-cells (see Figure 12.1).

c. T suppressor (T_S) cells

—are **CD8$^+$** cells.

—can modulate the extent of the immune response by suppressing the activity of T_H cells.

—may be important in **preventing autoimmune responses**.

Table 12.1. Biological Activity of Selected Cytokines in the Immune Response*

Cytokine	Secreted by	Activity
Interleukin 1 (IL-1)	Macrophages; various other cell types	Stimulates T_H-cell activation; promotes maturation and clonal expansion of B cells
Interleukin 2 (IL-2)	T_H cells	Stimulates proliferation of activated T_H, T_C, and B cells
Interleukin 4 (IL-4)	T_H cells	Stimulates B-cell activation; induces proliferation of mast cells and activated T and B cells; enhances phagocytic activity of macrophages; induces expression of HLA molecules by B cells and macrophages
Interleukin 5 (IL-5)	T_H cells	Stimulates proliferation and differentiation of B cells and eosinophils
Interleukin 6 (IL-6)	T_H cells; macrophages; fibroblasts	Increases antibody secretion by plasma cells; costimulates T_H-cell activation with IL-1
Interferon $\alpha^\dagger$ (INF-α)	T cells; NK cells; macrophages	Induces expression of class I HLA molecules on cells; activates NK cells
Interferon γ^+ (INF-γ)	T_H cells; NK cells; T_C cells	Induces expression of classd II HLA molecules by macrophages and other APCs; stimulates activation of macrophages, NK cells, and T_C cells, increasing their cytotoxic and/or phagocytic activity

*See Chapter 10 for discussion about cytokines involved in hematopoiesis.

$^+$Interferons also inhibit replication of viruses. Interferon γ is also called macrophage-activating factor (MAF).

d. T memory cells

- are **long-lived, committed, immunocompetent cells** that are formed during proliferation in response to an antigenic challenge.
- do not react against the antigen but remain in the circulation or in specific regions of the lymphoid system.
- increase the size of the original clone and thereby **provide a faster and greater secondary response** (anamnestic response) against a future challenge by the same antigen.

B. B lymphocytes *– Bursal derived (bone marrow in mammals)*

- originate and mature into **immunocompetent** cells within the bone marrow.
- have surface immunoglobulins (antibodies) attached to the external aspect of their plasma membrane; all the immunoglobulin (Ig) molecules on a given B cell recognize and bind to the **same antigenic determinant** (epitope).
- are responsible for the **humoral immune response**.
- proliferate and differentiate following an antigenic challenge to form plasma cells and B memory cells (see Figure 12.1).

1. Plasma cells *– Source of circulating AB*

- actively synthesize and secrete **antibody specific for the challenging antigen**.
- lack surface antibody.

2. B memory cells

- have the same properties as T memory cells.

C. Natural killer (NK) cells

- are one type of **null cell,** a small group of peripheral-blood lymphocytes that **lack the surface determinants** characteristic of T and B lymphocytes.
- exhibit an apparently **nonspecific cytotoxicity** against tumor cells and virus-infected cells; the mechanism by which NK cells recognize these target cells is not understood as yet.
- also can kill specific target cells that have antibodies bound to their surface antigens in a process known as **antibody-dependent cell-mediated cytotoxicity (ADCC);** macrophages, neutrophils, and eosinophils also exhibit ADCC.

D. Macrophages

- function both as **antigen-presenting cells** (APCs) and as **cytotoxic effector cells** in ADCC.
- produce **interleukin 1** (IL-1), which helps activate T_H cells, and several other cytokines that influence the immune response (see Table 12.1) or hematopoiesis (see Chapter 10 VII).
- also secrete **prostaglandin E_2** (PGE_2), which decreases certain immune responses.
- are activated by interferon γ (IFN-γ), which is also known as **macrophage-activating factor;** this activation increases the phagocytic and cytotoxic activity of macrophages.

III. Antigen Presentation and the Role of MHC Molecules

A. Major histocompatibility complex (MHC)

—is a large genetic complex with many loci that encode two main classes of membrane molecules: **class I molecules,** which are expressed by nearly all **nucleated cells,** and **class II molecules,** which are expressed by the various cells that function as **antigen-presenting cells.**

—is referred to as the **HLA (human leukocyte antigen) complex** in humans.

B. Immunogens

—are molecules that have the capability of **inducing an immune response**.

—are **antigens,** that is, molecules that can **react with an antibody or a T-cell receptor**.

1. Exogenous immunogens

—are endocytosed or phagocytosed by APCs and degraded intracellularly, yielding antigenic peptides (containing an epitope) that associate with **class II HLA molecules.** The epitope–class II HLA complexes are transported to the cell surface, where they are displayed (**presented**) to T lymphocytes.

2. Endogenous immunogens

—are produced **within** host cells (e.g., **viral proteins** synthesized within virus-infected cells and **tumor proteins** synthesized within cancerous cells).

—are degraded within host cells, yielding antigenic peptides that associate with **class I HLA molecules.** The peptide–class I HLA complexes are transported to the cell surface, where they are displayed (presented) to T lymphocytes.

C. HLA restriction—T lymphocytes

—of each type (except T memory cells) only recognize epitopes that are associated with either class I or class II HLA molecules as follows:

1. T_H cells and T_S cells recognize class II HLA molecules.

2. T_C cells recognize class I HLA molecules.

3. T memory cells recognize both class I and class II HLA molecules.

IV. Immunoglobulins

—are proteins that have **specific antibody activity** against one antigen, or a few closely related antigens.

—are synthesized and secreted by **plasma cells** and constitute the active agents of the humoral immune response.

—interact with antigens to form antigen–antibody complexes, which are cleared from the body by various mechanisms, some of which involve the **complement system**.

A. Structure—immunoglobulins

—are composed of monomers containing **two heavy chains** and **two light chains**.

—contain **constant regions,** which are identical in all immunoglobulin (Ig) molecules.

—also contain **variable regions,** which differ in antibody molecules that recognize different antigens. Thus the **variable regions determine the specificity** of an antibody molecule (i.e., its ability to bind to a particular antigenic determinant). Large antigens may have multiple antigenic determinants, thus inducing production of antibodies with different specificities.

B. Ig classes—immunoglobulins

—in human serum comprise five classes (**isotypes**), which differ in the amino acid composition of their **heavy-chain constant regions.**

—of different isotypes exhibit functional differences.

1. IgA

—contains α heavy chains and exists as **monomers** and **dimers.**

—constitutes 10%–15% of the total serum Ig.

—is located in **serum** (monomer) and glandular **secretions** (dimer).

—binds to antigens on the body surface, in lumina of ducts, and lumen of the gastrointestinal tract.

2. IgD

—contains δ heavy chains and exists as **monomers.**

—constitutes less than 1% of the total serum Ig.

—is located primarily on the **plasma membrane of B cells.**

—is thought to function in **activation of B cells** following an antigenic challenge.

3. IgE

—contains ϵ heavy chains and exists as **monomers.**

—is present at very low concentrations in serum.

—is located primarily on the **plasma membrane of mast cells** and **basophils.**

—mediates **degranulation** of mast cells and basophils in immediate hypersensitivity (allergic) reactions (see Chapter 6 III G).

—is also known as **reaginic antibody.**

4. IgG

—contains γ heavy chains and exists as **monomers.**

—is the **most abundant serum Ig,** constituting about 80% of the total serum Ig.

—can cross the placenta and thus functions to **protect the fetus,** which lacks a functional immune system.

—activates the **complement system.**

—functions as an **opsonin** by binding to antigens on bacteria, rendering them more susceptible to phagocytosis.

5. IgM

—contains μ heavy chains and exists as **monomers** and **pentamers.**

—constitutes 5%–10% of the total serum Ig.

—is located in **serum** (pentamer) and on the **plasma membrane of B cells** (monomer).

—is the first antibody isotype produced in a primary immune response and in the neonate.

—activates the **complement system.**

V. Diffuse Lymphoid Tissue

—is especially prominent in the loose connective tissue deep to the epithelia lining body cavities.

—is organized as scattered clusters of lymphoid cells or as lymphoid (lymphatic) nodules.

A. Lymphoid (lymphatic) nodules

—are transitory, dense, spherical **accumulations of lymphocytes** (mostly B cells). The dark, peripheral region of nodules (**corona**) is composed largely of small, newly formed lymphocytes.

1. **Secondary nodules** have a central, lightly staining area, called the **germinal center,** which is composed of medium and large lymphocytes (**lymphoblasts**) and is the site of B-cell proliferation and differentiation into plasma cells.

2. **Primary nodules** lack germinal centers.

B. Peyer's patches

—are aggregates of lymphoid nodules found in the ileum.

—are components of the **gut-associated lymphoid tissue (GALT).**

VI. Lymphoid Organs

A. Lymph nodes

—are small, encapsulated, ovoid to kidney-shaped organs.

—are scattered along lymphatic vessels throughout the body, but are concentrated in the neck, axilla, and groin.

—have a **convex** surface that receives afferent lymphatic vessels, and a **concave** surface (the hilum) where arterioles enter and efferent lymphatic vessels and venules exit.

—are surrounded by a **capsule** of dense irregular connective tissue, which sends **trabeculae** into the substance of the node.

—are divided into **cortical** and **medullary** regions.

—possess a **stroma** composed of a supportive framework rich in **reticular fibers** intimately associated with the capsule and trabeculae. This reticulum is more tightly woven in the cortex than in the medulla.

—**function** to filter lymph, to maintain and produce T and B cells, and to house memory cells (especially T memory cells).

1. Cortex of lymph nodes

—lies deep to the capsule from which it is separated by a subcapsular sinus.

—is incompletely subdivided into smaller, intercommunicating compartments by connective tissue septa derived from the capsule.

—contains three major components: lymphoid nodules, sinusoids, and the paracortex.

a. Lymphoid nodules

—are composed mainly of B cells but also consist of a few T cells, follicular dendritic cells, macrophages, and reticular cells.

—may possess a germinal center.

b. Sinusoids

—are endothelium-lined spaces that extend along the capsule and trabeculae and are known as **subcapsular** and **cortical sinusoids,** respectively.

c. Paracortex

—is located between the cortex (with nodules) and the medulla.

—is composed of a non-nodular arrangement of **mostly T lymphocytes** (the thymus-dependent area of the lymph node).

—is the region where circulating lymphocytes gain access to lymph nodes via **postcapillary (high endothelial) venules** recognized by their lining of cuboidal endothelium.

2. Medulla of lymph nodes

—lies deep to the paracortex and cortex except at the region of the hilum.

—is composed of medullary sinusoids and medullary cords.

a. Medullary sinusoids

—are endothelium-lined spaces supported by reticular fibers and reticular cells, whose processes frequently span the sinusoids, retarding and causing turbulence in the lymph flow; they also contain macrophages.

—receive lymph from the cortical sinuses.

b. Medullary cords

—are composed of lymphocytes and plasma cells, many of which migrated from the cortex to enter the medullary sinusoids.

B. Thymus

—is derived both from endoderm (epithelial reticular cells) and mesoderm (lymphocytes). Nonthymic lymphoid organs are derived exclusively from mesoderm.

—begins to involute near the time of puberty.

—is surrounded by a connective tissue **capsule** that sends septa into the parenchyma, dividing it into incomplete lobules, each of which contains a **cortical** and **medullary region**.

—does not contain lymphoid nodules.

1. Thymic cortex

—is supplied by arterioles in the septa; these provide capillary loops that enter the substance of the cortex.

—is the region in which **T-cell maturation** occurs.

a. Epithelial reticular cells

—form a meshwork in whose interstices T lymphocytes are tightly packed.

—are pale cells (derived from the third pharyngeal pouch), having a large, ovoid, lightly staining nucleus, which often displays a nucleolus.

—possess **long processes** that surround the thymic cortex, isolating it both from the connective tissue septa and from the medulla. These processes, which are filled with bundles of **tonofilaments,** form desmosomal contacts with each other.

—possess cytoplasmic granules that are believed to contain **thymosin, serum thymic factor,** and **thymopoietin;** these function in the transformation of immature T lymphocytes into immunocompetent T cells.

b. Thymocytes

—are **immature T lymphocytes** present in large numbers within the thymic cortex in different stages of differentiation.

—are surrounded by processes of epithelial reticular cells, which help to segregate thymocytes from antigens during their maturation.

—migrate toward the medulla as they mature; however **most T cells die in the cortex** and are phagocytosed by macrophages.

c. Blood–thymus barrier

—exists in the **cortex only,** making it an immunologically protected region.

—ensures that antigens escaping from the bloodstream do not reach developing T cells in the thymic cortex.

—consists of the following layers: **endothelium** of the thymic capillaries and the associated basal lamina; **perivascular connective tissue** and **cells** (e.g., pericytes and macrophages); and **epithelial reticular cells** and their basal laminae.

[margin handwritten notes: -capillaries not leaky; -reticulo epithelial cells]

2. Thymic medulla *- no blood-thymus barrier -does have reticuloepithelial cells*

—is continuous between adjacent lobules.

—contains large numbers **of epithelial reticular cells** and **mature T lymphocytes,** which are loosely packed, causing the medulla to stain lighter than the cortex.

—also contains whorl-like accretions of epithelial reticular cells, called **Hassall's corpuscles.** These structures display various stages of keratinization and increase in number with age. Their function is unknown.

[margin handwritten: 11-49 Thymate]

—is the region from which **mature T cells exit the thymus** via venules and efferent lymphatic vessels. The T cells then migrate to nonthymic lymphoid structures.

C. Spleen *- has nodules but no cortex & medulla*

—is the largest lymphoid organ of the body.

—is covered by a simple squamous epithelium (peritoneum).

—is surrounded by a **capsule** of dense irregular connective tissue, which sends out **trabeculae** and is interspersed with **elastic fibers** and **smooth muscle cells.** The capsule's internal aspect has a reticular fiber stroma that extends along the trabeculae and forms the framework of the spleen.

—possesses a hilum where nerves and blood vessels enter and leave the organ.

—differs from the thymus and lymph nodes in that it **lacks a cortex, medulla,** and **afferent lymphatics.**

[margin handwritten: Zones →] —is divided into regions designated as **red pulp** *[blood rich]* and **white pulp** *[full of lymphocytes]*; the latter contains lymphoid nodules. These regions are separated by the **marginal zone.**

—has the following **functions:** filtration of blood; storage of erythrocytes; phagocytosis of damaged and aged erythrocytes; proliferation of B and T lymphocytes; and production of antibodies by plasma cells.

[bottom margin handwritten: c/m - thymus; lymph node; tonsil]

1. Vascularization of the spleen

—is derived from the **splenic artery,** which enters the hilum and gives rise to trabecular arteries; these are distributed to the splenic pulp via trabeculae.

a. **Trabecular arteries** leave the trabeculae and become invested by a **periarterial lymphatic sheath** (PALS). The sheathed arteries are designated as central arteries.

b. **Central arteries** branch but maintain their lymphatic sheath until they leave the white pulp to form several straight penicillar arteries.

c. **Penicillar arteries** enter the red pulp and possess three regions: pulp arterioles, macrophage-sheathed arterioles, and **terminal arterial capillaries**. The latter either drain directly into the splenic sinusoids (**closed circulation**) or terminate as open-ended vessels within the red pulp (**open circulation**).

d. **Splenic sinusoids** are drained by pulp veins, which are tributaries of the trabecular veins; these, in turn, drain into the splenic vein, which exits the spleen at the hilum.

2. White pulp of the spleen

—includes **all** of the organ's lymphoid tissue (diffuse and nodular) such as **lymphoid nodules** (mostly B lymphocytes) and **periarterial lymphatic sheaths** (mostly T cells) around the central arteries.

—also contains macrophages and other antigen-presenting cells.

3. Marginal zone of the spleen

—is a **sinusoidal region** between the red and white pulp located at the periphery of the PALS.

—receives blood from capillary loops derived from the central artery and thus is the **first site where blood contacts the splenic parenchyma**.

—is richly supplied by avidly phagocytic macrophages and other antigen-presenting cells.

—is the region where T and B lymphocytes **enter the spleen** before becoming segregated to their specific locations within the organ.

4. Red pulp of the spleen

—is composed of an interconnected network of sinusoids supported by a loose type of reticular tissue (**splenic cords**).

a. **Sinusoids**

—are lined by long, fusiform endothelial cells separated by relatively large intercellular spaces.

—have a discontinuous basal lamina, underlying the endothelium, and circumferentially arranged "ribs" of reticular fibrils.

b. **Splenic cords**

—are also called **cords of Billroth**.

—contain plasma cells, reticular cells, and various types of blood cells (including macrophages) enmeshed within the spaces of the reticular fiber network. Processes of the macrophages enter the lumina of the sinusoids through the large intercellular spaces between the endothelial cells.

D. Tonsils

4 key charac:

1) lymphatic nodule

—are **aggregates of lymphoid tissue,** which sometimes lack a capsule.
—are all located in the upper section of the digestive tract, lying beneath, but in contact with, the epithelium.
—assist in combating antigens entering via the nasal and oral epithelia.

2) stratified sq. epith. — *adenoid* - *pseudostratified ciliated (pharyngeal)*

1. Palatine tonsils

3) salivary glands
11-10

—consist of **two** tonsillar structures located between the palatopharyngeal and palatoglossal arches.

4) striated 'skeletal'
muscle

—possess **crypts,** deep invaginations of the stratified squamous epithelium covering of the tonsils; the crypts frequently contain debris.
—possess lymphoid nodules, some with germinal centers (secondary nodules).
—are separated from subjacent structures by a connective tissue **capsule.**

Fig 7-45

p. 247

2. Pharyngeal tonsil

—is a **single** tonsil located in the posterior wall of the nasopharynx.
—is covered by a ciliated pseudostratified columnar epithelium.
—contains several folds. Seromucous secretions enter the folds from glands located in the connective tissue deep to the tonsillar capsule.

3. Lingual tonsils

—are located on the dorsum of the posterior one-third of the tongue.
—are covered by a stratified squamous nonkeratinized epithelium.
—possess deep **crypts,** which frequently contain cellular debris. Ducts of mucous glands often open into the base of these crypts.
—are smaller and more numerous than the palatine and pharyngeal tonsils.

VII. Clinical Considerations

A. X-linked hypogammaglobulinemia

—is a recessive, X-linked **immunodeficiency** affecting male newborns.
—is characterized by a depressed (or absent) B-cell population and decreased levels of immunoglobulins, resulting in **impaired humoral immunity**.
—does not involve T cells; thus cell-mediated immune responses are normal.
—may be associated with the absence of tonsils and of germinal centers in lymphoid nodules.
—results in extreme **susceptibility to infection with encapsulated bacteria** (e.g., streptococcus, pneumococcus).

B. DiGeorge's syndrome

—is also called **congenital thymic aplasia**.
—is characterized by the congenital absence of the thymus and parathyroid glands resulting from abnormal development of the third and fourth pharyngeal pouches. The length of the philtrum and clefts of the nose are reduced.
—is associated with **abnormal cell-mediated immunity** but relatively normal humoral immunity.
—usually results in **death** from **tetany** or uncontrollable **infection**.

C. Acquired immunodeficiency syndrome (AIDS)

–is caused by infection with **human immunodeficiency virus (HIV),** which preferentially invades T_H cells, causing a severe depression in their number and thus **suppressing both cell-mediated and humoral immune responses.**

–is characterized by **secondary infections** with opportunistic microorganisms that cause pneumonia, toxoplasmosis, candidiasis, and other diseases.

–is also characterized by development of certain malignancies such as **Kaposi's sarcoma** and **non-Hodgkin's lymphoma.**

D. Hyperactive malarious splenomegaly

–is characterized by an **enlarged spleen** and very **high serum levels** of **IgM** and **malaria antibody,** as well as liver sinusoidal lymphocytosis.

–is likely due to **recurring malaria,** which disturbs T-cell control of the humoral immune system.

E. Hodgkin's disease

–is a malignancy involving **neoplastic transformation of lymphocytes** (most probably).

–occurs mostly in young adults and affects twice as many men as women.

–is characterized by the presence of **Reed-Sternberg cells,** which are giant cells with two large, vacuolated nuclei, each with a dense nucleolus.

–is marked by the painless, progressive **enlargement of the lymph nodes, spleen,** and **liver.** Other symptoms include anemia, fever, weakness, anorexia, and weight loss.

Review Test

Directions: Each of the numbered items or incomplete statements in this section is followed by answers or by completions of the statement. Select the **one** lettered answer or completion that is **best** in each case.

1. Which of the following statements concerning T helper cells is TRUE?

(A) They possess membrane-bound antibodies
(B) They can recognize and interact with antigens in the blood
(C) They produce numerous cytokines
(D) They function only in cell-mediated immunity
(E) Their activation depends on interferon γ

2. Which of the following statements concerning T cytotoxic (T_C) cells is FALSE?

(A) They kill foreign cells
(B) They possess T-cell receptors
(C) They possess CD8 surface markers
(D) They possess CD4 surface markers
(E) They secrete interferon γ

3. Which of the following cell types is thought to function in preventing immune responses against self-antigens?

(A) T_S cells
(B) B cells
(C) T memory cells
(D) T_H cells
(E) Mast cells

4. Which of the following statements concerning interferon γ is TRUE?

(A) It is produced by macrophages
(B) It is produced by T suppressor cells
(C) It activates macrophages
(D) It inhibits macrophages
(E) It induces viral proliferation

5. Circulating antibodies are synthesized and secreted by which of the following cells?

(A) T memory cells
(B) B memory cells
(C) T helper cells
(D) Plasma cells
(E) T cytotoxic cells

6. Which of the following statements concerning epithelial reticular cells of the thymus is FALSE?

(A) They are located in the thymic cortex
(B) They are not located in the thymic medulla
(C) They have long processes
(D) They rest upon basal laminae
(E) They assist in forming the blood–thymus barrier

7. Which of the following statements concerning Hassall's corpuscles is TRUE?

(A) They are located in the thymic cortex of young individuals
(B) They are located in the thymic cortex of old individuals
(C) They are derived from mesoderm
(D) They are located in the thymic medulla
(E) They are derived from T memory cells

8. Following their maturation in the thymus and release into the circulation, T lymphocytes migrate preferentially to which of the following sites?

(A) Paracortex of lymph nodes
(B) Cortical lymphoid nodules of lymph nodes
(C) Hilus of lymph nodes
(D) Lymphoid nodules of the tonsils
(E) Lymphoid nodules of the spleen

9. In which of the following sites do lymphocytes become immunocompetent?

(A) Germinal center of secondary lymphoid nodules
(B) White pulp of the spleen
(C) Thymic cortex
(D) Red pulp of the spleen
(E) Paracortex of lymph nodes

10. Which of the following statements about IgG is TRUE?

(A) It is located in the serum and on the membrane of B cells
(B) It can cross the placenta
(C) It is involved in allergic reactions
(D) It exists as a pentamer
(E) It binds to antigens on the body surface and in the lumen of the gastrointestinal tract

Directions: Each group of items in this section consists of lettered options followed by a set of numbered items. For each item, select the **one** lettered option that is most closely associated with it. Each lettered option may be selected once, more than once, or not at all.

Questions 11–14

Match each description below with the corresponding lettered structure or region shown in the photomicrograph of a spleen section.

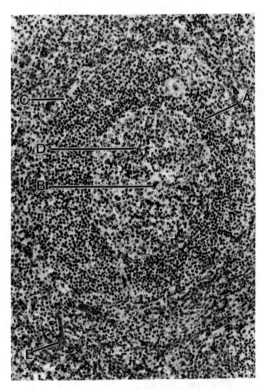

Reprinted from Gartner LP and Hiatt JL: *Atlas of Histology.* Baltimore, Williams & Wilkins, 1987, p 141.

11. Vessel surrounded by a periarterial lymphatic sheath (PALS)

12. Region containing small, newly formed B cells

13. Site of active B-cell proliferation

14. Region where T and B cells enter the spleen from the bloodstream

Questions 15–18

Match each description or term below with the corresponding structure shown in the light micrograph of a section of a lymph node.

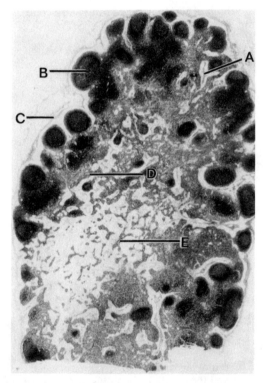

Reprinted from Gartner LP and Hiatt JL: *Atlas of Histology.* Baltimore, Williams & Wilkins, 1987, p 133.

15. Trabecula

16. Medullary cord

17. Thymic-dependent zone

18. Site of B-lymphocyte proliferation

Questions 19–20

Match each disorder with the sign or symptom that is most characteristic of it.

(A) Impaired cell-mediated immunity but normal humoral immunity
(B) Impaired humoral immunity but normal cell-mediated immunity
(C) Impaired cell-mediated and humoral immunity
(D) Presence of Reed–Sternberg cells
(E) Enlarged spleen and high serum levels of IgM

19. Hodgkin's disease

20. AIDS

Answers and Explanations

1–C. T helper cells produce a number of cytokines that affect other cells involved in both the cell-mediated and humoral immune responses. T helper cells possess antigen-specific T-cell receptors (not antibodies) on their membrane. These cells recognize and interact with antigenic determinants that are associated with class II HLA molecules on the surface of antigen-presenting cells. IL-1 is necessary for activation of T helper cells.

2–D. T cytotoxic cells are CD8$^+$. T helper cells and some T memory cells possess CD4.

3–A. The immune response is decreased by T suppressor (T_S) cells. Their activity is thought to help prevent autoimmune responses against self-antigens.

4–C. Interferon γ activates macrophages, as well as NK cells and T cytotoxic cells, enhancing their phagocytic and/or cytotoxic activity.

5–D. Plasma cells produce circulating antibodies. Following an antigenic challenge, proliferation and differentiation of B cells give rise to plasma cells and B memory cells.

6–B. Epithelial reticular cells are located in the medulla as well as in the cortex. Their long processes surround thymocytes, helping to form the blood–thymus barrier.

7–D. Hassall's corpuscles are concentrically arranged accretions of epithelial reticular cells (derived from endoderm) found only in the medulla of the thymus.

8–A. T lymphocytes are preferentially located in the paracortex of lymph nodes, whereas B lymphocytes are found in lymphoid nodules located in lymph nodes, tonsils, and the spleen.

9–C. T lymphocytes mature and become immunocompetent in the cortex of the thymus, whereas B lymphocytes do so in the bone marrow. Following an antigenic challenge, lymphocytes proliferate and differentiate in various lymphoid tissues.

10–B. IgG, the most abundant immunoglobulin isotype in the serum, can cross the placenta. It is present only in the serum, exists as a monomer, and functions to activate complement and as an opsonin.

11–B. Central artery is surrounded by a PALS.

12–A. The corona, the dark peripheral region of a lymphoid nodule, contains newly formed B cells.

13–D. The germinal center is the lightly stained, central region of a lymphoid nodule where B cells proliferate and differentiate.

14–C. The marginal zone of the spleen, which lies between the red and white pulp, receives blood from capillary loops and is the region where T and B cells enter the splenic parenchyma.

15–A. Trabecula.

16–E. Medullary cord.

17–D. The paracortex is composed mostly of T lymphocytes and thus is termed the thymic-dependent zone.

18–B. B-cell proliferation occurs in germinal centers of lymphoid nodules located in the cortex of lymph nodes.

19–D. A characteristic histologic sign of Hodgkin's disease is the presence of Reed-Sternberg cells.

20–C. AIDS is marked by deficiency of both the cell-mediated and humoral immune systems due to depletion of the population of T helper cells. With compromised immune systems, AIDS patients are susceptible to secondary infections and likely to develop certain malignancies including Kaposi's sarcoma and non-Hodgkin's lymphoma.

13

Endocrine System

−product -hormone → target tissue

Endocrine- DNES - diffuse neuroendocrine system cells throughout body unicell.
−sometimes paracrine

I. Overview—The Endocrine System

−comprises several **ductless glands, clusters of cells** located within certain organs, and isolated **endocrine cells** in the epithelial lining of the gastrointestinal and respiratory systems.

−includes the **pituitary, thyroid, parathyroid, adrenal,** and **pineal glands,** which are discussed in this chapter.

−is characterized by the secretion of endocrine **hormones** directly into the bloodstream.

−interacts with the nervous system to modulate and control the body's metabolic activities.

II. Hormones

−are **chemical messengers** that are carried via the bloodstream to distant **target cells**.

−include low-molecular-weight **water-soluble** proteins and polypeptides (e.g., insulin, glucagon, follicle-stimulating hormone) and **lipid-soluble** substances, principally the steroid hormones (e.g., progesterone, estradiol, testosterone).

A. Water-soluble hormones

−interact with specific **cell-surface receptors** on target cells (see Chapter 1 IV B).

1. G-protein–linked receptors

−are utilized by some hormones (e.g., epinephrine, thyroid-stimulating hormone, serotonin). Binding leads to production of a **second messenger** that affects metabolic activity.

2. Catalytic receptors

−are utilized by insulin and growth hormone. Binding activates protein kinases that phosphorylate target proteins.

B. Lipid-soluble hormones

−diffuse across the plasma membrane of target cells.

193

—bind to specific receptors in the cytosol or nucleus, forming hormone–receptor complexes that regulate transcription of DNA.

III. Pituitary Gland (Hypophysis)

—lies below the hypothalamus to which it is structurally and functionally connected.

—consists of two major subdivisions—the **adenohypophysis** and the **neurohypophysis**. Each subdivision is derived from a distinct embryonic analog, which is reflected in its unique cellular constituents and functions.

[handwritten: outpocketing of Surface ectoderm of oral cavity Rathke's pouch]

A. Adenohypophysis (Figure 13.1)

—constitutes the anterior portion of the pituitary gland.

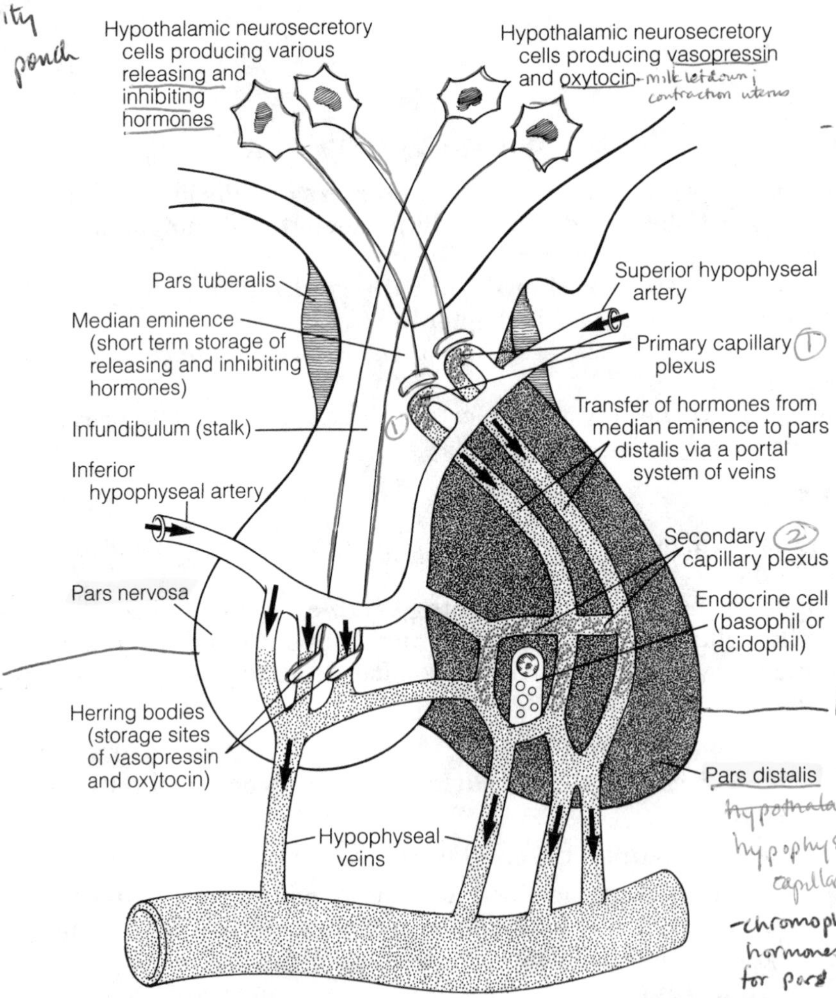

[handwritten annotations on figure: milk letdown; contraction uterus; – NERVOUS TISSUE VERY VASCULARIZED HERE; Anterior; POSTERIOR; – portal system – capillary bed for Hypothalam. input, 1 for output hypophyseal portal sys. capillary bed @ ___ connection Cap bed 2; –chromophils, probes make hormones to be released for pars distalis in Adeno]

Hypothalamic neurosecretory cells producing various releasing and inhibiting hormones

Hypothalamic neurosecretory cells producing vasopressin and oxytocin

Pars tuberalis

Median eminence (short term storage of releasing and inhibiting hormones)

Infundibulum (stalk)

Inferior hypophyseal artery

Pars nervosa

Herring bodies (storage sites of vasopressin and oxytocin)

Hypophyseal veins

Superior hypophyseal artery

Primary capillary plexus (1)

Transfer of hormones from median eminence to pars distalis via a portal system of veins

Secondary capillary plexus (2)

Endocrine cell (basophil or acidophil)

Pars distalis

Figure 13.1. Diagram of the pituitary gland showing its connections to the hypothalamus, sites of hormone synthesis and storage, and vascularization. The adenohypophysis lies to the right and consists of the pars distalis, pars tuberalis, and pars intermedia (not shown). The neurohypophysis consists of the infundibulum (stalk) and pars nervosa. Various releasing and inhibiting hormones stored in the median eminence are transferred, via the hypophyseal portal system, to the pars distalis. (Adapted from Junqueira LC et al: *Basic Histology,* 7th ed. Norwalk, CT, Appleton & Lange, 1992, p 395.)

[handwritten: PITUITARY – gonadotropins]

–originates from an ectodermal diverticulum of the stomodeum (**Rathke's pouch**).

–is subdivided into the **pars distalis, pars tuberalis,** and **pars intermedia**.

1. Pars distalis

–has an external collagenous capsule and an internal supportive reticular fiber network.

–consists of irregular cords of parenchymal cells lying adjacent to **fenestrated** capillaries.

a. Chromophils – pick up stains

–are parenchymal cells that **stain intensely**.

–are classified into two types, depending on the dyes they bind using special histologic stains. With hematoxylin–eosin stain, the distinction between the two cell types is much less obvious.

–synthesize, store, and release several hormones.

–are regulated by specific stimulatory and inhibitory hormones that are produced by neurosecretory cells in the **hypothalamus** and conveyed to the pars distalis via a system of portal blood vessels in the median eminence.

(1) Acidophils

–bind acid dyes and often stain **orange** or **red**.

–are most common in the center and posterolateral regions of the pars distalis.

–are small, spherical cells (15–20 μm in diameter) of two subtypes: somatotrophs and mammotrophs.

(a) Somatotrophs

–produce **somatotropin (growth hormone),** which is contained intracellularly within spherical, membrane-bounded granules (300–400 nm in diameter).

–possess a centrally positioned nucleus, moderate-sized Golgi complex, rod-shaped mitochondria, and abundant rough endoplasmic reticulum (RER).

–are stimulated by **somatotropin-releasing hormone (SRH)** and are inhibited by **somatostatin**.

(b) Mammotrophs

–produce **prolactin,** which is stored in small membrane-bounded granules (200 nm in diameter). These granules enlarge (up to 600 nm in diameter) in pregnant and lactating women; after suckling is complete, they undergo **crinophagy**.

–are stimulated by **prolactin-releasing hormone (PRH)** and are inhibited by **prolactin-inhibiting hormone (PIH)**.

(2) Basophils

–bind basic dyes and typically stain **blue**.

–are most common at the periphery of the pars distalis.

–include three subtypes: corticotrophs, thyrotrophs, and gonadotrophs.

(a) Corticotrophs

–produce **adrenocorticotropic hormone (ACTH)** and **lipotropic hormone (LPH),** a precursor of β-endorphin. Both hormones are stored in granules that are 250–400 nm in diameter.

–are spherical to ovoid in shape, with relatively few organelles and an eccentric nucleus.

–are stimulated by **corticotropin-releasing hormone (CRH).**

(b) Thyrotrophs

–produce **thyroid-stimulating hormone (TSH),** which is stored in small granules (150 nm in diameter), usually located just beneath the plasma membrane.

–are stimulated by **thyrotropin-releasing hormone (TRH).**

(c) Gonadotrophs

–produce **follicle-stimulating hormone (FSH)** and **luteinizing hormone (LH)** in both sexes, although the latter is often referred to as **interstitial cell–stimulating hormone (ICSH)** in men. These hormones are stored in large granules (200–400 nm in diameter).

–have abundant rough endoplasmic reticulum and mitochondria and a well-developed Golgi complex.

–are stimulated by **gonadotropin-releasing hormone (GnRH).**

b. Chromophobes

–are parenchymal cells that **do not stain** intensely and probably represent several cell types.

–in the light microscope appear as small cells, arranged close to one another in clusters, that lack (or have only a few) secretory granules.

–in the electron microscope sometimes resemble degranulated chromophils, suggesting that they may represent different stages in the life cycle of various acidophil and basophil populations.

–may also represent undifferentiated cells that are capable of differentiating into various types of chromophils.

2. Pars intermedia

–lies between the pars distalis and pars nervosa.

–is characterized by the presence of numerous **colloid-containing cysts (Rathke's cysts)** that are lined by cuboidal cells.

–possesses **basophilic cells,** which sometimes extend into the pars nervosa. In some vertebrates, these cells secrete **melanocyte-stimulating hormone (MSH),** although their function in humans is unknown.

3. Pars tuberalis

–surrounds the cranial part of the infundibulum (hypophyseal stalk).

–is composed of cuboidal **basophilic cells,** arranged in cords along an abundant capillary network.

–probably secretes **follicle-stimulating hormone** and **luteinizing hormone.**

B. Neurohypophysis (see Figure 13.1)

–constitutes the posterior portion of the pituitary gland.

–originates from an evagination of the hypothalamus.

—is divided into the **infundibulum,** which is continuous with the hypothalamus, and the **pars nervosa,** which is the main body of the neurohypophysis.

1. **Hypothalamohypophyseal tract**
   —contains the unmyelinated axons of **neurosecretory cells** whose somata are located in the **supraoptic** and **paraventricular nuclei** of the hypothalamus.
   —transports **oxytocin, vasopressin, neurophysin** (a binding protein specific for each hormone) and ATP to the pars nervosa.

2. **Pars nervosa** -has glial cells
   —contains the distal ends of the hypothalamohypophyseal axons and is the site where the neurosecretory granules in these axons are stored in accumulations known as **Herring bodies**.
   —releases oxytocin and vasopressin into fenestrated capillaries in response to nerve stimulation.

3. **Pituicytes**
   —occupy approximately 25% of the volume of the pars nervosa.
   —are glial-like cells that support the nervous tissue of the neurohypophysis.
   —possess numerous cytoplasmic processes and contain lipid droplets, intermediate filaments, and pigments.

C. **Physiologic effects of pituitary hormones**

   1. **Hormones of the pars distalis**

      a. **Somatotropin (growth hormone)**
         —increases metabolism in most cells.
         —indirectly stimulates epiphyseal plates and the growth of long bones via production of **somatomedins** in the liver.

      b. **Prolactin**
         —promotes development of the mammary gland during pregnancy.
         —stimulates milk production during lactation.

      c. **Adrenocorticotrophic hormone (ACTH)**
         —stimulates synthesis and release of hormones (particularly glucocorticoids) by the adrenal cortex.

      d. **Follicle-stimulating hormone (FSH)**
         —stimulates growth of secondary ovarian follicles and estrogen secretion in women.
         —stimulates Sertoli cells, which are located in seminiferous tubules and function in spermatogenesis.

      e. **Luteinizing hormone (LH)**
         —promotes ovulation, formation of the corpus luteum, and progesterone secretion in women.
         —stimulates testosterone synthesis by Leydig cells in the testes.

      f. **Thyroid-stimulating hormone (TSH)**
         —stimulates synthesis and release of thyroid hormones (T_3 and T_4) by thyroid follicular cells.

2. Hormones released by the pars nervosa

a. Oxytocin

—induces smooth muscle contraction in the wall of the uterus during copulation and at parturition.

—induces contraction of myoepithelial cells in the mammary gland during nursing.

b. Vasopressin

—is also called **antidiuretic hormone (ADH)**.

—renders the cells of renal collecting tubules permeable to water, which is resorbed to produce a concentrated urine.

—acts as a powerful vasoconstrictor of smooth muscle cells in the tunica media of arterioles.

D. Vascularization of the pituitary gland

1. Arterial supply

—is from two pairs of blood vessels derived from the internal carotid artery.

a. The right and left **superior** hypophyseal arteries serve the pars tuberalis, infundibulum, and median eminence.

b. The right and left **inferior** hypophyseal arteries serve mostly the pars nervosa.

2. Hypophyseal portal system (see Figure 13.1)

a. Primary capillary plexus

—consists of **fenestrated** capillaries coming off the superior hypophyseal arteries.

—is located in the **median eminence** where stored hypothalamic neurosecretory hormones enter the blood.

—is drained by **hypophyseal portal veins,** which descend through the infundibulum into the adenohypophysis.

b. Secondary capillary plexus

—consists of **fenestrated** capillaries coming off the hypophyseal portal veins.

—is located in the **pars distalis** where neurosecretory hormones leave the blood to stimulate or inhibit the parenchymal cells.

E. Regulation of the pars distalis

1. Neurosecretory cells in the hypothalamus synthesize specific hormones that enter the hypophyseal portal system and stimulate, or inhibit, the parenchymal cells of the pars distalis.

2. The hypothalamic neurosecretory cells in turn are regulated by the level of hormones in the blood (**negative feedback**) or by other physiologic (or psychologic) factors.

3. Some hormones (e.g., thyroid hormones, cortisol) exert negative feedback on the pars distalis directly.

IV. Thyroid Gland

—is composed of two lobes connected by an **isthmus**.

−has a dense irregular collagenous connective tissue capsule, in which the **parathyroid glands** are embedded.

−is subdivided by capsular septa into lobules containing **follicles**. These septa also serve as conduits for blood vessels, lymphatic vessels, and nerves.

A. Thyroid follicles

−are spherical structures filled with **colloid**, a viscous gel consisting mostly of **iodinated thyroglobulin**.

−are enveloped by a layer of epithelial cells, called **follicular cells**, which in turn are surrounded by **parafollicular cells**. These two parenchymal cell types rest on a basal lamina, which separates them from the abundant network of **fenestrated capillaries** in the connective tissue.

−function as synthesis and storage sites for thyroid hormones.

B. Follicular cells

1. Structure—follicular cells

−are normally cuboidal in shape but become columnar when stimulated and squamous when inactive.

−possess a distended rough endoplasmic reticulum with many ribosome-free regions, a supranuclear Golgi complex, numerous lysosomes, and rod-shaped mitochondria.

−contain many small **apical vesicles,** which are involved in the transport and release of thyroglobulin and enzymes into the colloid.

−possess short, blunt microvilli, which extend into the colloid.

2. Synthesis and release of thyroid hormones

−occurs by the sequence of events illustrated in Figure 13.2.

−is promoted by **thyroid-stimulating hormone,** which binds to G-protein–linked receptors on the basal surface of follicular cells.

C. Parafollicular cells

−are also called **clear (C) cells** because they stain less intensely than thyroid follicular cells.

−are present singly or in small clusters of cells located **between the follicular cells and basal lamina**.

−belong to the population of **APUD cells** (amine **p**recursor **u**ptake and **d**ecarboxylation cells), which are also known as enteroendocrine cells or diffuse endocrine cells.

−possess elongated mitochondria, substantial amounts of rough endoplasmic reticulum, a well-developed Golgi complex, and numerous membrane-bounded secretory granules (100–300 nm in diameter).

−synthesize and release **calcitonin,** a polypeptide hormone, which is also called **thyrocalcitonin**. Release of calcitonin is stimulated by high blood calcium levels.

D. Physiologic effects of thyroid hormones

1. T_4 and T_3

−act on a variety of target cells.

−**increase the basal metabolic rate** and thus promote heat production.

−have broad effects on gene expression and induction of protein synthesis.

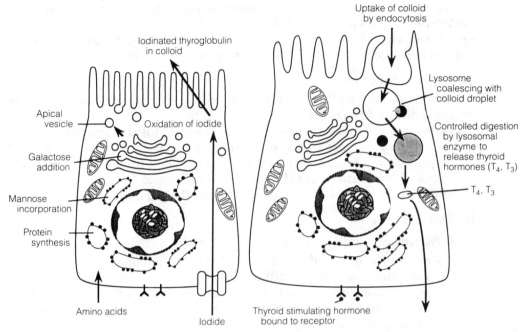

Figure 13.2. Synthesis and release of thyroxine (T_4) and triiodothyronine (T_3) by follicular cells of the thyroid gland. (*A*) Thyroglobulin is synthesized like other secretory proteins. Circulating iodide is actively transported into the cytosol, where a thyroid peroxidase oxidizes it and iodinates tyrosine residues on the thyroglobulin molecule; iodination occurs mostly at the apical plasma membrane. A rearrangement of the iodinated tyrosine residues of thyroglobulin in the colloid produces the iodothyronines T_4 and T_3. (*B*) Binding of thyroid-stimulating hormone to receptors on the basal surface stimulates follicular cells to become columnar and to form apical pseudopods, which engulf colloid by endocytosis. After the colloid droplets fuse with lysosomes, controlled hydrolysis of iodinated thyroglobulin liberates T_3 and T_4 into the cytosol. These hormones move basally and are released basally into the bloodstream and lymphatic vessels. (Adapted from Junqueira LC et al: *Basic Histology,* 7th ed. Norwalk, CT, Appleton & Lange, 1992, p 419, and from Fawcett DW: *Bloom and Fawcett: A Textbook of Histology,* 11th ed. Philadelphia, W.B. Saunders, 1986, p 507.)

2. Calcitonin

– functions primarily to **lower blood calcium levels** by inhibiting bone resorption.

V. Parathyroid Glands

– are four small glands that lie on the posterior surface of the thyroid gland, embedded in its capsule. (PRINCIPLE)

– have a parenchyma composed of two types of cells—**chief cells** and **oxyphil cells**.

– are individually contained within a **capsule** composed of slender strands of collagen. Septa derived from the capsule penetrate each gland to convey blood vessels into its interior and to support its parenchymal cells.

– become infiltrated with fat cells in older persons; the number of oxyphil cells also increases.

A. Chief cells

– are small, **basophilic** cells, arranged in clusters.

– form anastomosing cords, surrounded by a rich, fenestrated capillary network.

−possess a centrally located, spherical nucleus, a well-developed Golgi complex, abundant rough endoplasmic reticulum, small mitochondria, glycogen, and secretory granules of variable size (200–400 nm in diameter).
−synthesize and secrete **parathyroid hormone (PTH)**. High blood calcium levels **inhibit** production of parathyroid hormone.

B. Oxyphil cells

−are large, **eosinophilic** cells that are present singly or in small clusters within the parenchyma of the gland.
−possess many large, elongated mitochondria, a poorly developed Golgi complex, and only a limited amount of rough endoplasmic reticulum.
−have no known function.

C. Parathyroid hormone

−functions primarily to **increase blood calcium levels** by stimulating bone resorption by osteoclasts.
−along with calcitonin provides a dual mechanism for regulating blood calcium levels.

VI. Adrenal (Suprarenal) Glands

−are **glands** that lie embedded in fat at the superior pole of each kidney.
−are derived from two embryonic sources: the ectodermal neural crest, which gives rise to the **adrenal medulla,** and mesoderm, which gives rise to the **adrenal cortex**.
−are invested by their own **capsule** of dense, irregular, collagenous connective tissue. Septa derived from the capsule penetrate each gland, serving as conduits for a rich vascular supply.

A. Adrenal cortex

−contains parenchymal cells that synthesize and secrete, but **do not store,** various **steroid hormones**.
−is divided into three concentric, histologically recognizable regions: the zona glomerulosa, zona fasciculata, and zona reticularis.

1. Zona glomerulosa

SHAPE DIFFERENT FROM SPECIES →SPECIES

−is located just beneath the adrenal capsule and constitutes about 13% of the total cortical volume.
−synthesizes and secretes **mineralocorticoids,** mostly **aldosterone** and some **deoxycorticosterone**. Hormone production is stimulated by angiotensin II and ACTH.
−is composed of small cells arranged in arch-like cords and spherical clusters. These cells possess a few small lipid droplets; a dense, round nucleus with one or two nucleoli; an extensive network of smooth endoplasmic reticulum (SER); short mitochondria with **shelf-like cristae;** a well-developed Golgi complex; rough endoplasmic reticulum; and free ribosomes.

ZONA INTERMEDIA -IN HORSE + SOMEWHAT IN CARNIVORE

2. Zona fasciculata

−is the largest region of the adrenal cortex, constituting up to 80% of its volume.
−synthesizes and secretes **glucocorticoids,** namely **cortisol and corticosterone**. Hormone production is stimulated by ACTH.

—is composed of cells that are larger than those in the zona glomerulosa and are arranged in columns, one or two cells thick, oriented perpendicularly to the capsule.

—contains longitudinally oriented **sinusoidal capillaries** running between neighboring columns of cells.

—cells contain many lipid droplets, which are lost during histologic processing; hence the cells appear vacuolated and are referred to as **spongiocytes.** These cells also possess spherical mitochondria with **tubular** and **vesicular cristae,** a network of smooth endoplasmic reticulum, some rough endoplasmic reticulum, lysosomes, and **lipofuscin pigment granules.**

3. Zona reticularis

—is the deepest layer of the cortex and constitutes about 7% of its total volume.

—synthesizes and secretes **androgens** (mostly **dehydroepiandrosterone** and some **androstenedione**) and perhaps small amounts of glucocorticoids. Hormone production is stimulated by ACTH.

—is composed of cells, arranged in anastomosing cords, that are smaller in size and contain fewer lipid droplets than spongiocytes; numerous, **large lipofuscin pigment granules** are common in these cells.

—also contains some cells with **pyknotic nuclei.**

4. Physiologic effects of adrenocortical hormones

a. Mineralocorticoids

—regulate **electrolyte** and **water balance** via their effect on epithelial cells of the renal tubules.

b. Glucocorticoids

—primarily regulate **carbohydrate metabolism** by promoting gluconeogenesis, deposition of liver glycogen, and elevation of blood glucose levels.

—also influence fat and protein metabolism and have diverse other effects.

c. Dehydroepiandrosterone

—has typical androgenic effects, **promoting masculine characteristics.**

B. Adrenal medulla (NEURAL C.C. DERIVATIVES)

—is completely invested by the adrenal cortex.

— CLUSTERS OR CORDS OF CELLS

—contains two populations of parenchymal cells, called **chromaffin cells,** which synthesize, store, and secrete the catecholamines **epinephrine** and **norepinephrine** (but not both in the same cell).

— DOMESTIC ANIMALS HAVE 2 ZONES

—also contains **sympathetic ganglion cells** scattered in the connective tissue.

1. Chromaffin cells

— DARK, OUTER
— LIGHT INNER

—are large, polyhedral cells containing secretory granules that stain intensely with chromium salts (chromaffin reaction).

—are arranged in short, irregular cords surrounded by an extensive capillary network.

• OFTEN GANGLIA & FASCICULI NERVE FIBERS

−are innervated by **preganglionic sympathetic (cholinergic) fibers,** making these cells analogous in function to postganglionic sympathetic neurons.

−possess a well-developed Golgi complex, isolated regions of rough endoplasmic reticulum, and numerous mitochondria.

−also contain large numbers of membrane-bounded granules (100–300 nm in diameter) containing one of the catecholamines, ATP, enkephalins, and **chromogranins,** which may function as binding proteins for epinephrine and norepinephrine.

   a. Epinephrine-producing cells contain **homogeneous,** electron-dense granules that are smaller than granules in norepinephrine-producing cells.

   b. Norepinephrine-producing cells contain **heterogeneous** granules that have an extremely electron-dense core surrounded by an electron-lucent halo.

2. Catecholamine release

−occurs in response to intense emotional reactions.

−is mediated by the preganglionic sympathetic fibers that innervate chromaffin cells.

3. Physiologic effects of adrenal catecholamines

−constitute the **"fight-or-flight" response**.

−include increased blood pressure, faster heart rate, faster breathing rate, and elevated blood glucose levels.

C. Blood supply to the adrenal glands

−is derived from the superior, middle, and inferior adrenal arteries, which form three groups of vessels: to the capsule, to parenchymal cells of the cortex, and directly to the medulla.

1. Cortical blood supply

   a. A fenestrated capillary network bathes cells of the zona glomerulosa.

   b. Straight discontinuous fenestrated capillaries supply the zona fasciculata and zona reticularis.

2. Medullary blood supply

   a. Venous blood rich in hormones reaches the medulla via the discontinuous fenestrated capillaries that pass through the cortex.

   b. Arterial blood from direct branches of capsular arteries form an extensive fenestrated capillary network among the chromaffin cells of the medulla.

   c. Medullary veins join to form the suprarenal vein, which exits the gland.

VII. Pineal Gland — IN EPITHALAMUS — DORSAL TO 3ʳᵈ VENTRICLE

−is also known as the **pineal body** or **epiphysis**.

−is a flattened, conical structure that **projects from the roof of the diencephalon**.

"BRAIN SAND" — CONCRETIONS; USUALLY MORE PRONOUNCED IN OLDER ANIMALS

—has a capsule formed of the **pia mater,** from which septa and trabeculae extend to subdivide the pineal into incomplete lobules. The trabeculae convey vascular elements into the interior of the gland.

—contains calcified concretions (**brain sand**) in its interstitium. The function of these is unknown.

—is composed primarily of **pinealocytes** and **neuroglial cells**.

A. Pinealocytes

—synthesize and secrete **serotonin** (usually during the day) and **melatonin** (usually at night).

—may also produce **arginine vasotocin,** a peptide that may be an antagonist of luteinizing hormone and follicle-stimulating hormone.

—are pale-staining cells with numerous long processes, which end in dilatations near capillaries.

—possess a large nucleus, well-developed smooth endoplasmic reticulum, some rough endoplasmic reticulum, free ribosomes, a Golgi complex, and numerous secretory granules.

—also contain microtubules, microfilaments, and unusual structures called **synaptic ribbons**. The latter are composed of dense tubular elements surrounded by synaptic vesicle–like spheroids whose function remains unclear.

B. Neuroglial (interstitial) cells

—resemble astrocytes, with elongated processes and a small, densely staining, oval nucleus.

—contain rough endoplasmic reticulum, microtubules, and many microfilaments and intermediate filaments.

VIII. Clinical Considerations

A. Pituitary adenomas

—are **common tumors** of the anterior pituitary.

—may or may not produce excess pituitary hormones.

—enlarge and often suppress secretion by the remaining pars distalis cells.

—frequently destroy surrounding bone and neural tissues.

—are treated by **surgical removal**.

B. Graves' disease

—is characterized by a diffuse **enlargement of the thyroid gland** and **protrusion of the eyeballs** (exophthalmic goiter).

—is associated with the presence of columnar-shaped thyroid follicular cells, excessive production of thyroid hormones, and decreased amounts of follicular colloid.

—is caused by the binding of autoimmune IgG antibodies to TSH receptors, which stimulates the thyroid follicular cells.

C. Simple goiter

—is enlargement of the thyroid gland caused by **insufficient iodine** (<10 μg/day) in the diet.

—is usually not associated with either hyperthyroidism or hypothyroidism.

—is treated by administration of **dietary iodine**.

D. Hyperparathyroidism

—is overactivity of the parathyroid glands, resulting in excess secretion of parathyroid hormone and subsequent bone resorption (see Chapter 7 II J 1).

—is associated with **high blood calcium levels,** which may lead to deposition of calcium salts in the kidney and walls of blood vessels.

—may be caused by a benign tumor of the parathyroid glands.

E. Addison's disease

—is characterized by secretion of **inadequate amounts of adrenocortical hormones** due to destruction of the adrenal cortex.

—is most often caused by an autoimmune disease or can be a sequel of tuberculosis.

—is life-threatening and requires **steroid treatment.**

F. Diabetes insipidus

—results from inadequate amounts of vasopressin (antidiuretic hormone); it is discussed in Chapter 18 VII D.

SEE PANCREAS p. 260, 261

(ENDOCRINE SECTION — ISLETS OF LANGERHAANS)

Review Test

Directions: Each of the numbered items or incomplete statements in this section is followed by answers or by completions of the statement. Select the **one** lettered answer or completion that is **best** in each case.

1. Protein hormones act initially on target cells by

(A) attaching to receptors on the nuclear membrane
(B) attaching to receptors in the nucleolus
(C) diffusing through the plasma membrane
(D) attaching to receptors on the plasma membrane

2. Which of the following statements concerning adrenal parenchymal cells is TRUE?

(A) Those of the zona fasciculata produce androgens
(B) Those of the adrenal medulla produce epinephrine and norepinephrine
(C) Those of the zona glomerulosa produce glucocorticoids
(D) Those of the cortex contain numerous secretory granules

3. Characteristics of pinealocytes include which one of the following?

(A) They produce melatonin and serotonin
(B) They are astrocyte-like cells
(C) They contain calcified concretions of unknown function
(D) They are postganglionic sympathetic somata

4. Which of the following statements about thyroid follicles is FALSE?

(A) They contain thyroglobulin
(B) They are lined by epithelial cells that vary in appearance with the activity of the gland
(C) Their lining includes parafollicular cells that release calcitonin into the colloid
(D) They are surrounded by a fenestrated capillary network

5. A 40-year-old woman is diagnosed with Graves' disease. Which of the following characteristics would NOT be associated with her condition?

(A) Excessive production of thyroid hormones
(B) Protrusion of the eyeballs
(C) Columnar-shaped thyroid follicular cells
(D) Inadequate levels of iodine in her diet
(E) Decreased amounts of follicular colloid

6. Prolactin is synthesized and secreted by which of the following cells?

(A) Acidophils in the pars distalis
(B) Basophils in the pars tuberalis
(C) Somatotrophs in the pars distalis
(D) Basophils in the pars intermedia

7. Adrenocorticotropic hormone (ACTH) is produced by which of the following cells?

(A) Chromophobes in the pars distalis
(B) Neurosecretory cells in the median eminence
(C) Acidophils in the pars distalis
(D) Neurons of the paraventricular nucleus in the hypothalamus
(E) None of the above

Directions: Each group of items in this section consists of lettered options followed by a set of numbered items. For each item, select the **one** lettered option that is most closely associated with it. Each lettered option may be selected once, more than once, or not at all.

Questions 8–11

Match each item below with the corresponding lettered structure shown in the photomicrograph.

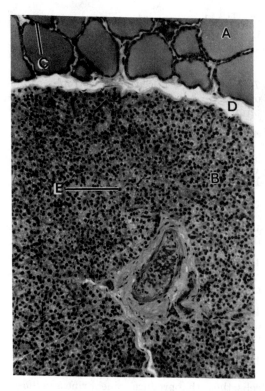

Reprinted from Gartner LP and Hiatt JL: *Atlas of Histology*. Baltimore, Williams & Wilkins, 1987, p 154.

8. Parathyroid gland

9. Colloid

10. Follicular cell

11. Capsule

Questions 12–17

Match each description below with the corresponding lettered region shown in the photomicrograph of a section of the adrenal gland.

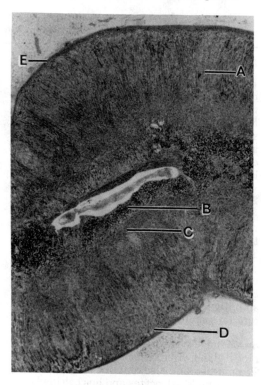

Reprinted from Gartner LP and Hiatt JL: *Atlas of Histology*. Baltimore, Williams & Wilkins, 1987, p 155.

12. Region containing ganglion cells

13. Region containing spongiocytes

14. Region where aldosterone is released

15. Region containing chromaffin cells

16. Region containing cells that produce cortisol

17. Region containing cells that produce dehydroepiandrosterone

Questions 18–20

Match each physiologic effect with the hormone that exerts the effect.

(A) Calcitonin
(B) Parathyroid hormone
(C) Triiodothyronine (T$_3$)

18. Acts upon osteoclasts to stimulate calcium mobilization from bone

19. Lowers blood calcium levels by inhibiting bone resorption

20. Helps to regulate body temperature

Answers and Explanations

1–D. Protein hormones initiate their action by binding externally to transmembrane receptor proteins in the target-cell plasma membrane. Receptors for some hormones (e.g., thyroid-stimulating hormone, serotonin, epinephrine) are linked to G proteins; other receptors, including those for insulin and growth hormone, have protein kinase activity.

2–B. Chromaffin cells in the adrenal medulla synthesize and store epinephrine and norepinephrine in secretory granules, which also contain ATP, chromogranins, and enkephalins. The cortical parenchymal cells of the zona fasciculata produce glucocorticoids, and those of the zona glomerulosa produce mineralocorticoids. The cortical parenchymal cells do not store their secretory products and thus do not contain secretory granules.

3–A. Pinealocytes, the parenchymal cells of the pineal gland, produce melatonin and serotonin. The pineal gland also contains neuroglial cells that resemble astrocytes and has calcified concretions in its interstitium.

4–C. Thyroid parafollicular cells produce calcitonin but release it into adjacent fenestrated capillaries. These cells lie between the follicular cells and the basal lamina enveloping the follicle and thus do not abut the colloid.

5–D. Graves' disease (exophthalmic goiter) results in an enlarged thyroid gland due to stimulation of the follicular cells by binding of autoimmune antibodies to TSH receptors. It is not caused by iodine deficiency.

6–A. Prolactin is produced by mammotrophs, one of the two types of acidophils located in the pars distalis of the pituitary gland. As their name implies, these cells produce a hormone that regulates the development of the mammary gland during pregnancy and lactation.

7–E. ACTH is produced by corticotrophs, a type of basophil, present in the pars distalis of the pituitary gland.

8–B. Parathyroid gland.

9–A. Colloid.

10–C. Follicular cell.

11–D. Connective tissue capsule.

12–B. Sympathetic ganglion cells are located in the adrenal medulla.

13–A. Spongiocytes are present in the zona fasciculata of the adrenal cortex. These lipid-rich cells produce mostly glucocorticoids.

14–D. Aldosterone, a mineralocorticoid, is produced and released by cells in the zona glomerulosa of the adrenal cortex.

15–B. Chromaffin cells are located in the adrenal medulla. There are two types of chromaffin cells: those that produce epinephrine and those that produce norepinephrine.

16–A. Cortisol, a glucocorticoid, is produced and secreted by cells in the zona fasciculata of the adrenal cortex.

17–C. Dehydroepiandrosterone, a weak androgen, is synthesized and released by cells in the zona reticularis of the adrenal cortex.

18–B. Parathyroid hormone increases blood calcium levels by stimulating osteoclasts to mobilize calcium from bone tissue.

19–A. Calcitonin lowers blood calcium levels and thus has an effect antagonistic to that of parathyroid hormone. It is produced by parafollicular cells of the thyroid gland.

20–C. Triiodothyronine and thyroxine both increase the basal metabolic rate, which affects heat production and body temperature. These thyroid hormones have numerous other effects.

14

Skin

I. Overview—The Skin

−is the relatively elastic covering of the body.

−constitutes about 16% of the total body weight and is the heaviest single organ.

−is composed of two layers, the **epidermis** and **dermis,** which **interdigitate** with each other to form an **irregular contour**.

−is underlain by a deeper superficial fascia layer, the **hypodermis,** which is **not** considered part of the skin. This layer of loose connective tissue binds skin loosely to the subjacent tissue.

−contains several **appendages** (sweat glands, hair follicles, sebaceous glands, and nails). The skin plus its appendages is called the **integument**.

−serves the following important **functions:** protection of the body against injury, desiccation, and infection; regulation of body temperature; absorption of ultraviolet (UV) radiation, which is necessary for synthesis of vitamin D; and reception of touch, temperature, and pain stimuli from the external environment.

II. Epidermis

−is the **superficial layer** of the skin.

−is classified as **stratified squamous keratinized epithelium** and is primarily of **ectodermal origin**.

−is composed of **keratinocytes,** the predominant cell type, and three different types of nonkeratinocytes: **melanocytes, Langerhans' cells,** and **Merkel's cells**.

−is constantly being regenerated (about every 30 days) by the mitotic activity of it keratinocytes, which typically divide at night.

−overlies projections of the dermis (**dermal papillae**), forming a series of **epidermal ridges**. On the fingertips, these surface features are visible as **fingerprints,** whose configuration is genetically determined and thus unique to each individual.

A. Layers of the epidermis

1. Stratum basale (germinativum)

−is the **deepest layer** of the epidermis.

—contains keratinocytes that are cuboidal to columnar in shape, **mitotically active,** and attached directly to the basal lamina by **hemidesmosomes** (see Chapter 5 IV B).

—also contains **melanocytes** and **Merkel's cells.**

2. Stratum spinosum

—consists of a few layers of polyhedral keratinocytes (**prickle cells**); these cells have extensions termed "intercellular bridges" by early histologists; they are now known to terminate in **desmosomes** (see Chapter 5 III A 3).

—also contains **Langerhans' cells**.

a. Keratinocytes in the deeper aspects of the stratum spinosum are **mitotically active.**

b. **Malpighian layer** (stratum malpighii) consists of the stratum spinosum and stratum basale. Nearly all the mitotic activity in the epidermis occurs in this region.

c. In the upper aspects of the stratum spinosum are keratinocytes that contain **membrane-coating granules.** The contents of these granules are released into the intercellular spaces in the form of lipid-containing sheets that are **impermeable to water and many foreign substances.**

3. Stratum granulosum

—comprises three to five layers of flattened keratinocytes that contain **keratohyalin granules, bundles of keratin filaments** (tonofilaments), and occasional **membrane-coating granules**.

—is the most superficial layer in which nuclei are observed.

a. Keratohyalin granules (not membrane-bounded) contain histidine- and cystine-rich proteins, which appear to bind the keratin filaments together.

b. The cytoplasmic aspect of the plasma membrane of keratinocytes in the stratum granulosum is reinforced by an electron-dense layer 10–12 nm thick.

4. Stratum lucidum

—is a clear, homogeneous layer just superficial to the stratum granulosum, which is often difficult to distinguish in histologic sections.

—is found only in **palmar** and **plantar skin**.

—consists of keratinocytes that **lack nuclei** and organelles but contain **eleidin,** a substance thought to be a transformation product of keratohyalin.

5. Stratum corneum

—is the **superficial layer** of the epidermis.

—consists of 15–20 layers of flattened, nonnucleated, **keratinized** "cells" filled with tonofilaments. These **nonviable, scale-like** structures (termed **horny cells,** or **squames**) have the shape of a 14-sided polygon.

—continuously sheds cells of the outermost layer (sometimes called the **stratum disjunctum**), by a process termed **desquamation.**

B. Nonkeratinocytes in the epidermis

1. Melanocytes

–are present in the **stratum basale** and originate from the neural crest.

–synthesize a **dark brown pigment (melanin)** in oval-shaped organelles (**melanosomes**), which contain **tyrosinase,** an UV-sensitive enzyme directly involved in melanin synthesis.

–possess long, **melanosome-containing processes** that extend between the cells of the stratum spinosum. Melanin is transferred, via a unique mechanism known as **cytocrine secretion,** from these melanosome-filled tips into keratinocytes of the stratum spinosum.

  a. Melanosome number, size, rate of transfer, and aggregation pattern in keratinocytes varies with race.

  b. Melanin protects against tissue damage caused by UV radiation.

2. Langerhans' cells

–are **dendritic cells** (so named because of their long processes), which originate in the bone marrow.

–are located primarily in the **stratum spinosum**.

–contain characteristic paddle-shaped **Birbeck granules**.

–function as **antigen-presenting cells** in immune responses to contact antigens (contact allergies) and some grafts (see Chapter 12 III).

3. Merkel's cells

–are present in small numbers in the **stratum basale,** near areas of well-vascularized, richly innervated connective tissue.

–possess desmosomes and keratin filaments, suggesting that they are of epithelial origin.

–have a pale cytoplasm containing **small, dense-cored granules** that are similar in appearance to those in cells of the adrenal medulla.

–receive afferent nerve terminals and are believed to function as **sensory mechanoreceptors**.

C. Thick and thin skin

–are distinguished based on the **thickness of the epidermis**.

1. Thick skin

–has an epidermis that is **400–600 µm thick**.

–is characterized by a prominent stratum corneum, well-developed stratum granulosum, and distinct stratum lucidum.

–lines the palms of the hands and soles of the feet.

–**lacks** hair follicles, sebaceous glands, and arrector pili muscles.

2. Thin skin

–has an epidermis that is **75–150 µm thick**.

–has a less prominent stratum corneum than thick skin and generally lacks a stratum granulosum and stratum lucidum, although it contains individual cells that are similar to the cells of these layers.

–is present over most of the body surface.

–contains hair follicles, sebaceous glands, and arrector pili muscles.

III. Dermis

–is also called the **corium**.

–is the layer of the skin underlying the epidermis.

–is of **mesodermal origin**.

–is composed of dense, irregular connective tissue that contains many **type I collagen fibers** and networks of thick **elastic fibers**.

–is divided into a **superficial** papillary layer and a **deeper** more extensive reticular layer, but there is no distinct boundary between them.

A. Dermal papillary layer

–is uneven and forms dermal papillae, which interdigitate with the basal surface of the epidermis.

–is composed of thin, loosely arranged fibers and cells.

–contains capillary loops and **Meissner's corpuscles,** which are fine-touch receptors.

B. Dermal reticular layer

–constitutes the major portion of the dermis.

–is composed of dense **bundles of collagen fibers** and thick **elastic fibers**.

–may contain **pacinian corpuscles** (pressure receptors) and **Krause's end-bulbs** (cold and pressure receptors) in its deeper aspects.

IV. Glands in the Skin

A. Eccrine sweat glands

–are coiled, **simple tubular glands** consisting of a secretory unit and duct.

–are present in skin throughout the body.

–are innervated by cholinergic fibers.

1. Secretory unit of eccrine sweat glands

–is composed of three cell types: dark cells, clear cells, and myoepithelial cells.

–is embedded in the dermis.

a. Dark cells

–line the lumen of the gland.

–contain numerous secretory granules.

–secrete a mucus-rich material.

b. Clear cells

–underlie the dark cells and are rich in mitochondria and glycogen.

–contain intercellular canaliculi, which extend to the lumen of the gland.

–secrete a watery, electrolyte-rich material.

c. Myoepithelial cells

–lie scattered in an incomplete layer beneath the clear cells, adjacent to the basal lamina.

–contract and aid in expressing the gland's secretions into the duct.

2. Duct of eccrine sweat glands

–is narrow and lined by **stratified cuboidal epithelial cells,** which contain many keratin filaments and have a prominent terminal web. The cells forming the external (basal) layer of the duct have many mitochondria and a prominent nucleus.

—leads from the secretory unit through the superficial portions of the dermis, to penetrate the interpapillary peg of the epidermis and spiral through all of its layers to deliver sweat to the outside.

—**modifies the secreted material** as it passes through by absorbing electrolytes and excreting ions, urea, lactic acid, drugs, and so on.

B. Apocrine sweat glands

—include the **large, specialized sweat glands** located in various areas of the body (e.g., axilla, areola of the nipple, perianal region) and the **ceruminous (wax) glands** of the external auditory canal.

—do not begin to function until puberty and are **responsive to hormonal influences**.

—have a large coiled secretory unit enveloped by scattered myoepithelial cells.

—empty their viscous secretion into hair follicles at a location superficial to the entry of sebaceous gland ducts. Although the term **apocrine** implies that a portion of the cytoplasm becomes part of the secretion, electron micrographs have shown that this **does not occur**.

—are innervated by **adrenergic fibers**.

C. Sebaceous glands

—are **branched acinar glands** with a short duct.

—empty into the neck of a hair follicle.

—are embedded in the dermis over most of the body surface with the exception of the palms and soles. They are most abundant on the face, forehead, and scalp.

—are **holocrine glands,** which produce their secretion (called **sebum**) as follows:

1. The mitotically active cells at the periphery of the secretory sacs (acini) divide, giving rise to daughter cells.

2. As these cells mature, they synthesize a lipid secretory product, which accumulates in their cytoplasm.

3. Eventually the nuclei shrink and the cells disintegrate, releasing the secretory product and cellular debris.

V. Hair Follicle and Arrector Pili Muscle

A. Hair follicle

—is an **invagination of the epidermis** extending deep into the dermis.

—comprises the following structural components:

1. Hair shaft

—is a long, slender filament that is located in the center of the follicle and extends above the surface of the epidermis.

—consists of an inner **medulla, cortex,** and outer **cuticle**.

—at its deep end is continuous with the **hair root**.

2. Hair bulb

—is the terminal expanded region of the hair follicle in which the hair is rooted.

—is deeply indented by a **dermal papilla,** which contains a capillary network necessary for sustaining the follicle.

—contains cells that form the internal root sheath and medulla of the hair shaft.

3. Internal root sheath
   —lies deep to the entrance of the sebaceous gland.
   —is composed of **Henle's layer, Huxley's layer,** and the **cuticle.**

4. External root sheath
   —is a direct continuation of the stratum malpighii of the epidermis.

5. Glassy membrane
   —is a **noncellular layer,** representing a thickening of the basement membrane.
   —separates the hair follicle from the surrounding dermal sheath.

B. Arrector pili muscle
   —attaches at an **oblique angle to the dermal sheath** surrounding a hair follicle.
   —extends superficially to underlie sebaceous glands, passing through the reticular layer of the dermis and **inserting into the papillary layer** of the dermis.
   —is a **smooth muscle** whose contraction elevates the hair and is responsible for formation of "goose bumps," caused by depressions of the skin where the muscles attach to the dermis.

VI. Nails
—are located on the distal phalanx of each finger and toe.
—are hard keratinized plates that rest on a bed of epidermis.
—are covered at their proximal end by a fold of skin, called the **cuticle,** or **eponychium,** which corresponds to the stratum corneum. This overlies the crescent-shaped, whitish **lunula.**
—are underlain at their distal (free) edge by the **hyponychium,** which also is composed of stratum corneum.
—grow as the result of mitoses of cells in the matrix of the **nail root.**

VII. Clinical Considerations

A. Epidermolysis bullosa
   —is a group of **hereditary** diseases of the skin characterized by **blister formation** following minor trauma.
   —is caused by **defects in the intermediate filaments** that provide mechanical stability and in the **anchoring fibrils** that attach the epidermis to the dermis.

1. Epidermolysis bullosa simplex
   —involves a separation within the basal epidermal cells.
   —is caused by a mutation in genes encoding intermediate filaments.
   —is associated with blisters that heal with little or no scarring.

2. Junctional epidermolysis bullosa
   —involves a separation within the lamina lucida of the basal lamina.
   —is associated with blisters that heal with scarring.

3. Distrophic epidermolysis bullosa

—involves a separation in the upper dermis.

—is caused by a mutation in the gene encoding type VII collagen, the major component of anchoring fibrils.

—is associated with blisters that heal with scarring.

B. Warts (verrucae)

—are common **skin lesions** caused by a **virus**.

—may occur anywhere on the skin (or oral mucosa) but are most frequent on the dorsal surfaces of the hands, often **close to the nails**.

—have the following histologic features: marked epidermal hyperplasia, eosinophilic cytoplasmic inclusions, and deeply basophilic nuclei. By electron microscopy, many intranuclear viral particles are observed in the keratinocytes.

C. Skin cancers

—commonly originate from cells in the epidermis.

—usually can be treated successfully if they are diagnosed early and surgically removed.

1. Basal cell carcinoma

—arises from basal keratinocytes.

2. Squamous cell carcinoma

—arises from cells of the stratum spinosum.

D. Malignant melanoma

—is a form of skin cancer than can be life-threatening.

—originates from **melanocytes** that divide, transform, and invade the dermis and then enter the lymphatic and circulatory systems, **metastasizing** to a wide variety of organs.

—is treated by **surgical removal** of the skin lesion and regional lymph nodes. Subsequent **chemotherapy** also is required because of the extensive metastases.

—represents about 3% of all cancers.

E. Psoriasis

—is a relatively common skin disease characterized by visible, patchy lesions.

—results from a **decrease in the turnover time of keratinocytes** (less than 1 week) and an **increase in the number of mitotic cells** in the stratum basale and stratum spinosum. These changes cause the epidermis to become thickened.

—is often treated by **corticosteroids** and other antimitotic drugs.

Review Test

Directions: Each of the numbered items or incomplete statements in this section is followed by answers or by completions of the statement. Select the **one** lettered answer or completion that is **best** in each case.

1. Intercellular bridges are characteristic of which of the following layers of the epidermis?

(A) Stratum granulosum
(B) Stratum lucidum
(C) Stratum corneum
(D) Statum spinosum
(E) Stratum basale

2. Which of the following statements about eccrine sweat glands is FALSE?

(A) They are lacking in thick skin
(B) They are holocrine glands
(C) They have a narrow duct lined by a stratified cuboidal epithelium
(D) They have a secretory portion that includes myoepithelial cells

3. Which of the following statements about hair follicles is FALSE?

(A) They are always associated with an eccrine sweat gland
(B) They are present in thin skin but not thick skin
(C) They are lined by epithelial cells
(D) They are attached to an arrector pili muscle
(E) They extend into the dermis

4. Which of the following statements concerning skin melanocytes is FALSE?

(A) They synthesize a pigment that protects against damage caused by UV radiation
(B) They are located in the epidermis
(C) They transfer melanosomes to keratinocytes
(D) They give rise to keratinocytes

5. Which of the following statements concerning the stratum granulosum is TRUE?

(A) It contains melanosomes
(B) It lies superficial to the stratum lucidum
(C) It is the thickest layer of the epidermis in thick skin
(D) It contains keratohyalin granules

6. Which of the following statements concerning sebaceous glands is FALSE?

(A) They are located near arrector pili muscles
(B) They employ the mechanism of holocrine secretion
(C) They produce a secretion called sebum
(D) They are present in thick skin
(E) None of the above

7. Nails possess all of the following structures EXCEPT

(A) an eponychium, or cuticle
(B) a whitish crescent-shaped lunula
(C) a body, or nail plate, composed of soft keratin
(D) a nail root, whose matrix is responsible for nail growth

8. Skin has all of the following components EXCEPT

(A) an epidermis
(B) a dermis
(C) a hypodermis
(D) a stratified squamous keratinized epithelium
(E) Merkel's cells

9. Which of the following statements about Langerhans' cells is TRUE?

(A) They commonly are found in the dermis
(B) They function as sensory mechanoreceptors
(C) They function as receptors for cold
(D) They play an immunologic role in the skin
(E) They are of epithelial origin

10. Meissner's corpuscles are present in which of the following regions of the skin?

(A) Dermal reticular layer
(B) Dermal papillary layer
(C) Hypodermis
(D) Stratum basale
(E) Epidermal ridges

11. Which of the following statements concerning thin skin is TRUE?

(A) It does not contain sweat glands
(B) It lacks a stratum corneum
(C) It is less abundant than thick skin
(D) It contains hair follicles

Directions: Each group of items in this section consists of lettered options followed by a set of numbered items. For each item, select the **one** lettered option that is most closely associated with it. Each lettered option may be selected once, more than once, or not at all.

Questions 12–15

Match each description below with the corresponding lettered structure shown in the light micrograph of skin.

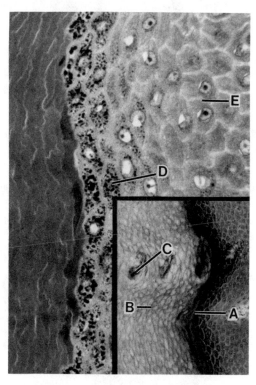

Reprinted from Gartner LP and Hiatt JL: *Atlas of Histology.* Baltimore, Williams & Wilkins, 1987, p 165.

12. A layer not present in thin skin

13. The stratum containing keratohyalin granules

14. The layer containing mitotically active cells

15. A structure that originates in the dermis and courses through the stratum corneum

Questions 16–17

Match each property below with the appropriate cell type.

(A) Merkel's cell
(B) Langerhans' cell
(C) Melanocyte

16. Contains a UV-sensitive enzyme called tyrosinase

17. Is thought to function as a sensory mechanoreceptor

Questions 18–20

Match each description below with the corresponding lettered structure in the electron micrograph of epidermal cells.

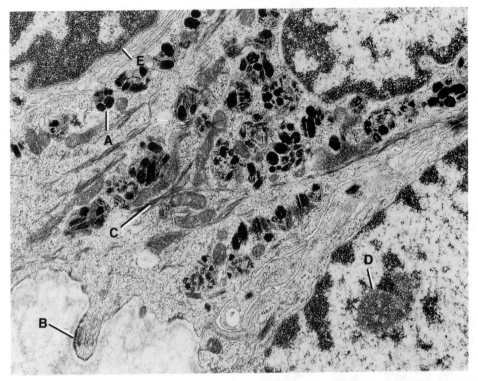

Reproduced from Strum J: *A Study Atlas of Electron Micrographs,* 3rd ed. Baltimore, University of Maryland School of Medicine, 1992, p 56.

18. A melanin-containing organelle

19. A structure that attaches keratinocytes to the basal lamina

20. A type of intermediate filament abundant in epidermal cells

Answers and Explanations

1–D. Observations with the electron microscope show that "intercellular bridges" are associated with desmosomes (maculae adherentes), linking the processes of adjacent cells in the stratum spinosum. Desmosomes also link cells within the other epidermal layers, but these cells do not form processes characteristic of "bridges." The keratinocytes of the stratum basale also contain hemidesmosomes, which attach the cells to the underlying basal lamina.

2–B. Eccrine sweat glands are merocrine glands; that is, only the contents of their secretory granules are released. They are found in both thick and thin skin.

3–A. Hair follicles are associated with sebaceous glands, not with eccrine sweat glands. They are located only in thin skin, are lined by epithelial cells, and are connected indirectly to an arrector pili muscle via the dermal sheath surrounding each follicle.

4–D. Melanocytes do not give rise to keratinocytes. Early in development, melanocytes migrate into the epidermis from the neural crest and form part of the stratum basale. They produce melanin pigment and transfer it to keratinocytes.

5–D. The stratum granulosum contains a number of dense keratohyalin granules but not melanosomes. It lies just deep to the stratum lucidum and is a relatively thin layer in the epidermis of thick skin.

6–D. Sebaceous glands are found only in thin skin.

7–C. The body of a nail is composed of hard, not soft, keratin.

8–C. The hypodermis is adipose connective tissue (superficial fascia) lying deep to the reticular layer of the dermis. It is not considered to be part of the skin.

9–D. Langerhans' cells in the epidermis function as antigen-presenting cells by trapping antigens that penetrate the epidermis and transporting them to regional lymph nodes where they are displayed to T lymphocytes. In this way, these cells assist in the immune defense of the body. They originate in the bone marrow and do not arise from epithelium.

10–B. Meissner's corpuscles are encapsulated nerve endings present in dermal papillae, which are part of the papillary layer of the dermis. They function as receptors for fine touch.

11–D. In contrast to thick skin, which lacks hair follicles, thin skin contains many of them.

12–A. The stratum lucidum.

13–D. The stratum granulosum.

14–E. The stratum spinosum.

15–C. Duct of an eccrine sweat gland.

16–C. Tyrosinase is an enzyme directly involved in melanin synthesis, which is the primary function of melanocytes.

17–A. Merkel's cells in the epidermis receive afferent nerve endings and are believed to function as sensory mechanoreceptors.

18–A. Melanosome present in a melanocyte.

19–B. Hemidesmosome in keratinocyte located in the stratum basale.

20–C. Bundle of keratin filaments (tonofilaments) in keratinocytes of the stratum basale.

15

Respiratory System

I. Overview—The Respiratory System

—includes the **lungs** and a series of **airways** that connect them to the external environment.

—is divided into two major divisions: a **conducting portion,** consisting of airways that deliver air to the lungs, and a **respiratory portion,** consisting of structures within the lungs where oxygen in inspired air is exchanged for carbon dioxide in the blood.

—possesses lining epithelia, supporting structures, glands, and other features that are characteristic of each part. These are summarized in Table 15.1.

II. Conducting Portion of the Respiratory System

—includes the nose, nasopharynx, larynx, trachea, bronchi, and bronchioles down to and including the terminal bronchioles.

—**warms, moistens,** and **filters the air** before it reaches the sites where gas exchange occurs.

A. Nasal cavity

1. Nares

—are the nostrils (paranasal openings), whose outer portions are lined by **thin skin**.

2. Vestibule

—is the first portion of the nasal cavity where the epithelial lining becomes **nonkeratinized**. Posteriorly, the lining changes to **respiratory epithelium** (pseudostratified ciliated columnar epithelium with goblet cells).

—contains **vibrissae** (thick, short hairs), which filter out large particles from the inspired air.

—has a lamina propria that is **vascular** (many venous plexuses) and contains **seromucous glands**.

3. Olfactory epithelium

—is located in the roof of the nasal cavity, on either side of the nasal septum and on the superior nasal conchae.

Table 15.1. Comparison of Respiratory System Components

Division	Skeleton	Glands	Epithelium	Ciliated Cells	Goblet Cells	Special Features
Nasal cavity: Vestibule	Hyaline cartilage	Sebaceous and sweat glands	Stratified squamous keratinized	No	No	Vibrissae
Respiratory	Bone and hyaline cartilage	Seromucous	Pseudostratified ciliated columnar	Yes	Yes	Large venous plexuses
Olfactory	Nasal conchae (bone)	Bowman's glands	Pseudostratified ciliated columnar (tall)	Yes	No	Bipolar olfactory cells; sustentacular cells; basal cells; nerve fibers
Nasopharynx	Muscle	Seromucous	Pseudostratified ciliated columnar	Yes	Yes	Pharyngeal tonsil; entrance of eustachian tube
Larynx	Hyaline and elastic cartilage	Mucous and seromucous	Stratified squamous nonkeratinized and pseudostratified ciliated columnar	Yes	Yes	Vocal cords; striated muscle (vocalis); epiglottis
Trachea and primary bronchi	C-shaped hyaline cartilage "rings"	Mucous and seromucous	Pseudostratified ciliated columnar	Yes	Yes	Trachealis (smooth) muscle; elastic lamina; two mucous cell types; short cells; diffuse endocrine cells
Intrapulmonary bronchi	Plates of hyaline cartilage	Seromucous	Pseudostratified ciliated columnar	Yes	Yes	Two helically oriented ribbons of smooth muscle
Primary bronchioles	Smooth muscle	None	Simple ciliated columnar to simple cuboidal	Yes	Only in larger ones	Clara cells
Terminal bronchioles	Smooth muscle	None	Simple cuboidal	Some	None	Clara cells

Respiratory bronchioles	Some smooth muscle	None	Simple cuboidal except where interrupted by alveoli	Some	None	Occasional alveoli; Clara cells
Alveolar ducts	Smooth muscle at alveolar openings; some reticular fibers	None	Simple squamous	None	None	Linear structure formed of adjacent alveoli; type I and II pneumocytes; alveolar macrophages
Alveoli	Reticular fibers; elastic fibers at alveolar openings	None	Simple squamous	None	None	Type I and II pneumocytes; alveolar macrophages

Modified from Gartner LP and Hiatt JL: *Color Atlas of Histology.* Baltimore, Williams & Wilkins, 1990, p 181.

–is a tall **pseudostratified columnar epithelium** consisting of three cell types: olfactory cells, supporting (sustentacular) cells, and basal cells.

–has a lamina propria that contains many **veins** and **unmyelinated nerves** and houses **Bowman's glands**.

a. Olfactory cells

–are **bipolar nerve cells** characterized by a bulbous apical projection (**olfactory vesicle**) from which several modified cilia extend.

(1) Olfactory cilia (olfactory hairs) are very **long, nonmotile cilia,** which extend over the surface of the olfactory epithelium. Their proximal one-third contains a typical **9 + 2 axoneme pattern,** but their distal two-thirds is composed of 9 peripheral **singlet** microtubules surrounding a central pair of microtubules.

(2) Function of olfactory cilia is to act as receptors for odor.

b. Supporting (sustentacular) cells

–possess nuclei that are more apically located than those of the other two cell types.

–have many **microvilli** and a prominent **terminal web**.

c. Basal cells

–rest on the basal lamina but do not extend to the surface.

–form an incomplete layer of cells.

–are believed to be **regenerative** for all three cell types.

d. Bowman's glands

–produce a **thin, watery secretion,** which is released onto the olfactory epithelial surface via narrow ducts.

(1) Odorous substances dissolved in this watery material are detected by the olfactory cilia.

(2) The secretion also flushes the surface, preparing the receptors to receive new odorous stimuli.

B. Nasopharynx

–is the posterior continuation of the nasal cavities, and becomes the oropharynx at the level of the soft palate.

–is lined by **respiratory epithelium,** whereas the oropharynx and laryngopharynx are lined by **stratified squamous nonkeratinized epithelium**.

–possesses **mucous** and **serous glands** in the lamina propria beneath the respiratory epithelium, as well as an abundance of lymphoid tissue, including the **nasopharyngeal tonsil**. When diseased, this tonsil is referred to as **adenoids**.

C. Larynx

–connects the pharynx with the trachea.

–has a wall supported by **hyaline cartilages** (thyroid, cricoid, and lower part of arytenoids) and **elastic cartilages** (epiglottis, corniculate, and tips of arytenoids).

–also possesses **striated muscle,** connective tissue, and **glands** within its wall.

1. Vocal cords

—consist of skeletal muscle (the **vocalis**), the **vocal ligament** (formed by a band of elastic fibers), and a covering of **stratified squamous nonkeratinized epithelium**.

a. Contraction of the laryngeal muscles changes the size of the opening between the vocal cords, which affects the pitch of the sounds caused by air passing through the larynx.

b. Inferior to the vocal cords, the lining epithelium changes to **respiratory epithelium,** which lines air passages down through the trachea and primary bronchi.

2. Vestibular folds (false vocal cords)

—lie superior to the vocal cords.

—are folds of loose connective tissue containing glands, lymphoid aggregations, and fat cells.

—are covered by **stratified squamous nonkeratinized epithelium**.

D. Trachea and extrapulmonary (primary) bronchi

—have walls supported by **C-shaped hyaline cartilages** (C-rings), whose open ends face posteriorly. **Smooth muscle (trachealis)** extends between the open ends of these cartilages.

—contain dense **fibroelastic** connective tissue between adjacent C-rings, which permits the elongation of the trachea during inhalation.

1. Mucosa

—consists of a respiratory epithelium, a thick underlying basement membrane, and a lamina propria.

a. Respiratory epithelium

—in the trachea possesses the following cell types:

(1) Ciliated cells

—have **long, actively motile cilia,** which beat towards the mouth.

—move inhaled particulate matter trapped in mucus towards the oropharynx, thus protecting the delicate lung tissue from damage.

—also possess **microvilli**.

(2) Mature goblet cells

—have a goblet shape and are filled with **large mucous droplets,** which are secreted to trap inhaled particles.

(3) Small mucous granule cells

—contain varying numbers of **small mucous granules**.

—are sometimes called "brush" cells because of their many **microvilli**.

—actively **divide** and thus may be able to replace recently desquamated cells.

—also may represent goblet cells after they have secreted mucus.

(4) Diffuse endocrine cells

—are also known as **small granule cells** and APUD cells.

—contain many small granules concentrated in their basal cytoplasm.

—synthesize different **polypeptide hormones,** as well as **serotonin,** which often exert a local effect on nearby cells and structures (**paracrine regulation**). The peptide hormones may also enter the bloodstream and have an **endocrine effect** on distant cells and structures.

(5) Short (basal) cells

—rest on the basal lamina but do not extend to the lumen; thus this epithelium is pseudostratified.

—are able to **divide**.

b. Basement membrane

—is a very thick layer underlying the epithelium.

c. Lamina propria

—is a thin layer of connective tissue that lies beneath the basement membrane.

—contains longitudinally oriented **elastic fibers** separating the lamina propria from the submucosa.

2. Submucosa

—is a connective tissue layer containing many **seromucous glands**.

3. Adventitia

—contains **C-shaped hyaline cartilages**.

—forms the outer layer of the trachea.

E. Intrapulmonary bronchi

—arise from subdivisions of the primary bronchi.

—divide many times giving rise to **lobar** and **segmental bronchi**.

—have **irregular cartilage plates** in their walls.

—are lined by **respiratory epithelium**.

—possess **spiraling smooth muscle bundles** separating the lamina propria from the submucosa, which contains **seromucous glands**.

F. Primary and terminal bronchioles

—**lack glands** in their submucosa.

—contain **smooth muscle** rather than cartilage plates in their walls.

1. Primary bronchioles

—have a diameter of 1 mm or less.

—are lined by epithelium that varies from **ciliated columnar with goblet cells** in the larger airways to **ciliated columnar with Clara cells** in the smaller passages.

—divide to form several terminal bronchioles after entering the **pulmonary lobules**.

2. Terminal bronchioles (Figure 15.1)

—are the **most distal part of the conducting portion** of the respiratory system.

—have a diameter of less than 0.5 mm.

—are lined by a **simple cuboidal epithelium** that contains mostly **Clara cells,** some ciliated cells, and no goblet cells. Clara cells **divide,** and they have the following functions:

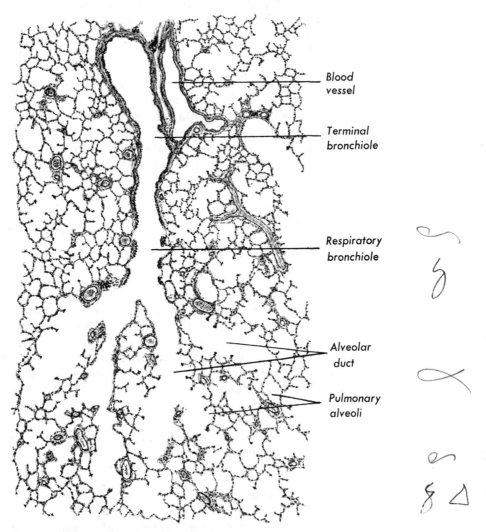

Figure 15.1. Drawing of a section of the lung. (Reprinted from Kelly DE et al: *Bailey's Textbook of Microscopic Anatomy,* 18th ed. Baltimore, Williams & Wilkins, 1984, p 631.)

> **a. Secretion of glycosaminoglycans,** which probably protect the bronchiolar lining
>
> **b. Metabolism of airborne toxins,** which is carried out by cytochrome P-450 enzymes present in the abundant smooth endoplasmic reticulum

III. Respiratory Portion of the Respiratory System

—includes the respiratory bronchioles, alveolar ducts, alveolar sacs, and alveoli, all of which are located within the lung (see Figure 15.1).

—is where **exchange of gases** takes place.

A. Respiratory bronchioles

—mark the transition from the conducting to the respiratory portion of the respiratory system.

—are lined by a **simple cuboidal epithelium** consisting of **Clara cells** and some **ciliated cells,** except where their walls are interrupted by **alveoli,** the sites where gas exchange occurs.

B. Alveolar ducts

—are **linear passageways** continuous with the respiratory bronchioles.

—have walls that consist of **adjacent alveoli,** which are separated from one another only by an **interalveolar septum**.

—are the most distal portion of the respiratory system to contain **smooth muscle,** which is present in their walls at the openings of adjacent alveoli.

—are lined by an **attenuated simple squamous epithelium** consisting of type I and type II **pneumocytes**.

C. Alveolar sacs

—are expanded outpouchings of numerous alveoli located at the distal ends of alveolar ducts.

D. Alveoli

—are pouch-like evaginations (about 200 μm in diameter) present in the wall of respiratory bronchioles, alveolar ducts, and alveolar sacs.

—have thin walls across which oxygen and carbon dioxide can diffuse between the air and blood.

—are separated from each other by **interalveolar septa,** which may contain one or more **alveolar pores** (pores of Kohn). These pores permit equalization of pressure between alveoli.

—are rimmed by **elastic fibers** at their openings and are supported by many **reticular fibers** in their walls.

—are lined by a **highly attenuated simple squamous epithelium** composed of type I and type II **pneumocytes**.

1. Type I pneumocytes

—are also called **type I alveolar cells**.

—cover about 95% of the alveolar surface.

—have an extremely **thin cytoplasm** and may be less than 80 nm in thickness.

—form **tight junctions** with adjacent cells.

—may have **phagocytic** capabilities.

—are **not** able to divide.

2. Type II pneumocytes

—are also called **type II alveolar cells, great alveolar cells, granular pneumocytes,** and **septal cells**.

—have a **cuboidal** shape and are most often **located near septal intersections**.

—bulge into the alveolus and have a free surface that contains short **microvilli** around its peripheral borders.

—are able to **divide** and **regenerate** both types of alveolar pneumocytes.

—form **tight junctions** with adjacent cells.

—synthesize **pulmonary surfactant,** which is stored in cytoplasmic **lamellar bodies**.

a. Structure—pulmonary surfactant

- –consists of **phospholipids** and at least four different **proteins** (SP-A, SP-B, SP-C, and SP-D).
- –forms **tubular myelin** (a network-like configuration) when it is first released from lamellar bodies. It then spreads to produce a **monomolecular film** over the alveolar surface, forming a **lower aqueous phase** and a **superficial lipid phase**.

b. Function—pulmonary surfactant

- –**reduces the surface tension** of the alveolar surface, permitting the alveoli to expand easily during inspiration and preventing alveolar collapse during expiration.

3. Alveolar macrophages

- –are also known as **alveolar phagocytes** and **dust cells**.
- –are the principal mononuclear **phagocytes** of the alveolar surface.
- –remove inhaled dust, bacteria, and other particulate matter trapped in the pulmonary surfactant, thus providing a vital line of defense in the lungs.
- –migrate to the bronchioles after becoming filled with debris. From there, they are carried via **ciliary action** to the upper airways, eventually reaching the pharynx where they are swallowed or expectorated.
- –may also exit by migrating into the interstitium and leaving via lymphatic vessels.

E. Interalveolar septum

- –is the wall, or partition, between two adjacent alveoli.
- –is bounded on its outer surfaces by the thin simple squamous epithelium lining the alveoli.
- –contains many **elastic and reticular fibers** in its thicker regions.
- –houses **continuous capillaries** in its central (interior) region.
- –accommodates the **blood–gas barrier,** which separates the alveolar airspace from the capillary lumen.

1. Structure of the blood–gas barrier

a. Thinnest regions of the barrier are 0.2 μm or less in thickness and consist of the following layers:

(1) **Type I pneumocytes and layer of surfactant** lining the alveolar airspace

(2) **Fused basal laminae of type I pneumocytes and capillary endothelial cells**

(3) **Endothelium of the continuous capillaries** within the interalveolar septum

b. Thicker regions of the barrier measure about 0.5 μm across and have an **interstitial space** interposed between the two basal laminae, which are not fused.

2. Function of the blood–gas barrier

- –is to permit the **diffusion of gases** between the alveolar airspace and the blood. **Oxygen** passes from the alveolus into the capillary, and **carbon dioxide** passes from the capillary blood into the alveolus.

IV. Lung Lobules

—vary greatly in size and shape, but each has an apex directed toward the pulmonary hilum and a wider base directed outward.

—contain a **single primary bronchiole,** which enters at the apex and branches to form five to seven terminal bronchioles. These in turn divide, ultimately giving rise to alveoli at the base of each lobule.

—are separated from adjacent lobules by an **incomplete septum**.

—possess lymphatic vessels running in their dense connective tissue, but these are absent from the interalveolar wall.

V. Pulmonary Vascular Supply

A. Pulmonary artery

—carries blood to the lungs to be **oxygenated**.

—enters the root of each lung and extends branches along the divisions of the bronchial tree.

—enters lung lobules where **its branches follow the bronchioles**.

B. Pulmonary veins

—in lung lobules run in the intersegmental connective tissue, **separated from the arteries**.

—after leaving the lobules, come close to divisions of the bronchial tree, and **run parallel with branches of the pulmonary artery** as they accompany bronchi to the root of the lung.

C. Bronchiole arteries and veins

—provide nutrients to and remove wastes from the nonrespiratory portions of the lung (bronchi, bronchioles, interstitium, and pleura).

—follow the branching pattern of the bronchial tree, and form anastomoses with the pulmonary vessels near capillary beds.

VI. Pulmonary Nerve Supply

—consists primarily of **autonomic fibers to the smooth muscle of bronchi and bronchioles**. Axons have also been demonstrated in the thicker parts of the interalveolar septae.

A. Parasympathetic stimulation

—causes **contraction** of pulmonary smooth muscle.

B. Sympathetic stimulation

—causes **relaxation** of pulmonary smooth muscle.

—can be mimicked by certain drugs that cause bronchial dilatation.

VII. Clinical Considerations

A. Hyaline membrane disease

—is frequently observed in **premature infants who lack adequate amounts of pulmonary surfactant**.

—is characterized by **labored breathing,** which results from difficulty expanding the alveoli due to a high alveolar surface tension (caused by inadequate levels of surfactant).

—can often be circumvented, if detected before birth, by the administration of **glucocorticoids,** which induce synthesis of surfactant.

B. Asthma

- —is marked by widespread **constriction of smooth muscle in the bronchioles,** causing a decrease in their diameter.
- —is associated with **extremely difficult expiration** of air, **accumulation of mucus** in the passageways, and **infiltration of inflammatory cells** into the region.
- —is often progressive and associated with **allergic reactions** during which vasoconstrictive substances are periodically released within the lungs.
- —is treated by administering drugs such as **epinephrine** and **isoproterenol,** which function to relax the bronchiolar smooth muscle, thus dilating the passageways.

C. Emphysema

- —results from **destruction of alveolar walls** and formation of large cyst-like sacs, reducing the surface available for gas exchange.
- —is marked by **decreased elasticity** of the lungs, which are unable to recoil adequately during expiration. In time, the lungs become expanded and enlarge the thoracic cavity ("barrel chest").
- —is associated with exposure to **cigarette smoke and other substances that inhibit α_1-antitrypsin,** a protein that normally protects the lungs from the action of **elastase** produced by alveolar macrophages.
- —can be a hereditary condition resulting from a defective α_1-antitrypsin. In such cases, gene therapy with recombinant α_1-antitrypsin is being used.

Review Test

Directions: Each of the numbered items or incomplete statements in this section is followed by answers or by completions of the statement. Select the **one** lettered answer or completion that is **best** in each case.

1. Characteristics of olfactory epithelium include which one of the following?

(A) It is located at the base of the nasal cavity
(B) It is classified as simple columnar
(C) It has an underlying lamina propria containing mucous glands
(D) It has modified cilia, which act as receptors for odor

2. Which of the following statements concerning terminal bronchioles is TRUE?

(A) They are part of the conducting portion of the respiratory system
(B) They function in gas exchange
(C) They do not contain ciliated cells
(D) They have cartilage plates present in their walls.

3. The larynx contains all of the following EXCEPT

(A) a stratified squamous nonkeratinized epithelium
(B) true vocal cords
(C) fibrocartilage
(D) skeletal muscle
(E) seromucous glands

4. The trachea possesses which one of the following components?

(A) Irregular cartilage plates in its wall
(B) Skeletal muscle in its wall
(C) An epithelium containing only two cell types
(D) A thick basement membrane underlying its epithelium

5. Intrapulmonary bronchi possess all of the following components EXCEPT

(A) seromucous glands in their submucosa
(B) cartilage in their walls
(C) an epithelium containing ciliated cells
(D) Clara cells within their lining epithelium
(E) spiraling smooth muscle bundles separating the lamina propria from the submucosa

6. Alveoli in alveolar sacs possess all of the following components EXCEPT

(A) elastic fibers in their walls
(B) a simple squamous lining epithelium
(C) reticular fibers in their walls
(D) smooth muscle in their walls
(E) pneumocytes

7. Which of the following statements concerning respiratory bronchioles is TRUE?

(A) No gas exchange occurs in them
(B) They do not have alveoli forming part of their wall
(C) They contain goblet cells in their lining epithelium
(D) They are included in the conducting portion of the respiratory system
(E) None of the above

8. The blood–gas barrier in alveoli may include each of the following EXCEPT

(A) cytoplasm of a type II pneumocyte
(B) basal lamina of type I alveolar cell
(C) cytoplasm of a type I pneumocyte
(D) endothelium of a continuous capillary
(E) basal lamina of an endothelial cell

9. Alveoli are present in all of the following parts of the respiratory tract EXCEPT

(A) alveolar sacs
(B) respiratory bronchioles
(C) alveolar ducts
(D) terminal bronchioles

10. True statements about asthma include which one of the following?

(A) It is due to a loss of lung elasticity
(B) It eventually causes the lungs to expand and leads to a "barrel chest"
(C) It is associated with difficulty expiring air from the lungs
(D) It may be helped by gene therapy, employing recombinant α_1-antitrypsin

11. Which of the following statements concerning alveolar macrophages is TRUE?

(A) They secrete α_1-antitrypsin
(B) They secrete elastase
(C) They originate from blood neutrophils
(D) They may play a role in causing hyaline membrane disease

Directions: Each group of items in this section consists of lettered options followed by a set of numbered items. For each item, select the **one** lettered option that is most closely associated with it. Each lettered option may be selected once, more than once, or not at all.

Questions 12–15

For each characteristic below, select the disorder to which it applies.

(A) Asthma
(B) Hyaline membrane disease
(C) Emphysema

12. May be treated with anti-elastase (α_1-antitrypsin) in some cases

13. Is caused by inadequate amounts of pulmonary surfactant

14. Is associated with a "barrel chest"

15. Is frequently treated with glucocorticoids

Questions 16–20

Match each structure below with the corresponding letter in the electron micrograph showing part of an alveolar wall.

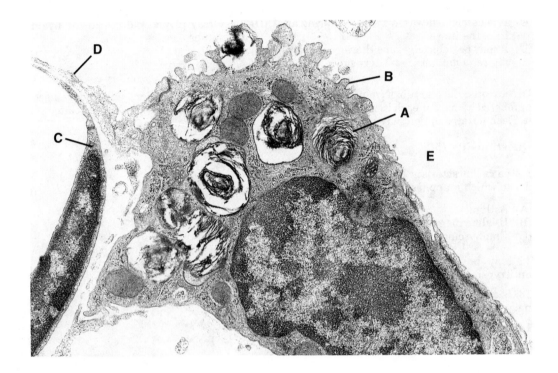

16. Microvillus

17. Portion of a type I pneumocyte

18. Lamellar body

19. Alveolar lumen

20. Capillary endothelium

Answers and Explanations

1–D. The olfactory epithelium possesses non-motile cilia, which act as receptors for odor. They are extensions of the bipolar nerve cells that form part of this tall pseudostratified epithelium located in the roof of the nasal cavity. Bowman's glands, which lie in the lamina propria beneath this epithelium, produce a watery secretion, which moistens the olfactory surface.

2–A. Terminal bronchioles are the most distal components of the conducting portion of the respiratory system. They lack alveoli and thus do not function in gas exchange. They are lined by an epithelium composed of two cell types: secretory (Clara) cells and ciliated cells. Cartilage is not present in bronchioles.

3–C. Hyaline and elastic cartilages, but not fibrocartilage, form part of the larynx.

4–D. The pseudostratified ciliated columnar epithelium lining the trachea rests upon a thick basement membrane and contains five cell types. The trachea possesses C-shaped cartilages with smooth muscle (the trachealis) extending between their ends.

5–D. Clara cells are found only in bronchioles.

6–D. The smallest airways to contain smooth muscle are alveolar ducts, where it forms a sphincter around the openings into adjacent alveoli. In alveolar sacs, the alveoli have openings rimmed only by elastic fibers.

7–E. Respiratory bronchioles have alveoli interrupting their walls, so some gas exchange takes place at this level. Their remaining walls are lined by a simple cuboidal epithelium consisting of Clara cells and ciliated cells. Respiratory bronchioles are categorized in the respiratory portion of the system.

8–A. Type II pneumocytes are located primarily near the intersections of interalveolar septa and do not form part of the blood–gas barrier.

9–D. Alveoli are not present in the terminal bronchioles, which constitute the last (most distal) component of the conducting portion of the respiratory system.

10–C. Asthma results from the constriction of smooth muscle in the bronchioles, which decreases their diameter and makes the expiration of air very difficult. Mucus accumulates in the airways and inflammatory cells invade the bronchiolar walls.

11–B. Alveolar macrophages secrete elastase. Normally, α_1-antitrypsin (a serum protein) interacts with elastase, thereby protecting the lung against damage that may lead to emphysema. Alveolar macrophages, like all macrophages, arise from blood monocytes, and they are unrelated to the pathogenesis of hyaline membrane disease.

12–C. Hereditary forms of emphysema are now being treated with recombinant α_1-antitrypsin, which has anti-elastase activity.

13–B. Hyaline membrane disease is caused by inadequate amounts of pulmonary surfactant.

14–C. A loss of lung elasticity in emphysema makes it difficult for the lungs to recoil normally during expiration. They enlarge and so does the thoracic cavity, producing a "barrel chest."

15–B. Glucocorticoids, which stimulate synthesis of pulmonary surfactant, are often used to prevent or alleviate hyaline membrane disease.

16–B. Microvillus extending from a type II pneumocyte.

17–D. A thin cytoplasmic process of a type I pneumocyte.

18–A. Lamellar body containing pulmonary surfactant.

19–E. Alveolar lumen.

20–C. Endothelial lining of continuous capillary in the interalveolar septum.

16

Digestive System: Oral Cavity and Alimentary Tract

I. Overview—The Digestive System

- comprises the **oral cavity** and **alimentary canal,** which includes the **esophagus, stomach,** and **small** and **large intestines,** and several **extrinsic glands**.
- consists of a hollow tube (highly modified in the oral cavity) whose lumen varies in diameter and is surrounded by the **mucosa, submucosa, muscularis externa,** and **serosa** (or **adventitia**).
- secretes various **enzymes** and **hormones** that function in the digestive process.

II. Oral Cavity

- includes the **lips, palate, teeth and associated structures,** and **tongue,** as well as the major salivary glands and lingual tonsils.
- is covered in most places by a **stratified squamous epithelium** whose **rete ridges** (pegs) interdigitate with tall **connective tissue papillae** of the subjacent connective tissue.

A. Lips

- are subdivided into an **external region,** a **vermillion zone,** and an **internal region**.
- have a dense irregular connective tissue core enveloping skeletal muscle.
- possess **sebaceous glands, sweat glands,** and **hair follicles** in the external region; **minor salivary glands** in the internal region; and occasional, nonfunctional sebaceous glands in the internal region and vermillion zone.

1. External region and vermillion zone

- are covered by a stratified squamous **keratinized** epithelium, which is underlain by **dermis**.

2. Internal region

- is covered by a stratified squamous **nonkeratinized** epithelium, which is underlain by dense irregular collagenous **connective tissue** (wet mucosa).

B. Palate

−is lined by **pseudostratified ciliated columnar epithelium** (respiratory epithelium) on its **nasal aspect**.

1. Hard palate

−is lined on its oral aspect by stratified squamous **parakeratinized to keratinized** epithelium.

−possesses a **bony shelf** to which bundles of collagen fibers firmly attach the mucosa.

−contains **adipose tissue** (anterolaterally) and **mucous minor salivary glands** (posterolaterally) in its mucosa.

2. Soft palate

−is lined on its oral aspect by stratified squamous **nonkeratinized** epithelium.

−has a core composed of thick bundles of **skeletal muscle** fibers.

−contains many **mucous minor salivary glands** in its mucosa.

C. Teeth

1. Structure—teeth

−are composed of an internal soft tissue, the **pulp,** and three calcified tissues: **enamel** and **cementum,** which form the surface layer, and **dentin,** which lies between the surface layer and pulp. As in bone, **calcium hydroxyapatite** is the mineralized material in the calcified dental tissues.

−contain an enamel-covered **crown,** a cementum-covered **root,** and a **cervix,** the region where the two surface materials meet.

a. Enamel

−has a highly calcified matrix composed of two fibrous proteins, **enamelin** and **amelogen,** which are elaborated by **ameloblasts** during formation of the crown.

−is **acellular after tooth eruption** and therefore cannot repair itself.

b. Dentin

−surrounds the central **pulp chamber** and **pulp (root) canal**.

−has a calcified matrix containing **collagen fibers**.

−is manufactured by odontoblasts, which persist and continue to elaborate dentin.

c. Cementum

−has a **collagen-containing** calcified matrix, which is produced by **cementoblasts**.

−is **continuously elaborated** after tooth eruption, compensating for the decrease in tooth length resulting from abrasion of the enamel.

d. Dental pulp

−is a gelatinous connective tissue containing **odontoblasts** in its peripheral layer (closest to the dentin); fibroblasts and mesenchymal cells; and thin collagen fibers.

−is richly **vascularized** and contains **afferent nerve fibers**. All sensations from the pulp are interpreted as **pain** in the central nervous system.

2. Crown formation

—begins 6–7 weeks after conception as a horseshoe-shaped band, the **dental lamina,** which is derived from the oral **epithelium**. A dental lamina develops in each jaw and projects into the underlying **ectomesenchyme**.

—occurs **before** root formation.

—proceeds via the following sequential stages: **bud, cap, bell,** and **appositional**.

3. Root formation

—occurs after the crown is mostly completed and is accompanied by tooth eruption.

D. Dental supporting structures

1. Periodontal ligament

—is composed of dense irregular connective tissue whose collagen fibers are arranged in **principal fiber bundles,** which extend from cementum to bone, suspending the tooth in its **alveolus**.

—has an abundant nerve and vascular supply.

2. Gingivae

—are covered by stratified squamous **parakeratinized to keratinized** epithelium with long rete ridges.

—contain dense irregular connective tissue in which the collagen is arranged in five **principal fiber bundles**.

3. Alveolar bone

—consists of an inner layer (**cribriform plate**) and an outer layer (**cortical plate**) of compact bone with an intervening layer of cancellous bone (**spongiosa**).

E. Tongue

—is divided into an anterior two-thirds and a posterior one-third by the **sulcus terminalis,** a V-shaped depression whose posteriorly pointing apex ends in a shallow pit, the **foramen cecum**.

—is covered on its dorsal surface by stratified squamous **parakeratinized to keratinized** epithelium, whereas its ventral surface is covered by stratified squamous **nonkeratinized** epithelium. Both surfaces are underlain by a **lamina propria** and **submucosa** of dense irregular collagenous connective tissue.

—possesses a **muscular core,** which constitutes the bulk of the lingual tissue.

1. Lingual papillae

—are located on the **dorsal surface of the anterior two-thirds** of the tongue.

—are classified into four types differing in shape and function.

a. Filiform papillae

—are short, narrow, **highly keratinized,** pointed structures that project above the tongue surface.

b. Fungiform papillae
   —are mushroom-shaped structures that project above the surface of the tongue.
   —are interspersed among the filiform papillae.
   —contain **occasional taste buds**.

c. Foliate papillae
   —are shallow, longitudinal furrows located on the lateral aspect of the tongue.

d. Circumvallate papillae
   —are large, circular papillae each of which is surrounded by a moat-like furrow.
   —are 10–15 in number, distributed in a V pattern just **anterior to the sulcus terminalis**.
   #### (1) Glands of von Ebner
      —deliver their **serous secretion** into the furrow surrounding each papilla, assisting the taste buds in perceiving stimuli.
   #### (2) Taste buds
      —are **intraepithelial structures** located on the **lateral surfaces** of circumvallate papillae and the **walls** of the surrounding furrows.
      —function in perceiving salt, sour, bitter, and sweet taste sensations.

2. Muscular core of the tongue
   —is composed of several bundles of **skeletal muscle fibers,** which cross each other in three planes.
   —possesses numerous **minor salivary glands** interspersed among the muscle fibers.

3. Lingual tonsils
   —are located on the dorsal surface of the **posterior one-third** of the tongue.

III. Histologic Layers of the Alimentary Canal

—are similar throughout the alimentary canal, but display characteristic regional specializations (Table 16.1).

A. Mucosa

1. Epithelium
   —**lines the lumen** of the alimentary canal.

2. Lamina propria
   —consists of loose connective tissue housing glands and **lymphoid accumulations**.

3. Muscularis mucosae
   —is composed of one to three layers of **smooth muscle**.
   —is responsible for the **motility of the mucosa,** which is controlled by Meissner's (submucosal) plexus and some paracrine hormones.

B. Submucosa

−consists of dense irregular collagenous or fibroelastic connective tissue.

−contains glands **only** in the esophagus and duodenum.

−houses **Meissner's (submucosal) nerve plexus**.

C. Muscularis externa

−is composed of two layers of **smooth muscle** (three layers in the stomach). In the esophagus, **skeletal muscle** also is present.

−houses **Auerbach's (myenteric) nerve plexus,** which is primarily responsible for its innervation.

−is responsible for **gut motility,** which is controlled by innervation, stretching of the muscle, and several paracrine hormones (see Table 16.2).

D. External layer

1. Serosa consists of a **mesothelial lining** (peritoneum) over a layer of loose connective tissue.

2. Adventitia consists only of loose connective tissue.

IV. Divisions of the Alimentary Canal

A. Esophagus

−is lined by a **stratified squamous nonkeratinized epithelium**.

−possesses mucus-secreting **esophageal cardiac glands** in the lamina propria and mucus-secreting **esophageal glands proper** in the submucosa.

−has a **muscularis mucosae** consisting of a **single** longitudinal layer of smooth muscle.

−contains **skeletal muscle** in the upper third of the **muscularis externa;** a combination of smooth and skeletal muscle in the middle third; and **smooth muscle** in the lower third.

−conveys small portions of masticated and moistened food (**bolus**) from the pharynx into the stomach by **peristaltic activity of the muscularis externa**. Two physiologic **sphincters** (the pharyngoesophageal and the gastroesophageal) in the muscularis externa assure that a bolus is transported in one direction only, towards the stomach.

B. Stomach

−**acidifies** and converts the bolus delivered by the esophagus into a thick, viscous fluid known as **chyme**.

−also produces **digestive enzymes** and **hormones**.

1. General structure—stomach

−exhibits longitudinal folds of the mucosa and submucosa (called **rugae**), which disappear in the distended stomach.

−possesses numerous **gastric pits** (foveolae), which are deepest in the pylorus and shallowest in the cardia.

a. Gastric mucosa
(1) Epithelium

−is **simple columnar,** composed of mucus-producing surface lining cells (not goblet cells).

Table 16.1. Selected Histologic Features of the Alimentary Canal

Region	Epithelium	Lamina Propria	Muscularis Mucosae*	Submucosa†	Muscularis Externa‡
Esophagus	Stratified squamous	Esophageal cardiac glands	One layer: longitudinal	Collagenous CT Esophageal glands proper	Two layers: inner circular and outer longitudinal
Stomach	Simple columnar with gastric pits	Gastric glands	Two (sometimes three layers): inner circular; outer longitudinal; and outermost circular in places	Collagenous CT No glands	Three layers: inner oblique; middle circular; and outer longitudinal
Small intestine	Simple columnar with goblet cells	Villi; crypts of Lieberkühn; Peyer's patches in ileum (extend into submucosa); lymphoid nodules	Two layers: inner circular and outer longitudinal	Fibroelastic CT Brunner's glands in duodenum	Two layers: inner circular and outer longitudinal
Large intestine: Cecum and colon	Simple columnar with goblet cells	Crypts of Lieberkühn (lack Paneth cells); lymphoid nodules	Two layers: inner circular and outer longitudinal	Fibroelastic CT No glands	Two layers: inner circular and outer longitudinal, which is modified to form teniae coli
Rectum	Simple columnar with goblet cells	Crypts of Lieberkühn (fewer but deeper than in colon); lymphoid nodules	Two layers: inner circular and outer longitudinal	Fibroelastic CT No glands	Two layers: inner circular and outer longitudinal

	Epithelium		Muscularis mucosae*		Muscularis externa‡
Anal canal	Simple columnar cuboidal (proximally); stratified squamous non-keratinized (distal to anal valves); stratified squamous keratinized (anus)	Sebaceous glands; circumanal glands; lymphoid nodules; rectal columns of Morgagni (involve entire mucosa); hair follicles (anus)	Two layers: inner circular and outer longitudinal	Fibroelastic CT with large veins; no glands	Two layers: inner circular (which forms internal anal sphincter) and outer longitudinal
Appendix	Simple columnar with goblet cells	Crypts of Lieberkühn (shallow); lymphoid nodules (large, numerous and may extend into the submucosa)	Two layers: inner circular and outer longitudinal	Fibroelastic CT; confluent lymphoid nodules; no glands; fat tissue (sometimes)	Two layers: inner circular and outer longitudinal

*The muscularis mucosae is composed entirely of smooth muscle throughout the alimentary canal.

†CT = connective tissue.

‡The muscularis externa is composed entirely of smooth muscle in all regions except the esophagus. The upper one-third of the esophageal muscularis externa is all skeletal muscle; the middle third is a mixture of skeletal and smooth muscle; and the lower one-third is all smooth muscle.

Rumen — papillae
- str. sq. epith — absorptive
- 13-44
- no lamina muscularis (also in esophagus)

Retic — folds
- 13-48
- honeycomb " (folds)
- muscularis @ tip of folds
- lamina

Omasum — folds — "plies"
- lamina muscularis complete

Abomasum — stomach
- gland

(2) Lamina propria
—is a loose connective tissue housing smooth muscle cells, lymphocytes, plasma cells, mast cells, and fibroblasts.
—contains **gastric glands**.

(3) Muscularis mucosae
—is composed of a poorly defined inner circular layer, an outer longitudinal layer of smooth muscle, and, occasionally, an outermost
circular layer.

b. Gastric submucosa
—is composed of dense irregular **collagenous** connective tissue.
—contains fibroblasts, mast cells, and lymphoid elements embedded in
the connective tissue.
—houses **Meissner's (submucosal) plexus**.
—possesses arterial and venous plexuses that supply and drain the vessels of the mucosa, respectively.

c. Gastric muscularis externa
—is composed of **three layers** of smooth muscle: an incomplete inner
oblique; a thick middle circular, which forms the **pyloric sphincter;**
and an outer longitudinal layer.
—is responsible for mixing of gastric contents and emptying of the
stomach.
—is affected by various characteristics of the chyme (e.g., lipid content,
viscosity, osmolality, caloric density, and pH), which influence the
emptying rate.

d. Gastric serosa
—covers the external surface of the stomach.

2. Gastric glands
—are simple branched tubular glands located in the lamina propria of the
cardia, fundus, and **pylorus**.
—consist of an **isthmus,** which connects a gland to the base of a gastric pit,
a **neck,** and a **base**. The base of pyloric glands is much more convoluted
than that of cardiac or fundic glands.
—are composed of **parietal, chief, mucous neck,** and **enteroendocrine
cells**.
—also possess **regenerative cells** (located primarily in the neck and isthmus), which replace all other cells in the gland, pit, and luminal surface.

a. Parietal (oxyntic) cells
—are pyramidal-shaped cells concentrated in the upper half of gastric
glands.
—secrete **hydrochloric acid** (HCl) and **gastric intrinsic factor**. The
latter is necessary for absorption of vitamin B_{12} in the ileum.
—possess a unique intracellular **tubulovesicular system,** many mitochondria, and secretory **intracellular canaliculi,** which are deep
invaginations of the apical plasma membrane lined by **microvilli**.
—undergo a transition, when stimulated to secrete HCl, in which the
number of microvilli increases and the complexity of the tubulovesicular system decreases (suggesting that tubulovesicles are involved in
secretion).

b. Chief (zymogenic) cells

- are pyramidal-shaped cells located in the lower half of **fundic glands only**.
- secrete **pepsinogen** (a precursor of the enzyme pepsin) and the precursors of two other enzymes, **rennin** and **lipase**.
- display an abundance of basally located rough endoplasmic reticulum (RER), a supranuclear Golgi complex, and many apical zymogen (secretory) granules.

c. Mucous neck cells

- are located in the neck of gastric glands (may be able to divide).
- possess short microvilli, apical mucous granules, a prominent Golgi complex, numerous mitochondria, and some basally located rough endoplasmic reticulum.

d. Enteroendocrine cells

- belong to the population of **diffuse endocrine cells** (DES).
- are also referred to as **APUD cells** (**a**mine **p**recursor **u**ptake and **d**ecarboxylation cells).
- include more than a dozen different types of cells that contain many small hormone-containing granules, usually concentrated in the **basal** cytoplasm. A given enteroendocrine cell secretes only one hormone (Table 16.2).
- possess an abundance of mitochondria and rough endoplasmic reticulum and a moderately well-developed Golgi complex.

Table 16.2. Selected Hormones Secreted in the Alimentary Canal*

Hormone	Site of Secretion	Physiologic Effect
Cholecystokinin (CCK)	Small intestine	Stimulates release of pancreatic enzymes and contraction of gall bladder (with release of bile)
Gastric inhibitory peptide (GIP)	Small intestine	Inhibits gastric HCl secretion
Gastrin	Pylorus and duodenum	Stimulates gastric secretion of HCl and pepsinogen
Glicentin	Stomach through colon	Stimulates hepatic glycogenolysis
Glucagon	Stomach and duodenum	Stimulates hepatic glycogenolysis
Motilin	Small intestine	Increases gut motility
Neurotensin	Small intestine	Inhibits gut motility; stimulates blood flow to the ileum
Secretin	Small intestine	Stimulates bicarbonate secretion by pancreas and biliary tract
Serotonin and substance P	Stomach through colon	Increase gut motility
Somatostatin	Pylorus and duodenum	Inhibits nearby enteroendocrine cells
Urogastrone	Duodenum (Brunner's glands)	Inhibits gastric HCl secretion; enhances epithelial cell division
Vasoactive intestinal peptide (VIP)	Stomach through colon	Increases gut motility; stimulates intestinal ion and water secretion

*Some of these hormones also are secreted in other parts of the body and have additional physiologic effects.

3. Gastric juices

–contain water, HCl, mucus, pepsin, lipase, rennin, and electrolytes, all of which aid in the digestion of food.

–are **very acidic** (pH 2.0), facilitating the activation of pepsinogen to pepsin, which catalyzes the partial hydrolysis of proteins.

4. Regulation of gastric secretion

–is effected by neural activity (vagus nerve) and by several hormones.

a. Gastrin, released by enteroendocrine cells in the gastric and duodenal mucosa, **stimulates HCl secretion.**

b. Somatostatin, produced by enteroendocrine cells of the pylorus and duodenum, inhibits the release of gastrin and thus **indirectly inhibits HCl secretion.**

c. Urogastrone, produced by Brunner's glands in the duodenum, and **gastric inhibitory peptide,** produced by enteroendocrine cells in the small intestine, **directly inhibit HCl secretion.**

C. Small intestine

–is approximately 7-feet long.

–has three regions: the **duodenum** (proximal), **jejunum** (medial), and **ileum** (distal).

–secretes several **hormones.**

–continues and largely completes the **digestion** of foodstuffs and **absorbs** the resulting metabolites.

1. Luminal surface modifications—small intestine

–possesses plicae circulares, intestinal villi, and microvilli, which collectively **increase the luminal surface area by a factor of 400–600.**

a. Plicae circulares (valves of Kerckring)

–are permanent spiral folds of the **mucosa** and **submucosa,** present in the distal half of the duodenum, the jejunum, and proximal half of the ileum.

–**increase the surface area twofold to threefold.**

b. Intestinal villi (Figure 16.1)

–are permanent evaginations that possess, in their lamina propria core, numerous plasma cells and lymphocytes, fibroblasts, mast cells, smooth muscle cells, capillary loops, and a **single lacteal** (blind-ended lymphatic channel).

–**increase the surface area tenfold.**

c. Microvilli

–of the **apical** surface of the epithelial cells of each villus, possess actin filaments that interact with myosin filaments in the terminal web.

–**increase the surface area about twentyfold.**

2. Mucosa of the small intestine

a. Epithelium

–is **simple columnar,** composed of goblet cells, surface absorptive cells, and some enteroendocrine cells (see Figure 16.1).

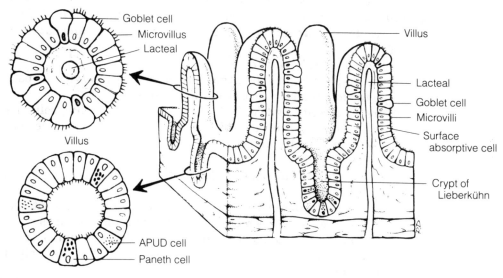

Figure 16.1. Three-dimensional drawing showing the spatial relationship of intestinal villi, crypts of Lieberkühn, and underlying muscularis mucosae in the small intestine. Note that intestinal villi are evaginations of the epithelium and lamina propria. Each villus contains a single blind-ended lacteal and capillary loops.

(1) Goblet cells

—are **unicellular glands** that produce **mucinogen,** which accumulates in membrane-bounded granules, distending the **apical** region (**theca**) of the cell. After being released, mucinogen is converted to **mucus,** a thick, viscous substance that acts as a **protective coating** of the small intestine lining.

—have their nucleus and other organelles in the basal region (**stem**) of the cell.

—increase in number from the duodenum to the ileum.

(2) Surface absorptive cells

—are tall columnar cells with numerous mitochondria, smooth and rough endoplasmic reticulum, and a Golgi complex.

—possess a layer of closely packed microvilli (**striated border**) on their free apical surface.

—have a **glycocalyx,** which overlies the microvilli and binds various enzymes.

—have well-developed **zonulae occludentes** (tight junctions) and **zonulae adherentes**.

(3) Enteroendocrine (APUD) cells

—include several types that produce and secrete **gastrin, cholecystokinin, gastric inhibitory peptide,** and other hormones (see Table 16.2).

b. Lamina propria

—occupies the cores of the villi and the interstices between the numerous **glands (crypts) of Lieberkühn.**

—consists of loose connective tissue with lymphoid cells, fibroblasts, mast cells, smooth muscle cells, nerve endings, and **lymphoid nodules**.

—also contains **lacteals** (blind-ended lymphatic vessels) and **capillary loops**.

(1) Crypts of Lieberkühn

—are simple tubular glands that extend from the intervillous spaces to the muscularis mucosae of the intestine.

—are composed of goblet cells (and oligomucous cells), columnar cells (similar to surface absorptive cells), enteroendocrine cells, regenerative cells, and Paneth cells.

(a) Paneth cells

—are pyramidal-shaped cells located at the base of the crypts of Lieberkühn.

—secrete **lysozyme,** an antibacterial enzyme, which is contained in large, apical, membrane-bounded secretory granules.

—also display extensive rough endoplasmic reticulum (basally), a large supranuclear Golgi complex, and many mitochondria.

(b) Regenerative cells

—are thin, tall, columnar **stem cells** located in the basal half of the crypts of Lieberkühn.

—divide to replace themselves and the other types of epithelial cells.

(2) Lymphoid nodules

—are usually small and solitary in the lamina propria of the duodenum and jejunum.

—increase in size and number in the ileum, where they form large contiguous aggregates, known as **Peyer's patches,** which extend through the muscularis mucosae into the submucosa.

(a) M (microfold) cells

—are highly specialized, have an unusual shape, and overlie lymphoid nodules.

—bind antigens and subsequently transport them to macrophages and lymphocytes also located in the lamina propria.

(b) Activated B lymphocytes

—respond to antigenic challenge by forming more B cells, which enter the lymph and blood circulation, then home back to their original locations, where they populate the lamina propria and differentiate into IgA-producing plasma cells.

(c) Plasma cells

—manufacture **IgA,** which is taken up by, and transported across, the intestinal epithelium (**transcytosis**) to the glycocalyx, where it remains as an immunologic defense against bacteria and antigens in the lumen.

c. Muscularis mucosae

—is composed of an inner circular and an outer longitudinal layer of smooth muscle.

3. Submucosa of the small intestine

—consists of **fibroelastic** connective tissue containing blood and lymphatic vessels, nerve fibers, and **Meissner's plexus**.

—also houses **Brunner's glands,** which are present only in the **duodenum**. These glands produce an **alkaline fluid** and **urogastrone**.

a. Alkaline mucin-containing fluid

—protects the duodenal epithelium from the acidic chyme entering from the stomach.

b. Urogastrone

—is a polypeptide hormone that enhances epithelial cell division and inhibits gastric HCl production.

4. Muscularis externa of the small intestine

—is composed of **two** layers of smooth muscle: an inner circular and an outer longitudinal layer. The inner layer participates in the formation of the **ileocecal sphincter**.

—houses **Auerbach's (myenteric) plexus** between its two layers.

5. External layer of the small intestine

a. Serosa covers all of the jejunum and ileum and part of the duodenum.

b. Adventitia covers the remainder of the duodenum.

D. Large intestine

—consists of the **cecum, colon** (ascending, transverse, descending, and sigmoid), **rectum, anal canal,** and **appendix**.

—contains some digestive enzymes received from the small intestine.

—houses bacteria that produce **vitamin B$_{12}$** and **vitamin K;** the former is necessary for hematopoiesis and the latter for coagulation.

—functions primarily in the **absorption of electrolytes and fluids,** thus compacting dead bacteria and indigestible remnants of the ingested material into **feces**.

—produces **abundant mucus,** which lubricates its lining and facilitates the passage and elimination of feces.

1. Cecum and colon

a. Mucosa of cecum and colon

—**lacks villi** and possesses no specialized folds.

(1) Epithelium

—is **simple columnar** with numerous goblet cells, absorptive cells, and occasional enteroendocrine cells.

(2) Lamina propria

—is similar to that of the small intestine, possessing lymphoid nodules, vascular elements, and closely packed crypts of Lieberkühn, which lack Paneth cells.

(3) Muscularis mucosa

—consists of an inner circular and outer longitudinal layer of smooth muscle cells.

b. Submucosa of cecum and colon

—is composed of **fibroelastic** connective tissue.

—contains blood and lymphatic vessels, nerves, and Meissner's (submucosal) plexus.

—has no glands.

c. Muscularis externa of cecum and colon

—is composed of an inner circular and a modified outer longitudinal layer of smooth muscle. The outer layer is gathered into three flat, longitudinal ribbons of smooth muscle that form the **teniae coli**. When continuously contracted, the teniae coli form sacculations of the wall known as **haustra coli**.

—contains Auerbach's (myenteric) plexus between its two layers.

d. External layer of the cecum and colon

(1) **Adventitia** covers the ascending and descending portions of the colon.

(2) **Serosa** covers the cecum and the remainder of the colon. Fat-filled outpocketings of the serosa (**appendices epiploicae**) are characteristic of the transverse and sigmoid colon.

2. Rectum

—is similar to the colon but contains fewer and deeper crypts of Lieberkühn (see Table 16.1).

3. Anal canal

—is the constricted continuation of the rectum.

a. Anal mucosa

—displays longitudinal folds called **anal columns** (or **rectal columns of Morgagni**), which join each other to form **anal valves**. The regions between adjacent valves are known as **anal sinuses**.

(1) **Epithelium**

—is simple columnar changing to **simple cuboidal** proximal to the anal valves.

—becomes **stratified squamous nonkeratinized** distal to the anal valves.

—changes to **stratified squamous keratinized** (epidermis) at the anus.

(2) **Lamina propria**

—is composed of **fibroelastic** connective tissue.

—contains sebaceous glands, circumanal glands, hair follicles, and large veins.

(3) **Muscularis mucosae**

—consists of an inner circular and an outer longitudinal layer of smooth muscle, both of which terminate at the anal valves.

b. Anal submucosa

—is composed of dense irregular **fibroelastic** connective tissue, which houses large veins.

c. Anal muscularis externa

—is composed of an inner circular and an outer longitudinal layer of smooth muscle. The inner circular layer forms the **internal anal sphincter.**

d. Anal adventitia

—attaches the anus to surrounding structures.

e. External anal sphincter

–is composed of **skeletal muscle** whose superficial and deep layers invest the anal canal.

–exhibits **continuous tonus,** thus maintaining a closed anal orifice. The degree of tonus is under **voluntary control,** so the retention or evacuation of feces normally can be controlled at will.

4. Appendix

–is a short **diverticulum** arising from the blind terminus of the cecum.

–has a narrow, stellate, or irregularly shaped lumen, which often contains debris.

–has a **thickened wall** due to the presence of large aggregates of **lymphoid nodules** in the mucosa and even in the submucosa (in middle-aged individuals).

a. Mucosa of the appendix

(1) Epithelium

–is **simple columnar** containing surface columnar cells and goblet cells.

(2) Lamina propria

–displays **numerous lymphoid nodules** (capped by M cells) and lymphoid cells.

–does not form villi but possesses **shallow crypts of Lieberkühn** with some goblet cells, surface columnar cells, regenerative cells, occasional Paneth cells, and numerous enteroendocrine cells (especially deep in the crypts).

(3) Muscularis mucosae

–is composed of an inner circular and outer longitudinal layer of smooth muscle.

b. Submucosa of the appendix

–is composed of **fibroelastic** connective tissue containing confluent lymphoid nodules and associated cell populations.

c. Muscularis externa of the appendix

–is composed of an inner circular and an outer longitudinal layer of smooth muscle.

d. Serosa of the appendix

–completely surrounds the appendix.

V. Digestion and Absorption

A. Carbohydrates

1. Salivary and pancreatic amylases hydrolyze carbohydrates to **disaccharides**. This process begins in the oral cavity, continues in the stomach, and is completed in the small intestine.

2. Disaccharidases present in the glycocalyx of the brush border cleave disaccharides into monosaccharides.

3. Monosaccharides are actively transported into surface absorptive cells and then discharged into the lamina propria, where they enter the circulation.

B. Proteins

1. **Pepsin** in the lumen of the stomach partially hydrolyzes proteins, forming a mixture of high-molecular-weight **polypeptides**. Pepsin activity is greatest at low pH.

2. **Pancreatic proteases** within the lumen of the small intestine hydrolyze the polypeptides received from the stomach into small peptides and amino acids.

3. **Small peptides and amino acids are actively transported** into surface absorptive cells where the remaining small peptides are degraded intracellularly into amino acids. Amino acids are discharged into the lamina propria where they enter the circulation.

C. Fats (Figure 16.2)

   —are degraded by **pancreatic lipase** into **monoglycerides, free fatty acids,** and **glycerol** in the lumen of the small intestine.

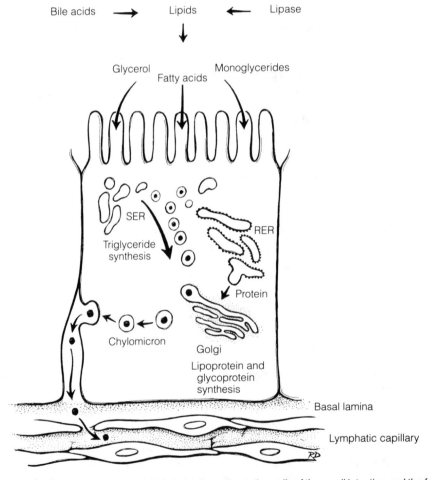

Figure 16.2. Diagram of the absorption of lipids by surface absorptive cells of the small intestine and the formation of chylomicrons. *RER* = rough endoplasmic reticulum; *SER* = smooth endoplasmic reticulum.

1. Absorption of lipid digestion products

–occurs primarily in the **duodenum** and **upper jejunum.**

a. Bile salts act on the free fatty acids and monoglycerides forming **water-soluble micelles.**

b. Micelles and **glycerol** then enter the surface absorptive cells.

2. Formation of chyle

a. Triglycerides are resynthesized from the digestion products within the smooth endoplasmic reticulum.

b. Chylomicrons are formed in the Golgi complex by complexing of the resynthesized triglycerides with proteins. Chylomicrons are transported to the lateral cell membrane and released by exocytosis; after crossing the basal lamina, they enter **lacteals** in the lamina propria to contribute to the formation of **chyle.**

c. Chyle enters the submucosal lymphatic plexus by contraction of smooth muscle cells in the intestinal villi.

3. Shorter-chain fatty acids

–of less than 10–12 carbon atoms are **not** reesterified but leave the surface absorptive cells directly and enter the circulation from the lamina propria.

D. Water and electrolytes

–are absorbed by surface absorptive cells of both the small and large intestine.

VI. Clinical Considerations

A. Disorders of the oral cavity

1. Aphthous ulcers (canker sores)

–are small, **painful,** shallow ulcers surrounded by a red border.
–are located on the mucous membranes of the lips, cheeks, alveolar mucosa, and gingivae.
–are caused by stress, fever, and a diet high in acidic foods (e.g., citrus fruits).
–are common and usually heal themselves.

2. Herpetic stomatitis

–is caused by herpes simplex virus type I. In the dormant state, this virus resides in the trigeminal ganglia.
–has a very high incidence and is transmitted by kissing.
–is characterized by **painful fever blisters** on the lips or in the vicinity of the nostrils. These blisters exude a clear fluid or are covered by a scab.

3. Oral leukoplakia

–is a white mucosal patch that usually appears on the lips, palate, tongue, and lining of the cheeks.
–results from **epithelial hyperkeratosis** (hypertrophy of the stratum corneum).
–is usually of unknown etiology, although the use of **smokeless tobacco** may precipitate its appearance.

4. Cancers of the oral region

—most commonly affect the lips, tongue, and floor of the mouth.
—initially resemble leukoplakia and are **asymptomatic**.
—have a high survival rate if recognized and treated in the early stages.

B. Disorders of the alimentary canal

1. Gastritis

—is inflammation of the gastric mucosa, a commonly occurring condition in individuals middle-aged and older, and is of unknown etiology.
—may be superficial and not very deleterious, or it may affect the entire thickness of the mucosa, causing it to atrophy.

2. Peptic ulcers

—are regions of the gastric or duodenal mucosa that have been destroyed by the action of gastric secretions.
—occur mostly in the cardiac and pyloric regions of the stomach as well as in the first portion of the duodenum.
—are caused by excess HCl secretions in about half the cases; other causes include nervous irritation, reduced vascular supply, reduced mucus secretion, and infection.
—are treated by stress management; elimination of alcohol, aspirin, and cigarette use; administration of antacids and/or drugs that inhibit gastric secretions; surgical sectioning of the vagus nerve; or removal of the affected region of the stomach.

3. Malabsorption disorders

—may lead to **malnutrition,** resulting in wasting diseases, if major nutrients (carbohydrates, amino acids, ions) cannot be assimilated.

a. Gluten enteropathy

—is also called **nontropical sprue**.
—results from the destructive effects of certain glutens (particularly of rye and wheat) on the intestinal villi, thus **reducing the surface area** available for absorption.
—is treated by eliminating wheat and rye products from the diet.

b. Idiopathic steatorrhea

—is characterized by **stools with a high fat content,** the result of malabsorption of digested fats.
—is of unknown etiology.

c. Malabsorption of vitamin B_{12}

—may cause **pernicious anemia** (see Chapter 10 VIII).
—results from inadequate production of **gastric intrinsic factor** by the parietal cells of the gastric mucosa.

d. Malabsorption of vitamin K

—may lead to clotting insufficiency.

4. Cholera-induced diarrhea

—is caused by action of cholera toxin, which blocks intestinal absorption of sodium ions and promotes excretion of water and electrolytes.

—causes death within a few days after onset unless the lost electrolytes and water are replaced.

5. Colorectal carcinoma

—is the second highest cause of cancer death in the United States.

—usually arises from **adenomatous polyps** and may be **asymptomatic** for many years.

—probably is **diet-related**. Diets high in fat and refined carbohydrates and low in fiber appear to be associated with colorectal carcinoma.

—most commonly affects individuals who are 50 years of age or older. The highest incidence is among those 60–70 years old.

—is usually treated by resection of the colon. The prognosis depends on the extent of the tumor. The highest survival rates are found among patients whose tumors do not extend beyond the muscularis mucosae.

6. Hemorrhoids

—present as **rectal bleeding during defecation**.

—are caused by the breakage of dilated, thin-walled vessels of venous plexuses either above (**internal hemorrhoids**) or below (**external hemorrhoids**) the anorectal line.

—are very common in individuals 50 years or older.

7. Appendicitis

—is usually associated with pain and/or discomfort in the lower right abdominal region, fever, nausea and vomiting, and an elevated white blood count.

—may lead to postoperative death from peritonitis if the appendix perforates.

Review Test

Directions: Each of the numbered items or incomplete statements in this section is followed by answers or by completions of the statement. Select the **one** lettered answer or completion that is **best** in each case.

1. The type of epithelium associated with the vermillion zone of the lips is

(A) stratified squamous nonkeratinized
(B) pseudostratified ciliated columnar
(C) stratified squamous keratinized
(D) stratified cuboidal

2. Which of the following cell types is present in the gastric glands of the pyloric stomach?

(A) Goblet cells
(B) Mucous neck cells
(C) Paneth cells
(D) Basal cells
(E) Chief cells

3. Secretin and cholecystokinin are produced and secreted by cells in the lining of the alimentary tract. Which of the following statements about these two substances is TRUE?

(A) They are produced by enteroendocrine cells in the lining of the stomach and small intestine
(B) They are digestive enzymes present within the lumen of the duodenum
(C) They are produced by Paneth cells
(D) They are hormones whose target cells are in the pancreas and biliary tract
(E) They are produced by Brunner's glands and released into the lumina of the crypts of Lieberkühn

4. Odontoblasts are responsible for the formation of which one of the following structures?

(A) Cementum
(B) Enamel
(C) Dentin
(D) Tooth crown only
(E) Tooth root only

5. Which of the following statements concerning the principal fiber bundles of the periodontal ligament is TRUE?

(A) They are composed of elastin
(B) They extend from the cementum to the enamel
(C) They extend from the dentin to the cementum
(D) They are composed of collagen
(E) They extend from one tooth to the next

6. Classic symptoms of appendicitis include all of the following EXCEPT

(A) fever
(B) nausea
(C) elevated white blood cell count
(D) rectal bleeding
(E) abdominal pain

7. Passage of a bolus through the esophagus into the stomach depends on all of the following EXCEPT

(A) skeletal muscle in the esophageal muscularis externa
(B) peristaltic activity of the esophageal muscularis mucosae
(C) pharyngoesophageal sphincter
(D) smooth muscle in the esophageal muscularis externa
(E) gastroesophageal sphincter

8. The small intestine has three histologically distinct regions. Which of the following statements concerning the histologic differences in the three regions is TRUE?

(A) Peyer's patches are present only in the ileum
(B) Goblet cells are present only in the epithelium of the duodenum
(C) Brunner's glands are located in the duodenum and jejunum but not the ileum
(D) Lacteals are present only in the lamina propria of the ileum
(E) The muscularis mucosae contains three layers of smooth muscle in the ileum and two layers in the duodenum and jejunum

9. All of the following materials can be absorbed directly by the surface absorptive cells of the intestines EXCEPT

(A) water
(B) monosaccharides
(C) free fatty acids
(D) ions
(E) small peptides

Directions: Each group of items in this section consists of lettered options followed by a set of numbered items. For each item, select the **one** lettered option that is most closely associated with it. Each lettered option may be selected once, more than once, or not at all.

Questions 10–13

Match each characteristic below with the condition associated with it.

(A) Herpetic stomatitis
(B) Oral leukoplakia
(C) Hemorrhoids
(D) Aphthous ulcers
(E) Idiopathic steatorrhea

10. May transform into squamous cell carcinoma

11. Usually transmitted by kissing

12. Usually accompanied by rectal bleeding

13. Often associated with a diet high in acid foods

Questions 14–17

Match each description below with the corresponding lettered structure in the photomicrograph of a histologic section of the jejunum.

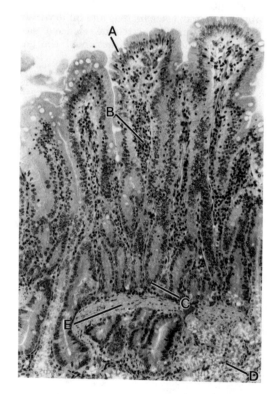

14. Cell that secretes mucinogen

15. Mucosal layer that contains lacteals

16. Mucosal layer consisting of two layers of smooth muscle fibers

17. Cell that secretes lysozyme

Answers and Explanations

1–C. The external aspect and vermillion zone of the lips are covered by thin skin, which contains a stratified squamous keratinized epithelium. The internal aspect of the lips is lined by a wet mucosa containing a stratified squamous nonkeratinized epithelium.

2–B. Mucous neck cells are located in the neck of gastric glands in all parts of the stomach, whereas only fundic glands contain chief (zymogenic) cells.

3–D. Secretin and cholecystokinin are hormones produced by enteroendocrine cells in the small intestine. Secretin stimulates bicarbonate secretion in the pancreas and biliary tract. Cholecystokinin stimulates the release of pancreatic enzymes and contraction of the gall bladder.

4–C. Dentin is manufactured by odontoblasts.

5–D. The principal fiber bundles of the periodontal ligament are composed of collagen fibers. They suspend a tooth in its alveolus, extending from the cribriform plate of the alveolar bone to the cementum on the root of the tooth. The fibers that extend from one tooth to the next are the transseptal fibers of the gingivae.

6–D. Rectal bleeding is not a sign of appendicitis; however, it often accompanies other gastrointestinal pathologies.

7–B. The smooth muscle in the muscularis mucosae plays no role in the movement of a bolus through the esophagus. This is accomplished by peristalsis of the esophageal muscularis externa, which contains both skeletal and smooth muscle. The sphincters at the proximal and distal end of the esophagus permit movement of food in only one direction, toward the stomach.

8–A. The primary histologic differences in the three regions of the small intestine are the presence of Peyer's patches in the lamina propria of the ileum and the presence of Brunner's glands in the submucosa of the duodenum. The duodenum and jejunum lack Peyer's patches, and the jejunum and ileum lack Brunner's glands. Goblet cells are present throughout the small intestine.

9–C. Free fatty acids and monoglycerides resulting from the action of lipase on ingested fats must be acted on by bile salts, forming micelles, before they can be absorbed. Monosaccharides, small peptides, and amino acids are absorbed in the small intestines. The large intestine, particularly the colon, is the primary site of water absorption, although absorption of ions and some water also occurs in the small intestine.

10–B. Oral leukoplakia is a white mucosal patch appearing on the surface of the oral cavity. It should be monitored carefully, as it may transform into squamous cell carcinoma.

11–A. Herpetic stomatitis, caused by herpes simplex virus type 1, is transmitted by kissing.

12–C. Rectal bleeding is commonly associated with hemorrhoids.

13–D. Aphthous ulcers (canker sores) are caused by stress, fever, and acidic foods such as citrus fruits.

14–A. Goblet cells secretes mucinogen, which is converted into mucus after being released.

15–B. The lamina propria, which occupies the cores of intestinal villi and interstices between crypts of Lieberkühn, contains lacteals (blind-ended lymphatic vessels).

16–E. The muscularis mucosae consists of two layers of smooth muscle and separates the lamina propria from the submucosa.

17–C. Paneth cells, which are located at the base of crypts of Lieberkühn, secrete lysozyme, an antibacterial enzyme.

17

Digestive System: Glands

I. Overview—Extrinsic Glands of the Digestive System

—are the **major salivary glands,** the **pancreas,** and the **liver** (with the associated **gallbladder**), all located outside of the digestive tract.

—produce enzymes, buffers, emulsifiers, and lubricants that are delivered to the lumen of the digestive tract via a system of ducts.

—also produce various hormones, blood proteins, and other products that are delivered to other parts of the body.

II. Major Salivary Glands

—consist of three **paired exocrine** glands: the **parotid, submandibular,** and **sublingual.**

—synthesize and secrete **salivary amylase, lysozyme, lactoferrin,** and a **secretory component,** which complexes with **IgA,** forming a complex that is resistant to enzymatic digestion in the saliva.

A. Structure—major salivary glands

—are classified as **compound tubuloacinar** (tubuloalveolar) glands.

—are further classified as **serous** or **mixed** (serous and mucous), depending on the type of secretory cells they contain.

—are surrounded by a **capsule** of dense irregular collagenous connective tissue whose septa subdivide each gland into lobes and lobules. The **excretory ducts** are housed in these septa, which also convey blood and lymphatic vessels and nerves to and from the parenchyma.

—contain **IgA-producing plasma cells** in the connective tissue.

1. Salivary gland acini

—consist of pyramidal-shaped serous or mucous cells arranged around a central lumen that connects with an **intercalated duct.** Mucous acini may be overlain with a crescent-shaped collection of serous cells called **serous demilunes.**

—possess **myoepithelial cells** in the acinar basal lamina.

—release a **primary secretion,** which resembles extracellular fluid. This is modified in the ducts to produce the **final secretion.**

a. **Parotid glands** consist of **serous** acini and are classified as serous.

b. **Sublingual glands** consist mostly of **mucous** acini capped with **serous demilunes**. They are classified as mixed.

c. **Submandibular glands** consist of both **serous** and **mucous** acini. They are classified as mixed.

2. **Salivary gland ducts**

a. **Intercalated ducts**

   —originate in the acini and join to form striated ducts.

   —may deliver **bicarbonate ions** into the primary secretion.

b. **Striated (intralobular) ducts**

   —are lined by **ion-transporting cells** that remove sodium and chloride ions from the luminal fluid (via a sodium pump) and actively pump potassium ions into it.

   —in each lobule converge and become the **interlobular (excretory) ducts,** which run in the connective tissue septa. These drain into the main duct of each gland, which empties into the oral cavity.

B. **Saliva**

   —is a **hypotonic** solution produced at the rate of about 1 L/day.

   —**lubricates** and **cleanses** the oral cavity by means of its water and glycoprotein content.

   —**controls bacterial flora** by the action of lysozyme, lactoferrin, and IgA, as well as by its cleansing action.

   —initiates **digestion of carbohydrates** by the action of salivary amylase.

   —acts as a solvent for substances that stimulate the taste buds.

III. Pancreas

   —is a retroperitoneal gland that produces **digestive enzymes** in its exocrine portion and several **hormones** in its endocrine portion (**islets of Langerhans**).

   —is surrounded by a **capsule** composed of delicate connective tissue that forms numerous septa, which subdivide the gland into lobules. These septa convey blood and lymphatic vessels into and out of the parenchyma and house the interlobular ducts.

A. **Exocrine pancreas**

   —is a **serous, compound tubuloacinar** gland.

1. **Pancreatic acinar cells**

   —are pyramidal-shaped serous cells arranged around a central lumen.

   —possess a round, basally located nucleus, abundant rough endoplasmic reticulum (RER), an extensive Golgi complex, numerous mitochondria, and many free ribosomes.

a. **Zymogen (secretory) granules**

   —are membrane-bounded and densely packed in the **apical** region of pancreatic acinar cells.

b. **Centroacinar cells**

   —form the initial intra-acinar portion of the intercalated ducts.

   —are low cuboidal with a pale cytoplasm.

2. Pancreatic ducts

a. Intercalated ducts begin within the acini (as centroacinar cells) and converge into a small number of **intralobular ducts,** which in turn empty into large interlobular ducts.

b. Interlobular ducts are located between lobules in the connective tissue septa; they empty into the main (or accessory) pancreatic duct.

c. Main pancreatic duct delivers secretions of the exocrine pancreas into the duodenum at the **papilla of Vater**. It fuses with the common bile duct just proximal to the papilla of Vater.

3. Exocrine pancreatic secretions

—are controlled by **cholecystokinin** (pancreozymin) and **secretin,** two hormones produced by enteroendocrine cells of the small intestine (see Table 16.2).

a. Enzyme-poor alkaline fluid

—is released in large quantities by **intercalated duct cells** stimulated by **secretin**.

—probably **neutralizes** the acidic chyme as it enters the duodenum.

b. Digestive enzymes

—are synthesized and stored in the pancreatic **acinar cells**. Their release is stimulated by **cholecystokinin**.

—are secreted as enzymes or **proenzymes** that must be activated in the intestinal lumen.

—include the following with those that are released as proenzymes indicated by (PE): trypsin (PE), chymotrypsin (PE), carboxypeptidase (PE), pancreatic amylase, pancreatic lipases, ribonuclease, deoxyribonuclease, and elastase (PE).

B. Islets of Langerhans (endocrine pancreas)

—are spherical clusters of epithelial (endocrine) cells surrounded by a fine network of **reticular fibers**.

—**are scattered among the acini of the exocrine pancreas** in an apparently random fashion, although they are slightly more abundant in the tail portion of the pancreas than in the remainder of the gland.

—are typically 100–200 μm in diameter and contain several hundred cells, but smaller islets also are found.

—are **richly vascularized** and occasionally are associated with autonomic nerve fibers.

1. Islet cells (Table 17.1)

—are of several types that can be differentiated from each other only by the use of special stains.

—produce several polypeptide hormones, but **each cell type produces only one hormone**.

2. Islet hormones

a. Glucagon is produced by alpha cells and acts to **elevate the blood glucose level**.

Table 17.1. Comparison of Secretory Cells in Islets of Langerhans

Cell Type	Granule Characteristics	Relative Numbers	Location	Hormone Synthesized
Alpha (A)	Spherical with a small halo between the membrane and electron-dense core	~ 20%	Positioned mostly at periphery of islets	Glucagon
Beta (B)	Small with an obvious halo between the membrane and irregular dense core	~ 70%	Concentrated in central region of islets but present throughout	Insulin
Delta (D)	Large and electron lucent	< 5%	Scattered throughout islets	Somatostatin
G	Small	Rare	Scattered throughout islets	Gastrin
PP	Small	Rare	Scattered throughout islets	Pancreatic polypeptide

   b. Insulin is produced by beta cells and acts to **decrease the blood glucose level**.

   c. Somatostatin is produced by delta cells and **inhibits release of hormones** by nearby secretory cells.

   d. Gastrin is produced by G cells and **stimulates gastric HCl secretion**.

   e. Pancreatic polypeptide is produced by PP cells and **inhibits release of exocrine pancreatic secretions**.

IV. Liver

   —is composed of a single type of parenchymal cell, the **hepatocyte**.
   —is surrounded by a capsule (**Glisson's capsule**) composed of thin connective tissue, whose septa subdivide the liver into lobes and lobules.
   —produces **bile** and plasma proteins and has numerous other functions including various metabolic activities.

   A. Liver lobules (Figure 17.1)

   1. Classic liver lobule

   —is a hexagonal mass of tissue primarily composed of **plates of hepatocytes,** which radiate like spokes from the region of the **central vein** toward the periphery (Figure 17.2).

   a. Portal areas (portal canals or portal triads)

   —are regions of the connective tissue between lobules that contain branches of the portal vein, hepatic artery, lymph vessel, and bile duct.
   —are present at each corner of a classic liver lobule.

   b. Liver sinusoids

   —are vessels that arise at the periphery of a lobule and run between adjacent plates of hepatocytes.
   —receive blood from the vessels in the portal areas and deliver it to the central vein.
   —are lined by **endothelial cells** that have large **fenestrations** and display discontinuities.

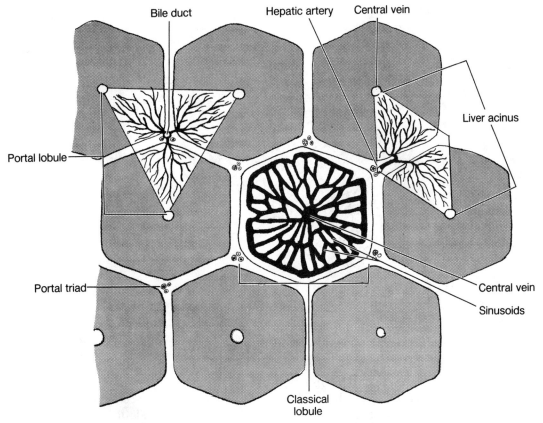

Figure 17.1. Diagram illustrating the defining characteristics of the classic liver lobule, portal lobule, and liver hepatic acinus of Rappaport. (Reprinted from Krause WJ and Cutts JH: *Concise Textbook of Histology,* 2nd ed. Baltimore, Williams & Wilkins, 1986, p 331.)

–also contain **Kupffer cells** in their endothelial lining. These are **phagocytic** cells derived from monocytes; they remove debris and cellular fragments from the bloodstream.

–**lack basal laminae.**

c. **Space of Disse**

–is the subendothelial space located between hepatocytes and liver sinusoids.

–contains short microvilli of the hepatocytes, reticular fibers (which maintain the architecture of the sinusoids), and occasional nonmyelinated nerve fibers.

–also contains stellate-shaped **fat-storing cells,** which preferentially store vitamin A.

–**functions** in the exchange of material between the bloodstream and hepatocytes, which do not directly contact the bloodstream.

2. **Portal lobule**

–is a **triangular region** whose three apices are neighboring central veins and whose center is located in a portal area (see Figure 17.1).

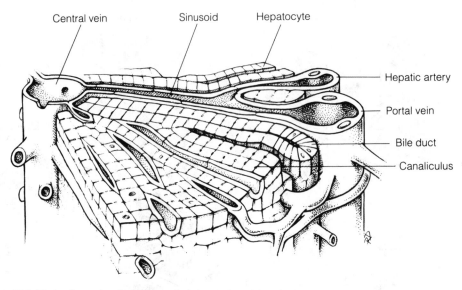

Figure 17.2. Three-dimensional representation of a portion of the classic liver lobule showing area served by one portal triad.

—contains portions of **three** adjacent classic liver lobules.

—is defined in terms of **bile flow** into a bile duct in the center of the lobule.

3. Hepatic acinus of Rappaport

—is a **diamond-shaped region** encompassing triangular sections of **two** adjacent classic liver lobules (whose apices are the central veins) and is divided by the common distributing vessels (see Figure 17.1).

—is defined in terms of **blood flow** from the distributing vessels in a single portal area.

—can be divided into **three zones** based on the proximity of the hepatocytes to the incoming blood.

—explains the histologic appearance of pathologic changes that occur in the liver.

B. Blood and bile flow (see Figure 17.2)

1. Blood flow into the liver

—is derived from two sources and is directed from the portal triads at the periphery of each classic liver lobule toward the central vein.

a. Hepatic artery brings oxygen-rich blood from the abdominal aorta and supplies 20%–30% of the liver's blood.

b. Portal vein brings nutrient-rich blood from the alimentary canal and spleen; it supplies 70%–80% of the liver's blood.

2. Blood flow out of the liver

—occurs via the **hepatic vein,** formed by the union of numerous **sublobular veins,** which collect blood from the central veins.

3. Bile flow

—is directed toward the periphery of the classic liver lobule (in the **opposite** direction of blood flow).

—is carried in a system of ducts that culminate in the left and right **hepatic ducts,** which leave the liver and carry bile to the gallbladder.

a. Bile canaliculi

—are narrow intercellular spaces between adjacent hepatocytes.

—receive the liver's **exocrine secretion** (bile) and carry it to the **canals of Hering** (bile ductules) located at the very periphery of classic liver lobules.

b. Bile ducts

—are located in the portal areas.

—receive bile from the canals of Hering.

—enlarge and fuse to form the hepatic ducts.

4. Porta hepatis

—is the fissure on the surface of the liver where the hepatic artery and portal vein enter the liver and the hepatic ducts leave.

C. Hepatocytes

—are large polyhedral cells (20–30 μm in diameter).

—usually contain one round, centrally located nucleus; some are binucleated. Occasionally nuclei are polyploid.

—possess abundant rough and smooth endoplasmic reticulum; numerous mitochondria, lysosomes, and peroxisomes; several Golgi complexes; and many lipid droplets and glycogen deposits.

—exhibit variations in their structural, histochemical, and biochemical characteristics depending on their distance from portal triads.

1. Hepatocyte surfaces facing space of Disse

—possess microvilli, which assist in the transfer of materials between the hepatocytes and the blood.

—are the sites where **endocrine secretions** leave hepatocytes and enter the liver sinusoids.

2. Abutting surfaces of adjacent hepatocytes

—frequently delineate **bile canaliculi,** small, tunnel-like expansions of the intercellular space. These are sealed off from the remaining intercellular space by **occluding junctions** located on each side of a canaliculus.

—possess microvilli that extend into the bile canaliculus.

—also have **gap junctions**.

D. Hepatic functions

1. Exocrine secretion involves the production and release of **bile,** a fluid composed of bilirubin glucuronide, bile acids, cholesterol, phospholipids, ions, and water. Hydrophobic bilirubin, a breakdown product of hemoglobin, is converted into water-soluble bilirubin glucuronide (a nontoxic compound) in the smooth endoplasmic reticulum (SER) of the liver.

2. Endocrine secretion involves the production and release of several **plasma proteins** (e.g., prothrombin, fibrinogen, albumin, factor III, and lipoproteins) as well as **urea**.

3. Metabolite storage occurs in the form of **glycogen** (stored glucose) and **triglycerides** (stored lipid).

4. **Gluconeogenesis** is the conversion of amino acids and lipids into glucose, a complex process catalyzed by a series of enzymes.

5. **Detoxification** involves the inactivation of various drugs, noxious chemicals, and toxins by enzymes, such as the **microsomal mixed-function oxidase** system, that catalyze the oxidation, methylation, or conjugation of such substances. These reactions usually occur in the smooth endoplasmic reticulum or, as in the case of alcohol, in peroxisomes.

6. **IgA transfer** involves the uptake of IgA across the space of Disse and its release into bile canaliculi. IgA is transported through the hepatic–biliary duct system to the intestine, where it serves a protective function.

V. Gallbladder

—communicates with the common hepatic duct via the **cystic duct,** which originates at the neck of the gallbladder.
—has a muscular wall whose contraction, stimulated by **cholecystokinin,** forces the bile from its lumen into the duodenum.
—functions in concentrating, storing, and releasing bile.
—contains the following four layers in its wall:

A. Mucosa
—is composed of a **simple columnar epithelium** and a richly vascularized lamina propria.
—presents highly convoluted folds when the gallbladder is empty.

B. Muscle layer
—is composed of a thin layer of **smooth muscle cells,** oriented in an oblique fashion.

C. Connective tissue layer
—consists of dense irregular collagenous connective tissue.
—houses nerves and blood vessels.

D. Serosa
—covers most of the gallbladder, but adventitia is present where the organ is attached to the liver.

VI. Clinical Considerations

A. Disorders of the salivary glands

1. Mumps
—is an infectious **viral disease** that affects the major salivary glands, primarily the **parotids**.
—is quite painful because the gland becomes markedly swollen.
—may cause orchitis; in severe cases (especially in adults), it may cause sterility.

2. Salivary gland tumors
—involve the parotids in 80% of cases; the submandibulars in less than 20%; and the sublinguals only rarely.

—are mostly benign (**benign pleomorphic adenoma**) when the parotids are involved, whereas half of those affecting the submandibulars are malignant.

—may be surgically removed. However, surgical removal of parotid tumors must be performed with great care because the **facial nerve,** which controls the muscles of facial expression, forms a plexus with the gland.

B. Pancreatic disorders

1. Type I (insulin-dependent) diabetes mellitus

—is characterized by **polyphagia** (insatiable hunger), **polydipsia** (unquenchable thirst), and **polyuria** (excessive urination).

—usually has a **sudden onset** before 20 years of age.

—results from a **low level of plasma insulin**.

—is treated with a combination of insulin therapy and diet.

2. Type II (non-insulin–dependent) diabetes mellitus

—commonly occurs in overweight individuals over 40 years of age.

—does **not** result from low levels of plasma insulin and is **insulin resistant,** which is a major factor in its pathogenesis.

—is usually controlled by diet.

C. Liver diseases

1. Cirrhosis

—is a **progressive** disease of the liver characterized by damage to hepatocytes, nodular degeneration, fibrosis, and pathologic alteration of the liver architecture.

—may result from alcohol abuse, biliary obstruction, or chronic poisoning.

2. Hepatitis

—is an inflammation of the liver, usually due to a viral infection but occasionally due to toxic materials.

a. Viral hepatitis A

—is also called **infectious hepatitis**.

—is caused by hepatitis A virus, which is frequently **transmitted by the fecal–oral route**.

—has a **short incubation period** (2–6 weeks).

—may be epidemic in populations of young adults.

—is usually nonfatal but may cause jaundice.

b. Viral hepatitis B

—is also called **serum hepatitis**.

—is caused by hepatitis B virus, which is **transmitted by blood** and its derivatives.

—has a **long incubation period** (6 weeks–5 months).

—has clinical symptoms similar to those associated with viral hepatitis A, but with more serious consequences including cirrhosis, jaundice, and death.

c. Viral hepatitis C

—is caused by hepatitis C virus.

—is responsible for about 80% of transfusion-related cases of hepatitis.

—is associated with **hepatocellular carcinoma.**

3. Jaundice (icterus)

—is characterized by **excess bilirubin** in the blood and deposition of **bile pigment** in the skin and sclera of the eyes, resulting in a yellowish appearance.

—may be hereditary or due to pathologic conditions such as excess destruction of red blood cells (**hemolytic jaundice**), liver dysfunction, and obstruction of the biliary passages (**obstructive jaundice**).

D. Gallstones (biliary calculi)

—are concretions, usually of fused crystals of **cholesterol,** that form in the gallbladder or bile duct.

—may accumulate to such an extent that the cystic duct is blocked, thus preventing emptying of the gallbladder.

—surgical removal may be necessary if less invasive methods fail to dissolve or pulverize them.

Review Test

Directions: Each of the numbered items or incomplete statements in this section is followed by answers or by completions of the statement. Select the **one** lettered answer or completion that is **best** in each case.

1. Which of the following statements concerning the pancreas is FALSE?

(A) It possesses islets of Langerhans
(B) It possesses serous demilunes
(C) Its acinar cells contain a round, basally located nucleus
(D) It contains more beta cells than alpha cells
(E) It exhibits large, electron-lucent granules in its D cells

2. An 18-year-old male presents subsequent to feeling faint. Symptoms include constant hunger, thirst, and excessive urination. The probable diagnosis is

(A) viral hepatitis A
(B) type I diabetes mellitus
(C) type II diabetes mellitus
(D) cirrhosis
(E) mumps

3. Which of the following statements concerning liver sinusoids is TRUE?

(A) They are continuous with bile canaliculi
(B) They are surrounded by a well-developed basal lamina
(C) They are lined by nonfenestrated endothelial cells
(D) They deliver blood to the central vein
(E) They deliver blood to the portal vein

4. A woman presents with yellow sclera and yellowish pallor. Blood tests indicate a low red blood cell count. The probable diagnosis is

(A) viral hepatitis A
(B) viral hepatitis B
(C) cirrhosis
(D) hemolytic jaundice
(E) type II diabetes mellitus

5. Which of the following statements concerning the gallbladder is FALSE?

(A) It concentrates bile
(B) It is lined by a simple squamous epithelium
(C) Bile leaves the gallbladder via the cystic duct
(D) It has smooth muscle cells in the walls
(E) It is affected by the hormone cholecystokinin

Directions: Each group of items in this section consists of lettered options followed by a set of numbered items. For each item, select the **one** lettered option that is most closely associated with it. Each lettered option may be selected once, more than once, or not at all.

Questions 6–10

Match the cell type with the secretory product that is associated with it.

(A) Glucagon
(B) Lysozyme
(C) Insulin
(D) Plasma proteins
(E) Proteases

6. Acinar cells of the exocrine pancreas

7. Pancreatic alpha cells

8. Pancreatic beta cells

9. Submandibular salivary gland

10. Hepatocytes

Questions 11–15

Match each structure below with the corresponding letter in the photomicrograph of a liver section showing portions of two hepatocytes.

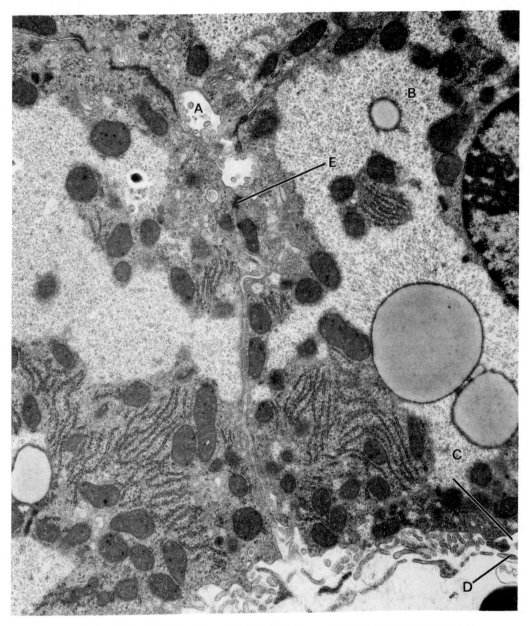

Reprinted from Gartner LP and Hiatt JL: *Atlas of Histology*. Baltimore, Williams & Wilkins, 1987, p 238.

11. Bile canaliculus

12. Sinusoidal endothelial cell

13. Glycogen deposits

14. Space of Disse

15. Occluding junction

Answers and Explanations

1–B. Serous demilunes are collections of serous cells that overlie mucous acini. Thus, they are present only in mixed glands. Since the exocrine pancreas produces only serous secretions, it contains no serous demilunes.

2–B. The three classic signs of type I diabetes (juvenile-onset) are polyphagia (excessive eating), polyuria (excessive urination), and polydipsia (excessive drinking). The condition occurs in young individuals, usually before the age of 20.

3–D. Liver sinusoids are lined by fenestrated endothelial cells, lack a basal lamina, and deliver blood directly to the central vein.

4–D. Yellow skin color and yellow sclera are indicative of jaundice. Since the patient had a low red blood cell count, the most probable diagnosis is hemolytic jaundice.

5–B. The gallbladder is lined by a simple columnar epithelium.

6–E. Several proteases are synthesized by pancreatic acinar cells and are delivered via the pancreatic duct to the duodenum.

7–A. Glucagon is produced by alpha cells of the islets of Langerhans. They are the second most abundant secretory cells of the endocrine pancreas.

8–C. Insulin is produced by pancreatic beta cells, which are the most abundant cell type of the islets of Langerhans.

9–B. Lysozyme, an enzyme with antibacterial activity, is produced primarily by the submandibular salivary glands.

10–D. Hepatocytes synthesize several plasma proteins including fibrinogen, prothrombin, and albumin.

11–A. Bile canaliculus.

12–D. Sinusoidal endothelial cell.

13–B. Glycogen deposits.

14–C. Space of Disse.

15–E. Occluding junction.

18

Urinary System

I. Overview—The Urinary System

−comprises the paired **kidneys** and **ureters** and the unpaired **bladder** and **urethra**.

−produces and excretes **urine,** thereby clearing the blood of waste products. The kidneys also regulate the electrolyte levels in the extracellular fluid and synthesize renin and erythropoietin.

II. General Renal Structure—Kidneys

−are paired, bean-shaped organs, enveloped by a thin **capsule** of connective tissue.

−can be divided into an outer **cortex** and inner **medulla**.

−each contain about 2 million **nephrons**. A nephron and the collecting tubule into which it drains form a **uriniferous tubule**.

A. Renal hilum

−is a concavity on the medial border of the kidney where arteries, veins, lymphatic vessels, and nerves are present, and where the renal pelvis is located.

B. Renal pelvis

−is a funnel-shaped expansion of the upper end of the **ureter**.

−is continuous with the **major renal calyces,** which in turn have several small branches, the **minor calyces**.

C. Renal medulla

−lies deep to the cortex but sends extensions (**medullary rays**) into the cortex.

1. Renal (medullary) pyramids

−are conical or pyramidal structures that compose the bulk of the renal medulla. Each kidney contains 10–18 renal pyramids.

−consist primarily of the thin limbs of **loops of Henle** and **collecting tubules**.

2. Renal papilla

–is located at the **apex** of each renal pyramid.

–has a perforated tip (**area cribrosa**) that projects into the lumen of a minor calyx.

D. Renal cortex

–is the superficial layer of the kidney.

–consists primarily of **renal corpuscles** and **convoluted tubules**.

1. Renal columns of Bertin

–are extensions of cortical tissue that run between adjacent renal pyramids.

2. Medullary rays

–are **groups of straight tubules** that extend from the **base of each renal pyramid** into the cortex.

E. Renal lobe

–consists of a renal pyramid and its closely associated cortical tissue.

F. Renal lobule

–consists of a central medullary ray and the closely associated cortical tissue on either side of it, extending as far as an **interlobular artery**.

–contains **many nephrons,** all of which drain into the same collecting tubule.

G. Renal interstitium

–is the connective tissue compartment of the kidney.

–consists primarily of **fibroblasts** and **mononuclear cells**. In the medulla, it consists of two additional cell types:

1. Pericytes are located along the blood vessels that supply the loops of Henle.

2. Interstitial cells produce **vasodepressor substances** (e.g., prostaglandins) and contain lipid droplets. Their long **processes** extend towards (and perhaps encircle) capillaries and tubules in the medulla.

III. Uriniferous Tubules

A. Nephrons

–consist of a **renal corpuscle, proximal convoluted tubule, loop of Henle,** and **distal convoluted tubule**.

–can be classified as **cortical** or **juxtamedullary,** depending on the location of the renal corpuscle. Juxtamedullary nephrons possess longer loops of Henle than do cortical nephrons and are responsible for establishing the interstitial concentration gradient in the medulla.

1. Renal corpuscle

–consists of the **glomerulus** (a tuft of capillaries) and **Bowman's capsule**.

–is the structure in which **filtration of the blood** occurs.

a. Bowman's capsule

(1) Parietal layer is the simple squamous epithelium that lines the outer wall of Bowman's capsule.

(2) Visceral layer (glomerular epithelium) is the modified simple squamous epithelium, composed of **podocytes,** that lines the inner wall of Bowman's capsule and envelops the glomerular capillaries.

(3) Bowman's space (capsular space) is the narrow, chalice-shaped cavity between the visceral and parietal layers into which the ultrafiltrate passes.

(4) Vascular pole is the site on Bowman's capsule where the afferent glomerular arteriole enters and the efferent glomerular arteriole leaves the glomerulus.

(5) Urinary pole is the site on Bowman's capsule where the capsular space becomes continuous with the lumen of the proximal convoluted tubule.

b. Podocytes

—are highly modified epithelial cells that form the **visceral layer** of Bowman's capsule.

—have complex shapes and possess several **primary processes** that give rise to many secondary processes called **pedicels**.

(1) Pedicles

—embrace the glomerular capillaries and interdigitate with pedicels arising from other primary processes.

—are coated, on their surfaces facing Bowman's space, with **podocalyxin,** a protein that is believed to maintain their organization and shape.

(2) Filtration slits

—are elongated spaces between adjacent pedicels. **Diaphragms,** a layer of **filamentous material,** bridge each slit.

c. Renal glomerulus

—is the **tuft of capillaries** that extends into Bowman's capsule.

(1) Glomerular endothelial cells

—form the inner layer of the capillary walls.

—have a thin cytoplasm that is thicker around the nucleus, where most organelles are located.

—possess large **fenestrae** (60–90 nm in diameter), but **lack the thin diaphragms** that typically span the openings in other fenestrated capillaries.

(2) Basal lamina

—is located **between** the podocytes and glomerular endothelial cells and is manufactured by **both** cell populations.

—is unusually **thick** (0.15–0.5 μm).

—contains three distinct **zones:**

(a) Lamina rara externa, an electron-lucent zone adjacent to the podocyte epithelium

(b) Lamina densa, a thicker, electron-dense intermediate zone of amorphous material

(c) Lamina rara interna, an electron-lucent zone adjacent to the capillary endothelium

(3) Mesangium

—is the interstitial tissue located between glomerular capillaries.

—is composed of mesangial cells and an amorphous extracellular matrix elaborated by these cells.

(a) Mesangial cells

—**phagocytose** large protein molecules and debris, which may accumulate during filtration or in certain disease states.

—also can **contract,** thereby decreasing the surface area available for filtration.

—possess **receptors for angiotensin II** and **atrial natriuretic peptide**.

(b) Mesangial matrix

—helps support glomerular capillaries.

d. Renal filtration barrier

—is composed of the **fenestrated endothelium** of the glomerular capillaries, the **basal lamina** (laminae rarae and lamina densa), and the **filtration slits** with diaphragms between pedicels.

—**permits passage** of water, ions, and small molecules from the bloodstream into the capsular space but **prevents passage** of large and/or most negatively charged proteins, thus forming an **ultrafiltrate of blood plasma** in Bowman's space.

(1) Laminae rarae contain **heparan sulfate,** which is polyanionic and **restricts passage of negatively charged proteins** into Bowman's space.

(2) Lamina densa contains **type IV collagen,** which acts as a **selective macromolecular filter** preventing passage of large protein molecules (MW > 69,000) into Bowman's space.

2. Proximal convoluted tubule (Figure 18.1)

—drains Bowman's space at the urinary pole of the renal corpuscle.

—**resorbs** from the glomerular filtrate all of the glucose, amino acids, and small proteins and at least 80% of the sodium chloride and water.

—**exchanges** hydrogen ions in the interstitium for bicarbonate ions in the filtrate.

—**secretes** into the filtrate organic acids (e.g., creatinine) and bases and certain foreign substances.

—is lined by a single layer of **irregularly shaped** (cuboidal to columnar) epithelial cells that have microvilli forming a prominent **brush border**. These cells exhibit the following structures:

a. Apical canaliculi, vesicles, and vacuoles (endocytic complex), which function in **protein absorption**

b. Prominent interdigitations along their lateral borders, which interlock adjacent cells with one another

c. Numerous **mitochondria,** compartmentalized in the basal region by extensive infoldings of the basal plasma membrane, which supply energy for the **active transport of sodium ions** out of the tubule

3. Loop of Henle (see Figure 18.1)

a. Descending thick limb of Henle's loop

—is also known as the straight portion (**pars recta**) of the proximal tubule.

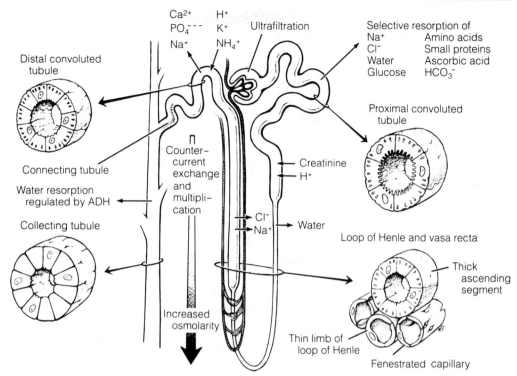

Figure 18.1. Schematic representation of a uriniferous tubule showing its major structural and functional features and its vascular associations. (Adapted from Williams PL and Warwick R (eds): *Gray's Anatomy,* 36th British ed. London, Churchill Livingstone, 1980, p 1393.)

—is lined by a simple **cuboidal** epithelium that has a prominent **brush border** and is similar to that lining the proximal convoluted tubule.

b. Thin limb of Henle's loop

—is composed of a descending segment, a loop, and an ascending segment, all of which are lined by simple **squamous** epithelial cells possessing a few short microvilli. The nuclei of these cells bulge into the lumen.

—in juxtamedullary nephrons, can be divided into **three** distinct portions based on the shape of the epithelial cells, their organelle content, the depth of their tight junctions, and their permeability to water.

c. Ascending thick limb of Henle's loop

—is also known as the straight portion of the distal tubule.

—is lined by **cuboidal** epithelial cells that possess only a few microvilli, an apically located nucleus, and mitochondria compartmentalized within basal plasma membrane infoldings.

—returns to the renal corpuscle of its origin, where it is in close association with the afferent and efferent glomerular arterioles. In this region, the wall of the tubule is modified, forming the **macula densa,** which is part of the juxtaglomerular apparatus.

4. Juxtaglomerular (JG) apparatus

—is located at the **vascular pole** of the renal corpuscle.

a. Components—the juxtaglomerular apparatus

(1) Juxtaglomerular cells

—are **modified smooth muscle cells** that exhibit some characteristics of protein-secreting cells.

—are located primarily in the wall of the **afferent arteriole,** but a few may be present in the efferent arteriole also.

—synthesize **renin** (a proteolytic enzyme) and store it in secretory granules.

(2) Macula densa cells

—are tall, narrow, closely packed epithelial cells of the **distal tubule**.

—have elongated, closely packed nuclei that appear as a "dense spot" by light microscopy.

—may **monitor the osmolarity and volume** of the fluid in the distal tubule and transmit this information to juxtaglomerular cells via the gap junctions existing between the two cell types.

(3) Extraglomerular mesangial cells

—are also known as **polkissen** (pole cushion) or **lacis cells**.

—lie between the afferent and efferent glomerular arterioles.

b. Function of the juxtaglomerular apparatus

—**maintains blood pressure** by the following mechanism:

(1) Decrease in extracellular fluid volume (perhaps detected by the macula densa) stimulates juxtaglomerular cells to release renin into the bloodstream.

(2) Renin acts on angiotensinogen in the plasma, converting it to **angiotensin I**. In capillaries of the lung (and elsewhere), angiotensin I is converted to **angiotensin II,** a potent vasoconstrictor that stimulates release of **aldosterone** in the adrenal cortex.

(3) Aldosterone stimulates the epithelial cells of the distal convoluted tubule to remove sodium and chloride ions. **Water follows the ions,** thereby **increasing the fluid volume** in the extracellular compartment, which leads to an increase in blood pressure.

5. Distal convoluted tubule (see Figure 18.1)

—is continuous with the macula densa and is similar histologically to the ascending thick limb of Henle's loop.

—is much shorter and has a wider lumen than the proximal convoluted tubule and **lacks a brush border**.

—**resorbs sodium ions** from the filtrate and actively transports them into the renal interstitium; this process is stimulated by **aldosterone**.

—also transfers potassium, ammonium, and hydrogen ions into the filtrate from the interstitium.

6. Connecting tubule

—is a short segment lying between the distal convoluted tubule and the collecting tubule into which it drains.

—is lined by the following two types of epithelial cells:

a. Principal cells

—have many infoldings of the basal plasma membrane.

—**remove sodium ions** from the filtrate and **secrete potassium ions** into it.

b. Intercalated cells

  −have many apical vesicles and mitochondria.

  −**remove potassium ions** from the filtrate and **secrete hydrogen ions** into it.

B. Collecting tubules (see Figure 18.1)

  −have a different embryologic origin than nephrons.

  −have segments in both the cortex and medulla.

  −converge to form larger and larger tubules.

1. Cortical collecting tubules

  −are located primarily within medullary rays, although a few are interspersed among the convoluted tubules in the cortex (**cortical labyrinth**).

  −are lined by a simple epithelium containing two types of **cuboidal** cells.

  a. Principal (light) cells possess a round, centrally located nucleus and a single, central **cilium**.

  b. Intercalated (dark) cells are less numerous than principal cells and possess **microplicae** (folds) on their apical surface and numerous apical cytoplasmic **vesicles**.

2. Medullary collecting tubules

  −in the **outer** medulla are similar in structure to cortical collecting tubules and contain both **principal** and **intercalated cells** in their lining epithelium.

  −in the **inner** medulla are lined only by **principal cells**.

3. Papillary collecting tubules (ducts of Bellini)

  −are large collecting tubules (200–300 μm in diameter) formed from converging smaller tubules.

  −are lined by a simple epithelium composed of **columnar** cells having a single central **cilium**.

  −empty at the **area cribrosa,** a region at the apex of each renal pyramid that has 10–25 openings through which the urine exits into a minor calyx.

IV. Renal Blood Circulation

  −is extensive, with total blood flow through both kidneys of about 1200 mL/min. At this rate, all the circulating blood in the body passes through the kidneys every 4–5 minutes.

A. Arterial supply to the kidney (Figure 18.2)

1. Branches of the renal artery

  −enter each kidney at the hilum and give rise to interlobar arteries.

2. Interlobar arteries

  −travel between the renal pyramids.

  −divide into several **arcuate arteries,** which run along the corticomedullary junction in a direction parallel to the kidney's surface.

3. Interlobular arteries

  −are smaller vessels that arise from the arcuate arteries.

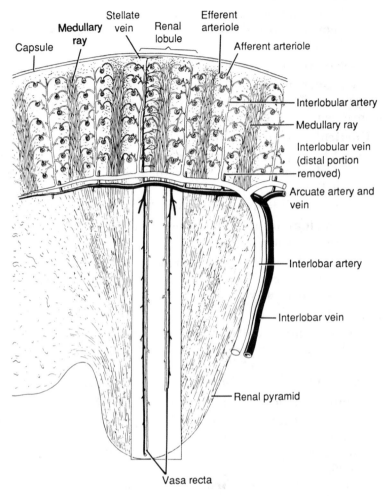

Figure 18.2. Blood circulation in the kidney. Arteries are shown in white and veins in black. Adjacent interlobular arteries, which extend outward from the arcuate artery, define the boundaries of a renal lobule. (Reprinted with permission from Junqueira LC et al: *Basic Histology,* 7th ed. Norwalk, CT, Appleton & Lange, 1992, p 388.)

—enter the cortical tissue and travel outward **between adjacent medullary rays**. Adjacent interlobular arteries delimit a renal lobule.

—give rise to **afferent (glomerular) arterioles** and also send branches to the interstitium just deep to the renal capsule.

4. Afferent arterioles

—are branches of the interlobular arteries.

—**supply the glomerular capillaries**.

5. Efferent arterioles

—arise from the glomerular capillaries.

—associated with cortical and midcortical nephrons leave the glomerulus and give rise to an extensive **peritubular capillary network** that supplies the cortical labyrinth.

6. Vasa recta

−arise from the efferent arterioles supplying **juxtamedullary nephrons**.

−are long, thin vessels that follow a straight path into the medulla and renal papilla (**arteriolae rectae**), where they form capillaries, and then loop back and increase in diameter toward the corticomedullary boundary (**venulae rectae**).

−are closely **associated with Henle's loops** to which they supply nutrients and oxygen.

−capillaries play a critical role in countercurrent exchanges with the interstitium.

B. Venous drainage of the kidney (see Figure 18.2)

1. Stellate veins

−are formed by convergence of **superficial cortical veins,** which drain the outermost layers of the cortex.

2. Deep cortical veins

−drain the deeper regions of the cortex.

3. Interlobular veins

−receive both stellate and deep cortical veins.

−join arcuate veins, which empty into interlobar veins. These then converge to form a branch of the renal vein, which exits the kidney at the hilum.

V. Regulation of Urine Concentration

−results in excretion of large amounts of dilute (**hypotonic**) urine when water intake is high (**diuresis**) and of concentrated (**hypertonic**) urine when body water needs to be conserved (**antidiuresis**).

−depends on events in the loops of Henle, collecting tubules, and vasa recta.

−is affected by the presence or absence of **antidiuretic hormone** (ADH), which is secreted from the pars nervosa of the pituitary gland when water must be conserved.

A. Countercurrent multiplier system (Figure 18.3)

−depends on the **increasing osmotic concentration gradient** in the renal interstitium in going from the outer medulla to the renal papillae.

−involves **ion and water exchanges** between the **renal interstitium** and the **filtrate in Henle's loop**.

1. In the descending limb of Henle's loop

a. The **isotonic** filtrate coming from the proximal convoluted tubules loses water to the interstitium and gains sodium and chloride ions.

b. The filtrate becomes **hypertonic**.

2. In the ascending thick limb of Henle's loop

a. No water is lost from the filtrate because this part of the nephron is **impermeable to water** in the presence or absence of ADH.

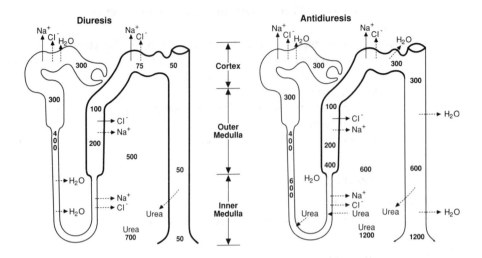

Figure 18.3. Summary of ion and water exchanges that occur in the uriniferous tubule in the absence (*left*) and presence (*right*) of antidiuretic hormone (ADH). The countercurrent multiplier system involving the loop of Henle produces an osmotic gradient in the medullary interstitium. Numbers refer to the local concentration in milliosmoles per liter (mosm/L). Segments of the tubule freely permeable to water are drawn with a thin line; impermeable segments are drawn with a thick line. In the distal convoluted tubule, some water follows sodium into the interstitium; sodium transport here is regulated by aldosterone. (Reprinted with permission from Weiss L: *Cell and Tissue Biology*, 6th ed. Baltimore, Urban & Schwarzenberg, 1988, p 840.)

    b. Chloride ions are **actively transported** from the filtrate into the interstitium, and sodium ions follow.

    c. An **osmotic gradient** thus is established in the **interstitium** of the outer medulla.

    d. The filtrate becomes **hypotonic**.

3. In the distal convoluted tubule

    –active resorption of sodium ions from the filtrate may occur (in response to aldosterone), resulting in some water loss as well.

B. Role of collecting tubules

    1. In the **absence of ADH,** the collecting tubules are **impermeable to water**. Thus the hypotonic filtrate coming from the ascending limb of Henle's loop is not changed, and **hypotonic** urine is excreted.

    2. In the **presence of ADH,** the collecting tubules become **permeable to water**. Thus the isotonic filtrate entering them from the distal convoluted tubule loses water, resulting in production of **hypertonic** (concentrated) urine.

C. Countercurrent exchange system (Figure 18.4)

    –involves passive ion and water exchanges between the renal interstitium and the blood in the vasa recta, the small straight vessels associated with the loops of Henle.

    –acts to maintain the interstitial osmotic gradient created by changes taking place in Henle's loop.

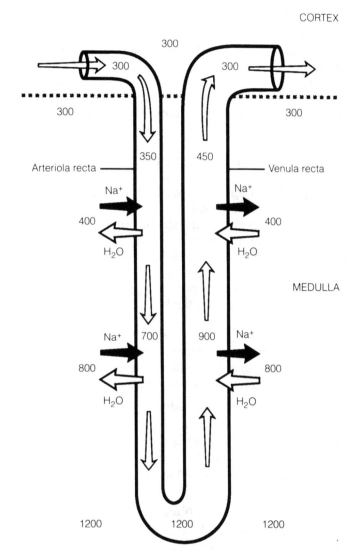

Figure 18.4. Summary of water and ion exchanges between the medullary interstitium and the blood in the vasa recta. These countercurrent exchanges are passive and do not disturb the osmotic gradient in the interstitial tissue. Numbers refer to the local osmolarity (mosm/L). (Adapted from Junqueira LC et al.: *Basic Histology*, 7th ed. Norwalk, CT, Appleton & Lange, 1992, p 390.)

   D. Effect of urea (see Figure 18.3)

   —is to aid in the production and maintenance of the interstitial osmotic gradient, mostly in the inner medulla.

   1. Urea concentrations in the filtrate progressively increase as water is lost from the medullary collecting tubules, causing the urea to diffuse out into the interstitium (thus contributing to the interstitial osmolarity).

   2. A high-protein diet increases urea levels in the filtrate and its subsequent entrapment in the interstitium, thus enhancing the kidney's ability to concentrate the urine.

VI. Excretory Passages

—include the minor and major **calyces** and the **renal pelvis,** located within each kidney, and the **ureter, urinary bladder,** and **urethra,** located outside the kidneys.

—generally possess a three-layered wall composed of a **mucosa of transitional epithelium** (except in the urethra) lying on a lamina propria of loose-to-dense connective tissue, a **muscularis (smooth muscle),** and an **adventitia**.

A. Ureter

—is the conduit between the renal pelvis of each kidney and the urinary bladder.

—has a **transitional epithelium** that is thicker and contains more cell layers than that of the renal calyces.

—possesses a **two-layered muscularis** (an inner longitudinal and outer circular layer of smooth muscle) in its upper two-thirds. The distal one-third of the ureter possesses an additional outer longitudinal layer of smooth muscle.

—contracts its muscle layers, producing **peristaltic waves** that propel the urine along so that it enters the bladder in spurts.

B. Urinary bladder

—possesses a **transitional epithelium** whose morphology differs in the relaxed (empty) and distended state; a thin lamina propria of **fibroelastic** connective tissue; and a **three-layered muscularis**.

1. Epithelium of relaxed bladder

—is five to six cell layers thick.

—has **rounded** superficial cells that bulge into the lumen, giving the lumen a scalloped contour. These cells contain unique **plaques** in their luminal plasma membrane and flattened, elliptical **vesicles** in their cytoplasm.

2. Epithelium of distended bladder

—is only three to four cell layers thick.

—has **squamous** superficial cells.

—has a larger luminal surface area than the relaxed bladder; however, the number of cell layers is reduced, resulting from insertion of the elliptical vesicles into the luminal plasma membrane of the surface cells.

C. Urethra

—moves urine from the bladder to the outside. In males, the urethra also carries semen during ejaculation.

—has a **two-layered muscularis** consisting of an inner longitudinal and outer circular layer of smooth muscle.

—is surrounded at some point by an **external sphincter of skeletal muscle,** which permits its voluntary closure.

1. Male urethra

—is about 20 cm long and is divided into **prostatic, membranous,** and **cavernous** portions.

–is lined by **transitional epithelium** in the prostatic portion and by **pseudostratified** or **stratified columnar epithelium** in the other two portions. The **fossa navicularis,** located at the distal end of the cavernous urethra, is lined by **stratified squamous epithelium**.

–contains mucus-secreting **glands of Littre** in the lamina propria.

2. Female urethra

–is much shorter (4–5 cm long) than the male urethra.

–is lined primarily by **stratified squamous epithelium,** although patches of pseudostratified columnar epithelium are present.

–may contain **glands of Littre** in the lamina propria.

VII. Clinical Considerations

A. Glomerulonephritis

–is a type of nephritis characterized by **inflammation of the glomeruli**.

–sometimes is marked by proliferation of podocytes, endothelial cells, and mesangial cells in the glomerular tuft; infiltration of leukocytes also is common.

–often occurs **secondary to a streptococcal infection** elsewhere in the body, which is thought to result in deposition of immune complexes in the glomerular basal lamina, damaging it and markedly reducing its filtering ability.

–also may result from **immune** or **autoimmune disorders**.

–is associated with production of urine containing blood (**hematuria**), protein (**proteinuria**), or both; in severe cases, decreased urine output (**oliguria**) is common.

–occurs in acute, subacute, and chronic forms. The chronic form, in which the destruction of glomeruli continues, leads eventually to renal failure and death.

B. Acute tubular necrosis

–involves the **destruction of epithelial cells** lining a specific portion of the nephron (e.g., the pars recta of the proximal tubule often is affected). As these cells die, they are sloughed off, forming casts that occlude the lumen.

–may be either **ischemic,** due to shock, extensive crush injuries, or severe bacterial infection, or **toxic,** due to ingestion of renal poisons (e.g., heavy metals such as mercury, organic solvents, antibacterial agents).

–results in severe **suppression of kidney function** (acute renal failure). If damage to the affected portion of the kidney is not too severe, recovery is possible; if damage is extensive, death results.

C. Chronic renal failure

–can result from a variety of diseases (e.g., diabetes mellitus, hypertension, atherosclerosis) in which blood flow to the kidneys is reduced, causing a decrease in glomerular filtration and tubular ischemia.

–is associated with pathologic changes (hyalinization) in the glomeruli and atrophy of the tubules, impairing virtually all aspects of renal function.

–is marked by **acidosis** and **hyperkalemia** because the acid–base balance cannot be maintained, and by **uremia** because of the inability to eliminate metabolic wastes.

–leads to neurologic problems, coma, and death if untreated.

D. Diabetes insipidus

—results from destruction of the paraventricular and supraoptic nuclei in the hypothalamus, which synthesize antidiuretic hormone, or vasopressin (see Figure 13.1).

—is associated with a **decreased ability of the kidney to concentrate urine** in the collecting tubules due to the reduced levels of ADH.

—is marked by dehydration, excessive thirst (**polydipsia**), and excretion of **high volumes of dilute urine**.

Review Test

Directions: Each of the numbered items or incomplete statements in this section is followed by answers or by completions of the statement. Select the **one** lettered answer or completion that is **best** in each case.

1. Which of the following statements concerning the structure of medullary rays is TRUE?

(A) They contain arched collecting tubules
(B) They contain proximal convoluted tubules
(C) They do not extend into the renal cortex
(D) They lie at the center of a renal lobule
(E) They contain thin limbs of loops of Henle

2. A nephron includes all of the following components EXCEPT

(A) a renal corpuscle
(B) a distal convoluted tubule
(C) a thin limb of the loop of Henle
(D) a collecting tubule
(E) pars recta of the proximal tubule

3. All of the following structures are located in the renal medulla EXCEPT

(A) vasa recta
(B) thin limbs of the loop of Henle
(C) afferent arterioles
(D) interlobar veins
(E) area cribrosa

4. Which of the following structures is present in the male urethra but is not present in the female urethra?

(A) Stratified squamous epithelium
(B) Transitional epithelium
(C) Glands of Littre
(D) External sphincter of skeletal muscle
(E) Connective tissue layer underlying the epithelium

5. Which of the following statements concerning cortical collecting tubules is always TRUE?

(A) They are lined by a simple epithelium containing two types of cells
(B) They are also known as the ducts of Bellini
(C) They empty on the area cribrosa
(D) They are permeable to water
(E) They are continuous with the ascending thick limb of Henle's loops

6. Shortly after suffering a severe β-hemolytic streptococcal infection, a medical student notices blood in her urine. Her physician orders several laboratory tests, which reveal oliguria and proteinuria. The most likely diagnosis is

(A) chronic renal failure
(B) diabetes insipidus
(C) glomerulonephritis
(D) acute tubular necrosis due to toxins
(E) acute tubular necrosis due to ischemia

7. The countercurrent multiplier system in the kidney involves the exchange of water and ions between the renal interstitium and

(A) the blood in the vasa recta
(B) the blood in the peritubular capillary network
(C) the filtrate in the proximal convoluted tubule
(D) the filtrate in the loop of Henle
(E) the filtrate in the medullary collecting tubule

Directions: Each group of items in this section consists of lettered options followed by a set of numbered items. For each item, select the **one** lettered option that is most closely associated with it. Each lettered option may be selected once, more than once, or not at all.

Questions 8–11

As the glomerular filtrate passes through the uriniferous tubule, ions and water are exchanged (actively or passively) with the renal interstitium. These exchanges result in the filtrate being isotonic, hypotonic, or hypertonic relative to blood plasma. For filtrate in each portion of the uriniferous tubule below, select the relative tonicity that applies in a condition of antidiuresis.

(A) Isotonic
(B) Hypotonic
(C) Hypertonic

8. Medullary collecting tubule

9. Bowman's (capsular) space

10. Distal portion of the ascending thick limb of the loop of Henle

11. Initial (thick portion) of the descending limb of the loop of Henle

Questions 12–16

Match each cell type with the appropriate description.

(A) Located within the tuft of glomerular capillaries where they phagocytose large protein molecules and debris
(B) Modified smooth muscle cells that secrete renin
(C) Form the visceral layer of Bowman's capsule
(D) Irregularly shaped epithelial cells whose microvilli form a prominent brush border
(E) Have long processes and secrete vasodepressor substances including prostaglandins
(F) Columnar epithelial cells possessing a single cilium

12. Podocytes

13. Medullary interstitial cells

14. Mesangial cells

15. Cells lining the proximal convoluted tubule

16. Juxtaglomerular cells

Questions 17–20

Match each structure below with the corresponding letter in the electron micrograph of a section of a renal corpuscle. *urinary space*

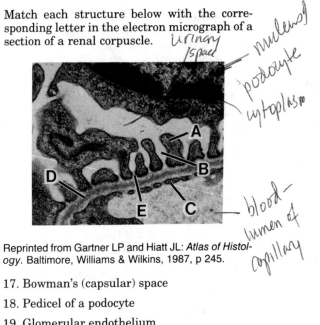

nucleus
podocyte
cytoplasm

blood - lumen of capillary

Reprinted from Gartner LP and Hiatt JL: *Atlas of Histology*. Baltimore, Williams & Wilkins, 1987, p 245.

17. Bowman's (capsular) space

18. Pedicel of a podocyte

19. Glomerular endothelium

20. Lamina densa

Answers and Explanations

1–D. A medullary ray contains the straight portions of tubules projecting from the medulla into the cortex, giving the appearance of striations or rays. A renal lobule consists of a centrally located medullary ray and its closely associated cortical tissue.

2–D. Collecting tubules originate embryologically from the ureteric bud, whereas nephrons originate from the metanephrogenic blastema. Because of this difference in origin, the collecting tubule is not considered to be part of the nephron. A nephron and its collecting tubule constitute a uriniferous tubule.

3–C. Afferent arterioles, which arise from interlobular arteries and supply the glomerular capillaries, are located in the renal cortex.

4–B. Only the male urethra contains transitional epithelium (in the prostatic portion). Stratified squamous epithelium lines most of the female urethra and the distal end of the cavernous urethra in males. Mucus-secreting glands of Littré are always present in the male urethra and may be present in the female urethra.

5–A. Cortical collecting tubules are lined by a simple epithelium containing principal (light) cells and intercalated (dark) cells. They are permeable to water only in the presence of antidiuretic hormone; in the absence of this hormone, they are impermeable to water. The large papillary collecting tubules, called ducts of Bellini, empty on the area cribrosa at the apex of each renal papilla.

6–C. The symptoms and findings are most characteristic of acute glomerulonephritis.

7–D. The countercurrent multiplier system in the loop of Henle involves ion and water exchanges between the filtrate and interstitium. It establishes an osmotic gradient in the interstitium of the medulla, which is greatest at the papilla.

8–C. The filtrate that enters the cortical collecting tubules is nearly isotonic. When antidiuretic hormone is present, water is removed from the filtrate in the collecting tubules, making the filtrate hypertonic by the time it reaches the medullary collecting tubules.

9–A. Filtration in the renal corpuscle yields an isotonic ultrafiltrate of blood plasma that enters Bowman's space.

10–B. The ascending thick limb of the loop of Henle is impermeable to water even in the presence of ADH but actively transports chloride ions from the filtrate into the interstitium (sodium ions follow passively). As a result, the filtrate becomes hypotonic as it approaches the distal convoluted tubule.

11–A. As the filtrate passes through the proximal convoluted tubule, the loss of ions is offset by the loss of water. As a result, the filtrate is still isotonic as it enters the thick descending limb of the loop of Henle (also called the pars recta of the proximal tubule).

12–C. The visceral layer of Bowman's capsule is composed of podocytes, whose processes envelop glomerular capillaries.

13–E. Interstitial cells in the renal medulla have long processes associated with capillaries and tubules. They also secrete vasodepressor substances.

14–A. Mesangial cells are present in the interstitium of the renal glomerulus. They are phagocytic and elaborate extracellular matrix that helps to support the walls of glomerular capillaries.

15–D. The proximal convoluted tubule is lined by a single layer of cuboidal to columnar epithelial cells that have microvilli forming a distinct brush border.

16–B. Juxtaglomerular cells are located primarily in the wall of the afferent arteriole at the vascular pole of the renal corpuscle. They are modified smooth muscle cells that release renin. This proteolytic enzyme triggers a mechanism that controls blood pressure.

289

17–B. Bowman's (capsular) space.

18–A. Pedicel of a podocyte.

19–C. Endothelium lining a glomerular capillary.

20–D. The lamina densa, located between the podocytes and glomerular endothelium, is part of the basal lamina and is the most important layer in the selective filtration of proteins.

19
Female Reproductive System

I. Overview—Female Reproductive System

–consists of the paired **ovaries** and **oviducts; uterus, vagina,** and **external genitalia;** and paired **mammary glands.**
–undergoes marked changes at the onset of puberty, which is initiated by **menarche.**
–exhibits monthly menstrual cycles and menses from puberty until the end of the reproductive years, which terminate at **menopause.**

II. Ovaries (Figure 19.1)

–are covered by a simple cuboidal (or squamous) surface epithelium, known as the **germinal epithelium.**
–possess a capsule (**tunica albuginea**) of dense irregular collagenous connective tissue.
–are each subdivided into a **cortex** and a **medulla,** which are not sharply delineated.

A. Ovarian cortex

–consists of **ovarian follicles** in various stages of development and a connective tissue **stroma** containing cells that respond in unique ways to hormonal stimuli.

1. Ovarian follicles (see Figure 19.1)

a. Primordial follicles

–are composed of a **primary oocyte** enveloped by a single layer of **squamous follicular cells.**

(1) Oocytes

–display a prominent, acentric, vesicular-appearing nucleus (**germinal vesicle**) that has a single nucleolus.
–possess a cytoplasm containing many Golgi regions, mitochondria, profiles of rough endoplasmic reticulum (RER), and well-developed annulate lamellae.
–become **arrested in prophase of meiosis I** during fetal life and may remain in this stage for years.

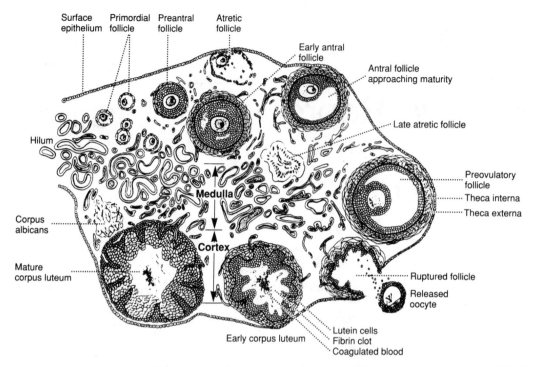

Figure 19.1. Diagram showing the structural features of the ovary. Note follicles and corpus luteum in different stages of development. (Reprinted from Weiss L: *Cell and Tissue Biology, A Textbook of Histology,* 6th ed. Baltimore, Urban & Schwarzenberg, 1988, p 855.)

    (2) Follicular cells
        —are attached to each other by **desmosomes**.
        —are separated from the surrounding stroma by a basal lamina.

  b. Growing follicles
    (1) Primary follicles
        —are **not** dependent on follicle-stimulating hormone (FSH) for their
          development.
        —possess an amorphous layer (**zona pellucida**) surrounding the
          oocyte. This layer, composed of three glycoproteins, is produced by
          the oocyte.
        —possess a basal lamina outside the follicular cells.
      (a) Unilaminar primary follicles
          —develop from primordial follicles.
          —possess a single layer of **cuboidal** follicular cells surrounding
            the oocyte.
      (b) Multilaminar primary follicles
          —develop from unilaminar follicles by proliferation of follicular
            cells.
          —possess several layers of follicular cells, which now are re-
            ferred to as **granulosa cells**.
          —are circumscribed by two layers of stromal cells: an inner cel-
            lular layer (**theca interna**) and an outer fibrous layer (**theca
            externa**).

−have a basal lamina separating the granulosa cells from the theca interna.

(2) Secondary (antral) follicles

−are established when fluid (**liquor folliculi**) begins to accumulate in the intercellular spaces between granulosa cells. The fluid-filled spaces coalesce to form a single large cavity called an **antrum**.

−are **dependent on FSH,** which stimulates the granulosa cells to convert androgens (produced by the theca interna cells) into **estrogens,** and to manufacture plasmalemma **receptors for luteinizing hormone (LH)**.

(a) Microvilli from the oocyte and narrow **processes** from the granulosa cells extend into the zona pellucida.

(b) Granulosa cells contact each other via gap junctions and also form gap junctions with the oocyte plasma membrane.

c. Graafian (mature) follicle

−is the one follicle, selected from a cohort of secondary follicles, that **will ovulate**.

−measures approximately 2.5 cm in diameter, and is evident as a large bulge on the surface of the ovary.

−has an acentrically positioned oocyte located on a small mound of granulosa cells (**cumulus oophorus**), which projects into the antrum. Granulosa cells surround the zona pellucida and line the antrum, forming an avascular layer.

(1) Theca interna cells manufacture **androgens,** which are transferred to granulosa cells where they are converted into **estrogens**.

(2) Theca externa is largely collagenous, with a few muscle cells, and contains many blood vessels, which provide nourishment to the theca interna.

(3) Primary oocyte completes its first meiotic division just prior to ovulation, forming a **secondary oocyte** and the first polar body. A second meiotic division begins but is blocked at metaphase.

(4) Ovulation occurs in response to a surge of LH as the secondary oocyte and its attendant cumulus cells (**corona radiata**) leave the ruptured follicle at the ovarian surface to enter the fimbriated end of the oviduct.

2. Corpus luteum

−is composed of **granulosa lutein cells** (modified granulosa cells) and **theca lutein cells** (modified theca interna cells).

−is a **temporary endocrine gland** whose formation depends upon LH.

−is richly supplied with blood vessels and capillaries.

a. Granulosa lutein cells

−are large (30 μm in diameter) and pale.

−possess an abundance of smooth endoplasmic reticulum and rough endoplasmic reticulum, many mitochondria, a well-developed Golgi complex, and lipid droplets.

−manufacture most of the body's **progesterone** and possibly convert small amounts of androgens to estrogens.

b. Theca lutein cells

–are small (15 μm in diameter) and are concentrated mainly along the periphery of the corpus luteum. These cells may represent a variant of the granulosa lutein cell population.

–manufacture **progesterone** and possibly androgens and small amounts of estrogen.

3. Corpus albicans

–is a small scar formed from the remnants of the corpus luteum after it ceases to function and degenerates.

4. Atretic follicles

–are follicles (in various stages of maturation) that are undergoing degeneration.

–are commonly present in the ovary; after a graafian follicle ovulates, the remaining secondary follicles degenerate.

–often show pyknotic changes in the nuclei of the granulosa cells as well as other degenerative changes.

B. Ovarian medulla

–contains large blood vessels, lymphatic vessels, and nerves in a loose connective tissue stroma.

C. Hormonal regulation of ovarian function (Figure 19.2)

1. Control of ovulation

a. Gonadotropin-releasing hormone (GnRH)

–from the hypothalamus causes the release of FSH and LH from the pars distalis of the pituitary gland.

b. FSH

–stimulates the growth and development of secondary ovarian follicles.

(1) Theca interna cells manufacture androgens, which are converted into estrogens by granulosa cells.

(2) Granulosa cells also secrete inhibin, folliostatin, and activin, all of which (in addition to estrogen) regulate FSH secretion.

(3) By approximately the 14th day of the menstrual cycle, estrogen blood levels become sufficiently high to facilitate a sudden brief surge of LH.

c. Surge of LH

–triggers the primary oocyte of the graafian follicle to complete meiosis I and to enter meiosis II, where it becomes blocked at metaphase.

–initiates ovulation of the secondary oocyte from the graafian follicle.

–promotes formation of the corpus luteum by theca cells and granulosa cells (which now have LH receptors).

2. Fate of the corpus luteum

a. Luteal hormones

(1) Progesterone, the major hormone secreted by the corpus luteum, inhibits the release of LH (by suppressing the release of GnRH) but promotes development of the uterine endometrium.

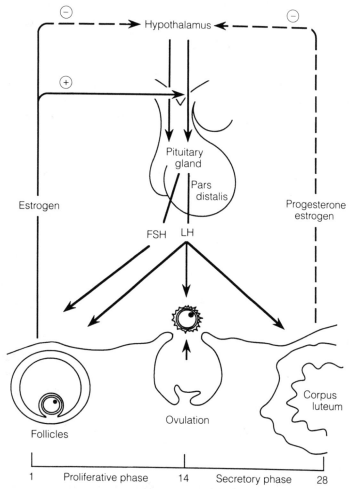

Figure 19.2. Diagram showing the hormonal relationships between the hypothalamus, pituitary gland, and ovary. Note that estrogen indirectly stimulates release of luteinizing hormone but inhibits release of follicle-stimulating hormone. *Solid lines* indicate stimulation; *broken lines* indicate inhibition. Scale at bottom refers to days in the menstrual cycle. (Adapted from Junqueira LC et al: *Basic Histology.* 7th ed. Norwalk, CT, Appleton & Lange, 1992, p 448.)

> **(2) Estrogen** inhibits the release of FSH (by suppressing the release of GnRH).
>
> **(3) Relaxin** facilitates parturition.

b. In the event of pregnancy

> **(1)** The syncytiotrophoblast of the developing placenta manufactures **human chorionic gonadotropin** (hCG) and **human chorionic somatomammotropin** (hCS).
>
> **(2) hCG maintains the corpus luteum of pregnancy for about 3 months,** at which time the placenta takes over the production of progesterone, estrogen, and relaxin.

c. In the absence of pregnancy

> **(1)** Neither LH nor hCG are present, and the **corpus luteum begins to atrophy**.

(2) Lack of estrogen and progesterone also permits the pituitary to release FSH, thus re-initiating the menstrual cycle.

III. Oviducts (Fallopian Tubes)

—are subdivided into four regions: the **infundibulum,** which has a **fimbriated end;** the **ampulla,** which is the most common site of fertilization; the **isthmus;** and the **intramural portion,** which traverses the wall of the uterus.
—have a wall consisting of **mucosa, muscularis,** and **serosa.**

A. Mucosa

—has extensive **longitudinal folds** in the infundibulum. The degree of folding progressively decreases in the remaining three regions of the oviduct.

1. Epithelium

—is **simple columnar** and consists of peg cells and ciliated cells.

a. Peg cells

—secrete a **nutrient-rich medium** that nourishes the spermatozoa (and preimplantation embryo).
—have a cytoplasm that contains abundant rough endoplasmic reticulum, a well-developed Golgi complex, and many apically located, electron-dense secretory granules.

b. Ciliated cells

—possess many cilia, which beat mostly towards the lumen of the uterus.
—may facilitate the transport of the developing embryo to the uterus.

2. Lamina propria

—consists of loose connective tissue containing reticular fibers, fibroblasts, mast cells, and lymphoid cells.

B. Muscularis

—is composed of an ill-defined inner circular and an outer longitudinal layer of smooth muscle.
—may assist in moving the preimplantation embryo towards the uterus by undergoing rhythmic contractions.

C. Serosa

—covers the outer surface of the oviduct.
—is composed of a simple squamous epithelium overlying a connective tissue layer.

IV. Uterus

—is subdivided into the **fundus,** body (**corpus**), and **cervix.**

A. Uterine wall

—consists of the **endometrium, myometrium,** and **adventitia** (or serosa).

1. Endometrium

—undergoes hormone-modulated cyclic alterations during different phases of the **menstrual cycle.**
—is lined by a **simple columnar** epithelium containing **secretory** and **ciliated cells.** The epithelium forms simple tubular glands.

−possesses a stroma resembling mesenchymal connective tissue, with **stellate-shaped cells** and an abundance of **reticular fibers**. Macrophages and leukocytes are also present.

−is divided into two indistinct layers: a thick superficial **functional layer** and a deeper **basal layer**.

a. Functional layer (functionalis)

−is the thick superficial layer of the endometrium that is sloughed and reestablished monthly due to hormonal changes during the menstrual cycle.

b. Basal layer (basalis)

−is the deeper layer of the endometrium that is preserved during menstruation.

−has endometrial **glands,** whose basal cells provide a source for **re-epithelialization** of the endometrium after the functional layer is sloughed.

c. Endometrial vascular supply

−consists of two types of arteries derived from those in the stratum vasculare of the myometrium.

(1) Coiled arteries extend into the functional layer and undergo pronounced changes during different stages of the menstrual cycle.

(2) Straight arteries do not undergo cyclic changes and terminate in the basal layer.

2. Myometrium

−is the thick smooth muscle tunic of the uterus.

−is composed of inner and outer longitudinal layers and a thick middle circular layer. The circular layer is richly **vascularized** and is often referred to as the **stratum vasculare**.

−thickens during pregnancy due to the hypertrophy and hyperplasia of individual smooth muscle cells.

−develops many gap junctions between its smooth muscle cells near the end of pregnancy. These junctions coordinate contraction of the muscle cells during parturition.

−undergoes powerful contractions at parturition, triggered by the hormone **oxytocin** and by **prostaglandins** (both of which are increased at term).

3. External covering

a. Serosa is present over surfaces of the uterus lying within the peritoneal cavity.

b. Adventitia is present along the retroperitoneal surfaces of the uterus.

B. Menstrual cycle

−**begins** on the day menstrual **bleeding appears**.

1. Menstrual phase (day 1–4)

−is characterized by a **hemorrhagic discharge (menses)** entering the lumen of the uterus from the endometrium.

−is triggered by **spasms of contraction** and **relaxation** by coiled arteries (caused by low levels of progesterone and estrogen). Long-term **vasoconstriction** of these arteries causes ischemia.

a. Sudden, intermittent **vasodilation** of the coiled arteries ruptures their walls, flooding the stroma with blood, detaching the epithelium, and dislodging the necrotic tissue.

b. The basal layer is not sloughed and does not become necrotic, since it is supplied by short straight vessels that do not undergo prolonged vasoconstriction.

2. Proliferative (follicular) phase (day 4–14)

–follows the menstrual phase and involves **renewal of the entire functional layer,** which was sloughed during menstruation.

–includes repair of glands, connective tissue, and vascular elements (coiled arteries).

a. The epithelium lining the luminal surface is renewed from mitotic activity of cells remaining in uterine glands of the basal layer.

b. Glands are straight and lined by a simple columnar epithelium.

c. Stromal cells divide, accumulate glycogen, and enlarge.

d. Coiled arteries extend approximately two-thirds of the way into the endometrium.

3. Secretory (luteal) phase (day 15–28)

–begins shortly after ovulation and is characterized by a **thickening of the endometrium,** resulting from edema and secretion by the endometrial glands.

a. Glands become coiled, their lumens contain secretory glycoprotein material, and their cells accumulate large amounts of glycogen basally.

b. Coiled arteries become more highly coiled and longer, extending into the superficial aspects of the functional layer.

V. Cervix

–does not participate in menstruation but alters its secretions during different stages of the menstrual cycle.

–has a wall composed mainly of dense collagenous connective tissue interspersed with numerous elastic fibers and a few smooth muscle cells.

–has a **simple columnar (mucus-secreting) epithelium** except for the inferior portion (continuous with the lining of the vagina), which is covered by a **stratified squamous nonkeratinized epithelium**.

–has branched **cervical glands,** which secrete a serous fluid near the time of ovulation that facilitates the entry of spermatozoa into the uterine lumen. During pregnancy, cervical glands produce a thick, viscous secretion that hinders the entry of spermatozoa (and microorganisms) into the uterus.

–prior to parturition, becomes dilated and softens due to the lysis of collagen in response to the hormone relaxin.

VI. Fertilization and Implantation

A. Fertilization

–usually takes place within the **ampulla of the oviduct**.

–occurs when a spermatozoon penetrates the corona radiata, zona pellucida, and the plasma membrane of a **secondary oocyte**.

    —triggers the resumption and completion of the second meiotic division, with the subsequent formation of an **ovum** and second polar body.

    —is completed when the male pronucleus (from the spermatozoon) and the female pronucleus (from the oocyte) fuse, forming a diploid ($2n$) cell known as a **zygote**.

B. Implantation

1. **Zygote** undergoes cell division (cleavage) and is transformed into a multicellular structure called a **morula,** which requires about 3 days to travel through the oviduct and enter the uterus.

2. **Conceptus** (the preimplantation embryo and its surrounding membranes) acquires a fluid-filled cavity and becomes known as a blastocyst.

3. **Blastocyst** attaches to the endometrium and becomes surrounded by an inner cellular layer, the **cytotrophoblast,** and an outer, multinucleated layer, the **syncytiotrophoblast.**

4. **Syncytiotrophoblast** invades the endometrium carrying along the blastocyst, which **implants** in the wall of the uterus. Formation of the placenta then begins.

VII. Placenta

—is a **transient** structure, consisting of a **maternal portion** and a **fetal portion**.

—permits the exchange of various materials between the maternal and fetal circulatory systems. This exchange occurs **without** mixing of the two separate blood supplies.

—secretes **progesterone, human chorionic gonadotrophin** (hCG), and **human chorionic somatomammotrophin,** which is a lactogenic and growth-promoting hormone.

—also produces **estrogen** with the assistance of the liver and adrenal cortex of the fetus.

A. Fetal portion of the placenta

—arises from the **chorion,** which envelops the embryo.

—consists of the **chorionic plate** from which primary, secondary, and tertiary chorionic villi will develop.

B. Maternal portion of the placenta

—consists of the **decidua basalis,** which provides an arterial supply and venous drainage for the lacunae.

—bathes the tertiary chorionic villi in maternal blood.

—contains stromal cells in the decidua basalis. Some of these cells enlarge and are transformed into **decidual cells** during the first half of pregnancy. Decidual cells produce **prolactin** and **prostaglandins**.

VIII. Vagina

—is a **fibromuscular** tube whose wall is composed of three layers: an inner **mucosa,** a middle **muscularis,** and an external **adventitia**.

—is circumscribed by **skeletal muscle** at its external orifice.

A. Mucosa

—is lined by a thick, **stratified squamous nonkeratinized epithelium**.

1. Epithelium

—contains **glycogen,** which is used by the vaginal bacterial flora to produce **lactic acid;** this acid lowers the pH during the follicular phase of the menstrual cycle and inhibits invasion by pathogens.

2. Lamina propria

—is composed of a fibroelastic connective tissue; is **highly vascular** in its deeper portion, which may be considered analogous to a submucosa.

—lacks glands throughout its length and is lubricated by cervical secretions.

B. Muscularis

—is composed of irregularly arranged layers of **smooth muscle** (thin inner layer circular and thicker outer longitudinal) interspersed with **elastic fibers**.

C. Adventitia

—is composed of **fibroelastic** connective tissue.

—fixes the vagina to the surrounding structures.

IX. External Genitalia (Vulva)

A. Labia majora

—are **fat-laden folds of skin** whose inner aspects are devoid of hair.

—contain **hair, sebaceous glands,** and **sweat glands** on their external surfaces.

—have a thin layer of smooth muscle in their wall.

B. Labia minora

—are folds of skin that possess a core of **highly vascular** connective tissue containing elastic fibers.

—lack hair follicles, but contain numerous **sebaceous glands,** which open directly onto the epithelial surface.

C. Vestibule

—is the space between the two labia minora that is partially occluded by the **hymen**.

1. Glands of Bartholin

—are two large **mucus-secreting glands** that open into the vestibule.

2. Minor vestibular glands

—are numerous, small mucus-secreting glands.

—are located around the urethra and clitoris.

—resemble the glands of Littré in the male.

D. Clitoris

—is composed of two small, cylindrical **erectile bodies,** which terminate in the prepuce-covered **glans clitoridis**.

—has a stroma that is rich in blood vessels and contains many sensory nerve fibers and specialized nerve endings (e.g., Meissner's corpuscles and Pacinian corpuscles).

X. Mammary Glands

–are each composed of about two dozen **compound tubuloalveolar glands;** each of these has its own lactiferous sinus and a duct that opens at the apex of the nipple.

–are identical in both sexes until puberty. Then, in females, hormonal changes lead to an increase in adipose tissue in the stroma, causing an enlargement of the glands and development of **lobules** and **terminal ductules**.

A. Resting mammary glands

–are characteristic of adult, nonpregnant females.

–are composed of **lactiferous sinuses** and **ducts** lined in most areas by a stratified cuboidal epithelium, whose lowest layer consists of **myoepithelial cells**.

–have a basal lamina separating the epithelial components from the underlying stroma.

B. Active (lactating) mammary glands

–are larger than resting mammary glands due to the development of **alveoli,** which occurs only during pregnancy as terminal ductules proliferate.

1. Alveolar cells (Figure 19.3)

–are cuboidal **secretory cells** lining the alveoli of active mammary glands.

–are surrounded by an incomplete layer of **myoepithelial cells**.

–are richly endowed with rough endoplasmic reticulum, several Golgi regions, numerous mitochondria, lipid droplets, and vesicles containing milk protein (caseins) and lactose.

2. Secretion by alveolar cells

–during lactation occurs by two different mechanisms.

a. Lipids are released via the **apocrine** mechanism, in which a portion of the cytoplasm is released into the alveolar lumen along with the lipid material.

b. Proteins are released into the alveolar lumen via the **merocrine** mechanism (exocytosis).

C. Nipple

–is a skin-covered conical protuberance composed of dense collagenous connective tissue interlaced with smooth muscle fibers, which act as a **sphincter**.

–contains the openings of the lactiferous ducts.

–is surrounded by a circular region of pigmented skin (**areola**), which becomes more deeply pigmented during pregnancy and contains the **areolar glands (of Montgomery)**.

D. Secretions of the mammary glands

1. Colostrum

–is a **protein-rich, yellowish fluid** produced during the first few days after birth.

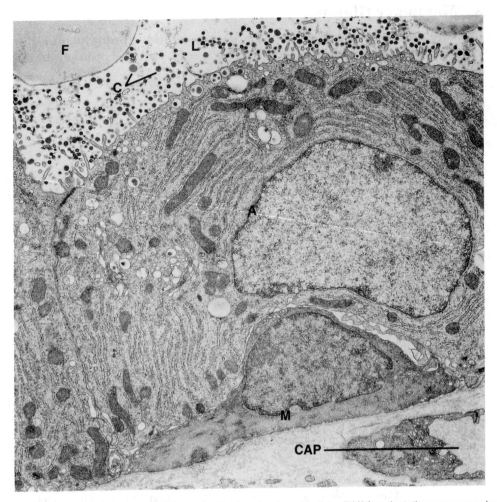

Figure 19.3. Transmission electron micrograph showing alveolar epithelial cell (*A*) from lactating mammary gland and an underlying myoepithelial cell (*M*). *CAP* = capillary; *L* = lumen of alveolus containing milk; *F* = fat droplet; *C* = casein. (Reprinted from Strum J: *A Study Atlas of Electron Micrographs,* 3rd ed. Baltimore, Univ. of Maryland School of Medicine, 1992, p 105.)

-is rich in cells (lymphocytes, monocytes), lactalbumin, fat-soluble vitamins, and minerals.

-also contains **antibodies (IgA),** which provide an immunologic defense for the newborn.

2. Milk

-begins to be secreted the third or fourth day after birth.

-consists of proteins (caseins, IgA, α-lactalbumin), many lipid droplets, and lactose.

-is released from the mammary glands, via the **milk ejection reflex,** in response to a variety of external stimuli related to suckling. The ejection reflex involves release of **oxytocin** (from axons in the pars nervosa of the pituitary gland). This hormone induces contraction of the **myoepithelial cells** located around the alveoli and ducts, forcing milk into the larger ducts and out of the breast.

XI. Clinical Considerations

A. In vitro fertilization (IVF)

1. **Gonadotropins** are administered to a women after her own hypo-thalmic–pituitary system has been suppressed. These hormones induce the maturation of ovarian follicles.

2. When follicles reach preovulatory size, **hCG** is administered to trigger ovulation.

3. The released eggs (oocytes) are retrieved and mixed with sperm from the male donor in vitro. One of the three following techniques is then employed:

   a. **Gamete intrafallopian transfer (GIFT)**
   - involves placing oocyte–sperm complexes into the oviduct **before fertilization** occurs so that fertilization itself may occur in vivo.
   - has the highest rate of success.

   b. **Zygote intrafallopian transfer (ZIFT)**
   - involves the next-day transfer of selected in vitro **fertilized zygotes** into the oviduct.
   - has a lower rate of success than GIFT.

   c. **Uterine embryo transfer (UET)**
   - involves in vitro incubation of oocyte–sperm complexes for 1–2 days and the subsequent transfer of **embryos in the 2-cell to 8-cell stage** into the fundus of the uterus.
   - is the most commonly used procedure, although it has the lowest rate of success.

B. Papanicolaou (Pap) smear

- epithelial cells are scraped from the lining of the cervix (or vagina) and are examined to detect cervical cancer.
- shows **variation in cell populations** with stages of the menstrual cycle.

C. Carcinoma of the cervix

- originates from stratified squamous nonkeratinized epithelial cells.
- may be contained within the epithelium and not invade the underlying stroma (**carcinoma-in-situ**), or it may penetrate the basal lamina and metastasize to other parts of the body (**invasive carcinoma**).
- occurs at a relatively high frequency but may be cured by surgery if discovered early (by Pap smears) before it becomes invasive.

D. Endometriosis

- is a condition in which uterine endometrial tissue exists in the pelvic peritoneal cavity.
- is associated with hormone-induced changes occurring in the ectopic endometrium during the menstrual cycle. As the endometrium is shed, bleeding occurs in the peritoneal cavity, causing severe pain and the formation of cysts and adhesions.
- may lead to **sterility** since the ovaries and oviducts become deformed and embedded in scar tissue.

E. Ectopic (tubal) pregnancy

—refers to **implantation** of the early **embryo** in the **wall of the oviduct** (or other abnormal site).

—results in the lamina propria forming decidual cells and undergoing other changes that would normally occur in the endometrium. Because the oviduct cannot support the developing embryo, it bursts, causing hemorrhaging into the peritoneal cavity.

—may result in shock; it **can be fatal** without immediate medical intervention.

F. Breast cancer

—may originate from the epithelium lining the ducts (**ductal carcinoma**) or the terminal ductules (**lobular carcinoma**).

1. If breast cancer is not treated early, the tumor cells **metastasize,** via lymphatic vessels, to the axillary nodes near the affected breast and later, via the bloodstream, to the lungs, bone, and brain.

2. **Early detection** by self-examination, mammography, or ultrasound has led to a reduction in the mortality rate associated with breast cancer.

Review Test

Directions: Each of the numbered items or incomplete statements in this section is followed by answers or by completions of the statement. Select the **one** lettered answer or completion that is **best** in each case.

1. Which of the following statements concerning secondary ovarian follicles is TRUE?

(A) They lack liquor folliculi
(B) They contain a secondary oocyte
(C) Their continued maturation requires FSH
(D) They lack a theca externa
(E) They have a single layer of cuboidal follicular cells surrounding the oocyte

2. Colostrum contains all of the following EXCEPT

(A) IgA antibodies
(B) IgG antibodies
(C) fat-soluble vitamins
(D) minerals
(E) lymphocytes

3. Which of the following statements concerning the corpus luteum is TRUE?

(A) It produces LH
(B) It produces FSH
(C) It derives its granulosa luteal cells from the theca externa
(D) It becomes the corpus albicans
(E) All of the above

4. LH exerts which one of the following physiologic effects?

(A) Triggers completion of the second meiotic division by secondary oocytes
(B) Triggers ovulation
(C) Suppresses release of estrogens
(D) Induces primary follicles to become secondary follicles

5. Which of the following statements concerning the oviduct is FALSE?

(A) It is lined by a simple columnar epithelium
(B) Its epithelium contains ciliated cells
(C) It functions in nourishing spermatozoa
(D) It possesses a fimbriated portion where fertilization most often occurs
(E) Its epithelium contains peg cells

6. The basal layer of the uterine endometrium

(A) becomes sloughed during menstruation
(B) has no glands
(C) is supplied by coiled arteries
(D) is supplied by straight arteries

7. One of the recognized phases of the menstrual cycle is termed the

(A) gestational phase
(B) active phase
(C) follicular phase
(D) resting phase

8. During the proliferative phase of the menstrual cycle, the functional layer of the endometrium undergoes which of the following changes?

(A) Blood vessels become ischemic
(B) The epithelium is renewed
(C) The stroma swells due to edema
(D) Glands become coiled

9. Which of the following statements concerning the vaginal mucosa is TRUE?

(A) It is lined by stratified columnar epithelium
(B) It is lined by stratified squamous keratinized epithelium
(C) It possesses no elastic fibers
(D) It is lubricated by glands located in the cervix
(E) Its cells secrete lactic acid

10. Which one of the following statements concerning the mammary gland is FALSE?

(A) It contains lactiferous ducts
(B) It produces and secretes colostrum
(C) It is identical in males and females prior to puberty
(D) It contains myoepithelial cells
(E) It forms alveoli during the proliferative phase of the menstrual cycle

Directions: Each group of items in this section consists of lettered options followed by a set of numbered items. For each item, select the **one** lettered option that is most closely associated with it. Each lettered option may be selected once, more than once, or not at all.

Questions 11–14

Match the characteristic with the appropriate clinical correlation.

(A) Breast cancer
(B) In vitro fertilization
(C) Endometriosis
(D) Ectopic tubal pregnancy
(E) Carcinoma of the cervix

11. Frequently is detected by a Pap smear

12. Is characterized by an ectopic endometrium hemorrhaging into the peritoneal cavity

13. May involve zygote intrafallopian transfer

14. May originate from terminal ductule epithelium

Questions 15–18

Match each structure or region with the corresponding letter in the photomicrograph of a section of a preovulatory graafian follicle.

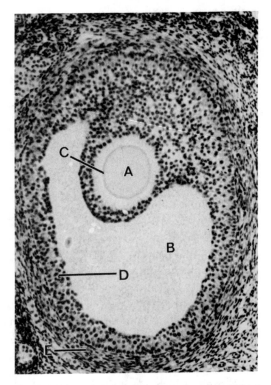

Reprinted from Gartner LP and Hiatt JL: *Atlas of Histology*. Baltimore, Williams & Wilkins, 1987, p 257.

15. Theca interna

16. Site where liquor folliculi is present

17. Granulosa cells

18. Antrum

Answers and Explanations

1–C. Secondary follicles are dependent on FSH for their continued development. They are established when liquor folliculi (an ultrafiltrate of plasma and granulosa-cell secretions) begins to accumulate among the granulosa cells. Secondary follicles contain a primary oocyte blocked in the prophase of meiosis I.

2–B. IgA antibodies are present in human colostrum and milk, whereas IgG antibodies are not. However, IgG antibodies are acquired by the fetus prior to birth by placental transfer from the mother.

3–D. A corpus albicans is formed from a corpus luteum that has ceased to function. LH and FSH are both produced in the anterior pituitary gland. Granulosa luteal cells are derived from the granulosa cells of an ovulated graafian follicle.

4–B. A sudden surge of luteinizing hormone near the middle of the menstrual cycle triggers ovulation.

5–D. Fertilization most often occurs in the ampulla of the oviduct, not in the infundibulum where fimbria are located.

6–D. The basal layer of the uterine endometrium is supplied by the straight arteries and contains the deeper portions of the uterine glands. Cells from these glands re-epithelialize the endometrial surface after the functional layer (supplied by the coiled arteries) has been sloughed.

7–C. The recognized phases of the menstrual cycle are the follicular (proliferative), secretory (luteal), and menstrual phases. The mammary glands are characterized by active (lactating) and resting phases. The term gestational phase refers to the period of pregnancy.

8–B. During the proliferative phase of the menstrual cycle, the entire functional layer of the endometrium is renewed, including the epithelium lining the surface and glands. Edema in the stroma and coiled glands are characteristic of the secretory phase of the cycle, and ischemia is responsible for the menstrual phase.

9–D. The vagina lacks glands throughout its length and is lubricated by secretions from cervical glands. It is lined by a stratified squamous nonkeratinized epithelium whose cells release glycogen, which is used by the normal bacterial flora of the vagina to manufacture lactic acid.

10–E. Alveoli develop in the mammary gland only during pregnancy.

11–E. Carcinoma of the cervix is frequently detected by Pap smears when epithelial cells scraped from this region are examined by light microscopy.

12–C. Endometriosis is characterized by bleeding into the peritoneal cavity from the ectopic endometrium. This occurs during the menstrual phase of the cycle.

13–B. Zygote intrafallopian transfer (ZIFT) is one method for achieving pregnancy using in vitro fertilization technology.

14–A. Breast cancer may originate from the epithelium lining the terminal ductules, leading to lobular carcinoma.

15–E. Theca interna.

16–B. Liquor folliculi is present in this area.

17–D. Granulosa cells.

18–B. Antrum.

20
Male Reproductive System

I. Overview—Male Reproductive System

–consists of the **testes, genital ducts,** accessory genital glands (**seminal vesicles, prostate gland,** and **bulbourethral glands**), and the **penis**.

–functions to produce **spermatozoa** (sperm), **testosterone,** and **seminal fluid,** which transports and nourishes the sperm as they pass through the excretory ducts. The penis delivers sperm to the exterior and also serves as the conduit for excretion of urine from the body.

II. Testes

–develop in the abdominal cavity and later descend into the scrotum, where they are suspended at the ends of the **spermatic cords**.

–are the sites where **spermatogenesis** occurs and the **male sex hormones,** primarily **testosterone,** are produced.

A. Testicular tunicae

1. Tunica vaginalis

–is a **serous sac,** derived from the peritoneum, which partially covers the anterior and lateral surfaces of each testis.

2. Tunica albuginea

–is the thick, fibrous connective tissue capsule of the testis.

–is lined by a highly vascular layer of loose connective tissue—the **tunica vasculosa**.

–is thickened posteriorly to form the **mediastinum testis** from which incomplete connective tissue septa arise to divide the organ into approximately 250 compartments (lobuli testis).

B. Lobuli testis

–are pyramidal-shaped compartments, which are separated by incomplete septa and can intercommunicate.

–contain one to four **seminiferous tubules** each. These tubules are embedded in a meshwork of loose connective tissue containing blood and lymphatic vessels, nerves, and interstitial cells of Leydig.

C. Interstitial cells of Leydig

—are round to polygonal cells located in the interstitial regions between seminiferous tubules.

—possess a large central nucleus, numerous mitochondria, a well-developed Golgi complex, and many lipid droplets. The latter contain cholesterol esters, which are precursors of testosterone.

—are richly supplied with capillaries and lymphatic vessels.

—are **endocrine cells** that produce and secrete **testosterone**. Secretion is stimulated by **luteinizing hormone** (interstitial cell–stimulating hormone) produced in the pituitary gland.

—mature and begin to secrete during puberty.

D. Seminiferous tubules

—are 30–70 cm long with a diameter of 150–250 μm.

—are enveloped by a fibrous connective tissue tunic composed of several layers of fibroblasts. **Myoid cells** are not present in humans.

—form tortuous pathways through the testicular lobules and then narrow into short, straight segments, the **tubuli recti,** which connect with the **rete testis**.

—are lined by a thick complex epithelium (**seminiferous,** or **germinal, epithelium**). This epithelium consists of four to eight cell layers and contains **spermatogenic cells,** from which the germ cells eventually develop (spermatogenesis), and **Sertoli cells,** which have several functions.

1. Sertoli cells (Figure 20.1)

a. Structure—Sertoli cells

—have a pale, oval nucleus, which displays frequent indentations and a large nucleolus.

—have a well-developed smooth endoplasmic reticulum (SER), some rough endoplasmic reticulum (RER), an abundance of mitochondria and lysosomes, and an extensive Golgi complex.

—possess **receptors for follicle-stimulating hormone (FSH)** on their plasma membrane.

—form **zonulae occludentes** (tight junctions) with adjacent Sertoli cells near their base, thus dividing the lumen of the seminiferous tubule into a **basal** and an **adluminal compartment**. These junctions are responsible for the blood–testis barrier, which protects developing sperm cells from autoimmune reactions.

b. Functions—Sertoli cells

—support, protect, and nourish the spermatogenic cells.

—phagocytose excess cytoplasm discarded by maturing spermatids.

—secrete a fluid into the lumen that transports spermatozoa through the seminiferous tubules to the genital ducts.

—synthesize **androgen-binding protein** (ABP) under the influence of FSH.

—secrete **inhibin,** a hormone that inhibits the synthesis and release of FSH by the anterior pituitary.

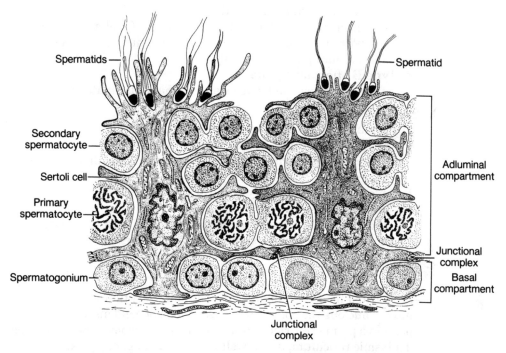

Figure 20.1. Drawing of the seminiferous (germinal) epithelium. Note the intercellular bridges between spermatocytes and the junctional complexes near the bases of adjacent Sertoli cells. These junctions divide the lumen into an adluminal and basal compartment. (Reprinted from Krause WJ and Cutts JH: *Concise Text of Histology*, 2nd ed. Baltimore, Williams & Wilkins, 1986, p 414.)

2. Spermatogenesis

- –does **not** occur simultaneously or synchronously in all seminiferous tubules but rather in wave-like sequences of maturation, referred to as **cycles of the seminiferous epithelium**.
- –is characterized by the daughter cells remaining connected to each other via **intercellular bridges** (see Figure 20.1). The resultant **syncytium** may be responsible for the **synchronous development** of germ cells along **any one** seminiferous tubule.
- –is divided into three phases: **spermatogenesis, meiosis,** and **spermiogenesis**.

3. Spermatogenic cells (see Figure 20.1)

a. Spermatogonia

- –are **diploid** germ cells located adjacent to the basal lamina of the seminiferous epithelium.

(1) Pale type A spermatogonia

- –possess a pale-staining nucleus, spherical mitochondria, a small Golgi complex, and abundant free ribosomes.
- –are **mitotically active** (starting at puberty) and give rise either to more cells of the same type (to maintain the supply) or to type B spermatogonia.

(2) Dark type A spermatogonia
  —represent mitotically **inactive** cells (in the G_0 phase of the cell cycle) with dark nuclei; they have the potential to resume mitosis and produce pale type A cells.

(3) Type B spermatogonia
  —undergo mitosis and give rise to primary spermatocytes.

b. Spermatocytes

(1) Primary spermatocytes
  —are large **diploid** cells with 4CDNA content.
  —undergo the **first meiotic division** (reductional division) to form secondary spermatocytes.

(2) Secondary spermatocytes
  —are **haploid** cells with 2CDNA.
  —quickly undergo the **second meiotic division** (equitorial division), without an intervening S phase, to form spermatids.

c. Spermatids
  —are small **haploid** cells containing only **1CDNA**.
  —are located near the lumen of the seminiferous tubule.
  —have a nucleus that often displays regions of condensed chromatin.
  —possess a pair of centrioles, mitochondria, free ribosomes, smooth endoplasmic reticulum, and a well-developed Golgi complex.

4. Spermiogenesis
  —is a unique process of **cytodifferentiation** whereby **spermatids are transformed into spermatozoa,** which are released into the lumen of the seminiferous tubule.
  —is divided into four phases:

a. Golgi phase
  —is characterized by the formation of an **acrosomal granule,** enclosed within an **acrosomal vesicle,** which becomes attached to the anterior end of the nuclear envelope of a spermatid.
  —is associated with migration of the centrioles away from the nucleus to form the **flagellar axoneme**. The centrioles then migrate back toward the nucleus to assist in forming the **connecting piece** associated with the tail.

b. Cap phase
  —is characterized by expansion of the acrosomal vesicle over much of the nucleus, forming the **acrosomal cap**.

c. Acrosomal phase

(1) The **nucleus** becomes condensed, flattened, and located in the head region.

(2) Mitochondria aggregate around the proximal portion of the flagellum, which develops into the middle piece of the tail.

(3) The **spermatid** elongates, a process that is aided by a temporary cylinder of microtubules called the **manchette**.

(4) By the end of the acrosomal phase, the spermatid is oriented with its acrosome pointing towards the base of the seminiferous tubule.

d. Maturation phase

—is characterized by the loss of excess cytoplasm and of the intercellular bridges connecting spermatids into a syncytium. The discarded material is phagocytosed by Sertoli cells.

—is completed when the **nonmotile** spermatozoa are released (tail first) into the lumen of the seminiferous tubule.

5. Spermatozoon

a. Head of the spermatozoon

—is flattened and houses a dense, homogeneous nucleus containing 23 chromosomes.

—also possesses the acrosome, which contains **hydrolytic enzymes** (e.g., acid phosphatase, neuraminidase, hyaluronidase, and proteases) that assist the sperm in penetrating the corona radiata and zona pellucida of the oocyte. Release of these enzymes is termed the **acrosomal reaction**.

b. Tail of the spermatozoon

(1) Neck

—houses the centrioles and the **connecting piece,** which is attached to the nine **outer dense fibers** of the remainder of the tail.

(2) Middle piece

—extends from the neck to the **annulus**.

—contains the axoneme, nine outer dense fibers, and a spirally arranged **sheath of mitochondria**.

(3) Principal piece

—extends from the annulus to the end piece.

—contains the axoneme with its surrounding dense fibers, which in turn are encircled by a **fibrous sheath** that has circumferentially oriented ribs.

(4) End piece

—consists of the axoneme and the surrounding plasma membrane.

E. Regulation of spermatogenesis

1. Critical testicular temperature

—is 35°C for spermatogenesis to occur.

2. Hormonal relationships (Figure 20.2)

a. Stimulation of testicular hormone production

—is effected by two pituitary gonadotropins.

(1) Luteinizing hormone (LH) stimulates the interstitial cells of Leydig to secrete **testosterone**.

(2) Follicle-stimulating hormone (FSH) promotes the synthesis of **androgen-binding protein** by Sertoli cells.

b. Testosterone

—is necessary for the normal development of male germ cells and secondary sex characteristics.

c. Androgen-binding protein

—binds testosterone and maintains a high concentration of it in the seminiferous tubule.

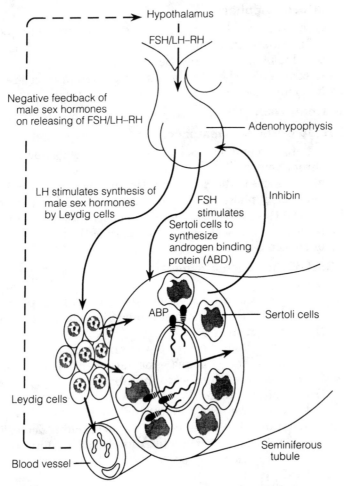

Figure 20.2. Diagram of the hormonal control of testicular function. Note the feedback inhibition of the pituitary by testicular hormones. FSH = follicle stimulating hormone; LH = luteinizing hormone; RH = releasing hormone. (Adapted from Fawcett DW: *Bloom and Fawcett's A Textbook of Histology,* 10th ed. Philadelphia, WB Saunders Company, 1975, p 839.)

—can also bind estrogens, thus inhibiting spermatogenesis.

d. Inhibition of FSH and LH release

(1) Excess levels of testosterone exert feedback inhibition on LH release.

(2) Inhibin, which is secreted by Sertoli cells, inhibits FSH release.

III. Genital Ducts

A. Intratesticular ducts

1. Tubuli recti

—are short, straight tubules lined by a **simple cuboidal epithelium** with **microvilli** and a single **flagellum**.

2. Rete testis

—is a labyrinthine plexus of anastomosing channels lined by a **simple cuboidal epithelium;** many of the cells possess a single luminal **flagellum**.

B. Extratesticular ducts

1. Ductuli efferentes

- —are a collection of 10–20 tubules leading from the rete testis to the ductus epididymidis.
- —have a thin circular layer of **smooth muscle** underlaying the basal lamina of the epithelium.
- —are lined by a **simple epithelium** composed of **alternating clusters** of **nonciliated cuboidal cells** and **ciliated columnar cells.**
- —reabsorb fluid from the semen.

2. Ductus epididymidis

- —together with the ductuli efferentes constitutes the **epididymis.**
- —is surrounded by **circular layers of smooth muscle,** which undergo **peristaltic contractions** that convey the sperm toward the ductus deferens.
- —is lined by a **pseudostratified columnar epithelium,** which is supported by a basal lamina and contains the following two cell types:

a. Basal cells

- —are round and appear undifferentiated, apparently serving as precursors of the principal cells.

b. Principal cells

- —are columnar in shape and possess nonmotile **stereocilia** (long, irregular microvilli) on their luminal surface.
- —contain in their cytoplasm endoplasmic reticulum, a large Golgi complex, lysosomes, and many apically located pinocytic and coated vesicles; the latter suggest that these cells function in **fluid resorption**.
- —secrete **glycerophosphocholine,** which probably inhibits **capacitation** (the process whereby a sperm becomes capable of fertilizing an oocyte).

3. Ductus (vas) deferens

- —has a **thick muscular wall** with inner and outer layers of longitudinally oriented smooth muscle, which are separated from one another by a middle circular layer.
- —has a narrow, irregular lumen lined by a **pseudostratified columnar epithelium** similar to that of the ductus epididymidis.

4. Ejaculatory duct

- —is the straight continuation of the ductus deferens beyond where it receives the duct of the seminal vesicle.
- —lacks a muscular wall.
- —enters the prostate gland and terminates in a slit on the **colliculus seminalis** in the prostatic urethra.

IV. Accessory Genital Glands

A. Seminal vesicles

1. Epithelium

- —is **pseudostratified columnar** tissue whose height varies with testosterone levels; it lines the **extensively folded mucosa.**

—contains many **yellow lipochrome pigment granules** and secretory granules, a large Golgi complex, many mitochondria, and an abundant rough endoplasmic reticulum.

2. Lamina propria

—consists of **fibroelastic** connective tissue surrounded by an inner circular and outer longitudinal layer of smooth muscle.

3. Adventitia

—is composed of **fibroelastic** connective tissue.

4. Secretory product

—is a yellow, viscous fluid containing substances that **activate sperm** (e.g., fructose, which is the energy source for sperm motility).

—constitutes about 70% of the human ejaculate.

B. Prostate gland

—surrounds the urethra as it exits the urinary bladder.

—consists of 30–50 discrete **branched tubuloalveolar glands,** which empty their contents, via excretory ducts, into the prostatic urethra. These glands are arranged in three concentric layers (**mucosal, submucosal,** and **main**) around the urethra.

—is covered by a **fibroelastic capsule** that contains smooth muscle. Septa from the capsule penetrate the gland and divide it into lobes.

1. Epithelium

—is **simple** or **pseudostratified columnar** and lines the individual glands making up the prostate.

—is composed of cells that contain abundant rough endoplasmic reticulum, a well-developed Golgi complex, numerous lysosomes, and many secretory granules.

2. Corpora amylacia

—are **concretions,** composed of glycoprotein, which may become calcified; their numbers increase with age.

3. Prostatic secretion

—is a whitish, thin fluid containing proteolytic enzymes, citric acid, acid phosphatase, and lipids.

—is regulated by dihydrotestosterone.

C. Bulbourethral (Cowper's) glands

—are located adjacent to the membranous urethra.

—empty their secretion into the lumen of the membranous urethra to lubricate it.

—are lined by a **simple cuboidal** or **columnar epithelium**.

—are surrounded by a **fibroelastic capsule** containing smooth and skeletal muscle.

V. Penis

A. Corpora cavernosa

—are **paired** masses of erectile tissue that contain **irregular vascular spaces** lined by a continuous layer of endothelial cells. These spaces are separated from each other by trabeculae of connective tissue and smooth muscle.

1. The vascular spaces decrease in size towards the periphery of the corpora cavernosa.

2. During erection, the vascular spaces become engorged with blood due to **parasympathetic impulses,** which constrict arteriovenous shunts and dilate the helicine arteries.

B. Corpus spongiosum

—is a single mass of **erectile tissue** that contains vascular spaces of uniform size throughout.

—has **trabeculae** that contain more elastic fibers and less smooth muscle than those of the corpora cavernosa.

C. Connective tissue and skin

1. Tunica albuginea

—is a thick fibrous connective sheath that surrounds the paired corpora cavernosa and the corpus spongiosa. The arrangement of dense collagen bundles permits extension of the penis during erection.

2. Glans penis

—is the dilated distal end of the corpus spongiosum.

—contains dense connective tissue and longitudinal muscle fibers.

—is covered by the **prepuce,** which is lined by stratified squamous nonkeratinized epithelium and is covered by skin.

3. Glands of Littre

—are mucus-secreting glands present throughout the length of the penile urethra.

VI. Clinical Considerations

A. Cryptorchidism

—is a developmental defect characterized by **failure of the testes to descend** into the scrotum.

—results in **sterility** because the temperature of the undescended testes (i.e., normal body temperature) inhibits spermatogenesis.

—does not affect testosterone production.

—is **associated with** a much higher incidence of **testicular malignancy** than in normally descended testes.

—can be **surgically corrected,** usually between 5 and 7 years of age. After corrective surgery, however, affected individuals produce abnormal sperm.

B. Klinefelter's syndrome

—is caused by an **excess number of X chromosomes**. The most common genotype is **XXY,** although other variants also occur.

—is characterized typically by a tall, thin stature, some mental retardation, small testes, elevated gonadotropin levels, and infertility.

C. Kartagener's syndrome

—is a hereditary disorder marked by **situs invertus** and **immotile cilia syndrome.**

—is characterized by **infertility in men** because their **spermatozoa are immotile.**

−is also characterized by **chronic respiratory infections** because affected individuals cannot clear microorganisms and debris from the respiratory tract.

D. Benign prostatic hypertrophy

−may result from hyperplasia of glandular and connective tissue elements of the prostate gland.

−most commonly involves only the **mucosal glands** (middle lobe of the prostate), which become enlarged.

−is frequently associated with an inability to begin and cease urination because the urethra is partially strangulated by the enlarged prostate.

−leads to **nocturia** (urination at night) and sensory urgency (the desire to urinate without having to void).

−is a common disease of older men, occurring in about 50% of men over 50 years of age and in 95% of those over 70.

E. Adenocarcinoma of the prostate gland

−is usually localized to the **posterior lobe**.

−can be diagnosed by palpation through the rectum.

−is hard and nodular to the touch rather than smooth and elastic (normal tissue).

−commonly **metastasizes to bone** via the circulatory system.

−occurs in about one-third of men over the age of 75 and is the second most common form of cancer in men.

Review Test

Directions: Each of the numbered items or incomplete statements in this section is followed by answers or by completions of the statement. Select the **one** lettered answer or completion that is **best** in each case.

1. Seminiferous tubules contain all of the following components EXCEPT

(A) Sertoli cells
(B) spermatocytes
(C) a fibrous connective tissue tunic
(D) skeletal muscle in their tunica propria
(E) a germinal epithelium

2. Which of the following statements concerning the interstitial cells of Leydig is TRUE?

(A) They become functional at puberty
(B) They are located within the seminiferous tubules
(C) They are stimulated by FSH
(D) They secrete much of the fluid portion of semen
(E) They respond to inhibin

3. Type A spermatogonia are germ cells that

(A) develop from secondary spermatocytes
(B) undergo meiotic activity subsequent to sexual maturity
(C) develop through meiotic divisions
(D) give rise to primary spermatids
(E) may be dark or pale

4. Which of the following statements concerning the ductus epididymidis is FALSE?

(A) It begins at the distal terminals of the ductuli efferentes
(B) It is lined by a pseudostratified columnar epithelium
(C) It reabsorbs fluid from its lumen
(D) It possesses motile cilia
(E) It secretes glycerophosphocholine

5. Testosterone is produced by which of the following structures?

(A) Interstitial cells of Leydig
(B) Sertoli cells
(C) Spermatogonia
(D) Spermatids
(E) Spermatocytes

6. Spermatozoa are conveyed from the seminiferous tubules to the rete testis via the

(A) ductus epididymidis
(B) tubuli recti
(C) ductuli efferentes
(D) ductus deferens

7. Spermiogenesis involves all of the following phases EXCEPT

(A) meiotic phase
(B) maturation phase
(C) Golgi phase
(D) cap phase
(E) acrosomal phase

8. The structural feature that best distinguishes the ductus deferens from the other genital ducts is its

(A) smooth-bore lumen
(B) thick muscular wall containing three muscle layers
(C) lining of transitional epithelium
(D) flattened mucosa
(E) nonmotile stereocilia

9. Androgen-binding protein is manufactured by which of the following structures?

(A) Prostate gland
(B) Sertoli cells
(C) Seminal vesicles
(D) Bulbourethral glands
(E) Leydig cells

10. A 55-year-old man presents with urinary complications. His complaints include difficulty in urinating and reduced urinary flow. Rectal palpation indicates enlargement of the middle lobe of the prostate without hard nodules at its posterior extent. The possible diagnosis is

(A) Kartagener's syndrome
(B) benign prostatic hyperplasia
(C) Klinefelter's syndrome
(D) adenocarcinoma of the prostate

Directions: Each group of items in this section consists of lettered options followed by a set of numbered items. For each item, select the **one** lettered option that is most closely associated with it. Each lettered option may be selected once, more than once, or not at all.

Questions 11–14

Match each disorder with the sign or symptom associated with it.

(A) Sterility due to high testicular temperature
(B) XXY chromosomal complement
(C) Infertility due to immotile sperm
(D) Glandular hyperplasia
(E) Metastasis to bone

11. Kartagener's syndrome

12. Cryptorchidism

13. Klinefelter's syndrome

14. Adenocarcinoma of the prostate

Questions 15–19

Match each of the following characteristics with the hormone associated with it.

(A) Follicle-stimulating hormone (FSH)
(B) Luteinizing hormone (LH)
(C) Inhibin
(D) Testosterone

15. Exerts negative feedback on the hypothalamus

16. Stimulates release of testosterone

17. Is produced by Sertoli cells

18. Promotes the synthesis of androgen-binding protein

19. Is necessary for development of secondary sex characteristics in males

Answers and Explanations

1–D. The wall of seminiferous tubule does not contain skeletal muscle. Although some species possess myoid cells, which resemble smooth muscle cells, humans do not.

2–A. Interstitial cells of Leydig become functional at puberty due to the action of LH produced in the pituitary gland.

3–E. Type A spermatogonia, which may be pale or dark, are primitive germ cells. Pale type A spermatogonia become mitotically active at puberty and give rise to type B spermatogonia, which undergo mitoses giving rise to primary spermatocytes.

4–D. The ductus epididymidis is lined by pseudostratified columnar epithelium, whose cells possess stereocilia, which are nonmotile.

5–A. The hormone testosterone is produced by the interstitial cells of Leydig.

6–B. The seminiferous tubules are connected to the rete testis by the tubuli recti.

7–A. Spermiogenesis is the process of cytodifferentiation by which spermatids are transformed into spermatozoa. It does not involve any cell division either by meiosis or mitosis.

8–B. The ductus (vas) deferens possesses three layers of smooth muscle in its wall, whereas the other genital ducts do not. Like the ductus epididymidis, the ductus deferens is lined by a pseudostratified columnar epithelium whose principal cells possess nonmotile stereocilia.

9–B. When stimulated by FSH, Sertoli cells manufacture androgen-binding protein.

10–B. The indicated symptoms are typically associated with benign prostatic hypertrophy. Rectal palpation indicating enlargement of the middle lobe without hard nodules distinguishes benign prostatic hypertrophy from prostatic adenocarcinoma.

11–C. Kartagener's syndrome is associated with situs inversus and immotile sperm; the latter causes affected men to be infertile.

12–A. The testicular temperature of cryptorchid patients, whose testes are undescended, is higher than that optimal for spermatogenesis and causes these men to be sterile.

13–B. Individuals with Klinefelter's syndrome have an extra X chromosome and commonly have an XXY chromosomal complement. These patients are infertile.

14–E. Adenocarcinoma of the prostate usually metastasizes to bone.

15–D. Testosterone exerts negative feedback on the hypothalamus resulting in inhibition of the release of follicle-stimulating hormone and luteinizing hormone from the pituitary.

16–B. LH stimulates secretion of testosterone by the interstitial cells of Leydig.

17–C. Inhibin is produced by Sertoli cells and exerts an inhibitory effect on the anterior pituitary.

18–A. Follicle-stimulating hormone promotes synthesis of androgen-binding protein by Sertoli cells.

19–D. Testosterone, the male sex hormone, is required for development of the secondary sex characteristics in males.

21

Special Senses

I. Overview—Special Sense Receptors

–are responsible for the five special senses of **taste, smell, seeing, hearing, and feeling** (touch, pressure, temperature, pain, and proprioception).
–**transduce stimuli from the environment into electrical impulses.**

II. Specialized Diffuse Receptors

–are **dendritic nerve endings** located in the skin, fascia, muscles, joints, and tendons.
–respond to stimuli related to **touch, pressure, temperature, pain,** and **proprioception**.
–are specialized to receive only **one** type of sensory stimulus, although they will respond to other types of stimuli if they are intense enough.
–are divided morphologically into **free nerve terminals** and **encapsulated nerve endings,** which are ensheathed in a connective tissue capsule.

A. Touch and pressure receptors

1. Pacinian corpuscles

–are large, ellipsoid **encapsulated** receptors located in the dermis and hypodermis and in the connective tissue of the mesenteries and joints.
–are especially abundant in the digits and breasts.
–are composed of a **multilayered capsule** of fibroblasts, collagen, and tissue fluid, surrounding an **inner unmyelinated nerve terminal.**
–resemble a sliced onion in histologic section.
–function in perceiving **pressure, touch,** and **vibration**.

2. Ruffini's endings

–are **encapsulated** receptors located in the dermis and joints.
–are composed of groups of **branched terminals,** from myelinated nerve fibers, surrounded by a thin connective tissue capsule.
–function in **pressure** and **touch** reception.

3. Meissner's corpuscles

–are ellipsoid, **encapsulated** receptors located in the dermal papillae of thick skin, eyelids, lips, and nipples.

–have a connective tissue capsule that envelops the nerve terminal and its associated Schwann cell.

–function in **fine-touch** perception.

4. Free nerve endings

–are **unencapsulated, unmyelinated** terminations located in the skin in longitudinal and circular arrays around most of the hair follicles.

–function in **touch** perception.

B. Temperature and pain receptors

1. Cold receptors

–respond to temperatures **below 25°–30°C.**

2. Heat receptors

–respond to temperatures above **40°–42°C.**

3. Nociceptors

–are sensitive to **pain stimuli.**

–are delicate myelinated fibers that lose their myelin before entering the epidermis.

III. Eye

Fig 18.1 –Embryo

–is the photosensitive organ responsible for vision.

–is composed of three layers: the **tunica fibrosa** (outer layer), **tunica vasculosa** (middle layer), and **retina.** *Fig 18.3 – Embryo | 18.7*

–receives **light** through the **cornea.** The light is focused by the **lens** on the **retina,** which contains specialized cells that encode the various patterns of the image for transmission to the brain via the **optic nerve.**

–possesses **intrinsic muscles** that adjust the aperture of the iris and alter the lens diameter, permitting accommodation for close vision.

–possesses **extrinsic muscles,** attached to the external aspect of the orb (eyeball), which move the eyes in a coordinated manner to access the desired visual fields.

–is moistened on its anterior surface with **lacrimal fluid** (tears) secreted by the **lacrimal gland.**

–is covered by the upper and lower eyelids, which protect the anterior surface.

A. Tunica fibrosa *Fig 18.7 embryo - mesoderm*

1. Sclera

–is the **opaque,** relatively avascular, fibrous connective tissue layer that covers the posterior five-sixths of the orb, which receives insertions of extrinsic ocular muscles.

2. Cornea *if becomes vascularized becomes cloudy*

–is the **transparent,** highly innervated, **avascular** anterior one-sixth of the tunica fibrosa.

–joins the sclera in a region called the **limbus,** which is highly vascularized.

–is composed of the following five layers:

a. Corneal epithelium *– derived from surface ect.*

–lines the **anterior aspect** of the cornea.

⌐ 3 TUNICS   - Adnexa
- Lens   - extrinsic eye muscles

–is a **stratified squamous nonkeratinized epithelium**.

–possesses **microvilli** in its superficial layer; these trap moisture, protecting the cornea from dehydration.

b. Bowman's membrane - in Human

–is a homogeneous **noncellular** layer that functions to provide form, stability, and strength to the cornea.

c. Corneal stroma

–is the thickest corneal layer.

–has channels, located in the region of the limbus, that are lined by **endothelium,** forming the **canal of Schlemm.** This canal drains fluid from the anterior chamber of the eye into the venous system.

d. Descemet's membrane

–is a thick (5–10 μm) basal lamina separating the stroma from the endothelium lining the cornea.

e. Corneal endothelium

–lines the **posterior aspect** of the cornea.

–is a **simple squamous epithelium** whose cells exhibit numerous **pinocytic vesicles**.

–**resorbs fluid** from the stroma, thus contributing to the transparency of the cornea.

B. Tunica vasculosa (uvea) mesoderm

1. Choroid

–is the **highly vascular, pigmented layer** of the eye whose loose connective tissue contains many **melanocytes**.

–is loosely attached to the tunica fibrosa.

–has a deep **choriocapillary layer** and **Bruch's membrane** (basement membrane), which extends from the **optic disk** to the **ora serrata**.

2. Ciliary body

–is the wedge-shaped **anterior expansion of the choroid**.

–completely encircles the lens and separates the ora serrata from the iris.

–is lined on its inner surface by two layers of cells: an **outer, pigmented columnar epithelium** rich in melanin and an **inner, nonpigmented simple columnar** epithelium.

a. Ciliary processes

–are radially arranged extensions (about 70) of the ciliary body.

–have a connective tissue core containing many **fenestrated capillaries.**

–are covered by two epithelial layers. The **unpigmented inner layer** transports components from the plasma filtrate in the posterior chamber and thus forms the **aqueous humor,** which flows to the anterior chamber via the pupillary aperature.

–possess **suspensory ligaments** (zonule) that arise from the processes and insert into the capsule of the lens, serving to anchor it in place.

b. Ciliary muscle *- mesoderm*

—is attached to the sclera and ciliary body in such a manner that its contractions stretch the choroid body and release tension on the suspensory ligament and lens. Contraction permits the lens to become more convex, allowing the eye to focus on nearby objects (**accommodation**). With advancing age, the lens loses its elasticity, thereby gradually losing the ability to accommodate.

—is innervated via **parasympathetic fibers** of the oculomotor nerve (CN III).

Contraction ⟹ see near

3. Iris *- made of vasculosa + retina*

—is the most anterior extension of the choroid, separating the anterior and posterior chambers.

—incompletely covers the anterior surface of the lens, forming an adjustable opening called the **pupil**.

—is covered by an incomplete layer of pigmented cells and fibroblasts on its anterior surface.

—has a wall composed of loose, vascular connective tissue containing melanocytes and fibroblasts.

—is covered on its deep surface by a two-layered epithelium with pigmented cells, which blocks light from entering the interior of the eye except via the pupil.

- Anterior surface is CT - bathed in AH

a. Eye color

—is blue if only a few melanocytes are present. Increasing amounts of pigment impart darker colors to the eye.

b. Dilator pupillae muscle

—is a **smooth muscle** whose fibers radiate from the periphery of the iris towards the pupil.

—contracts upon stimulation by **sympathetic** nerve fibers, **dilating the pupil**.

c. Sphincter pupillae muscle

—is **smooth muscle** arranged in concentric rings around the pupillary orifice.

—contracts upon stimulation by **parasympathetic** nerve fibers, **constricting the pupil**.

C. Refractive media of the eye

1. Aqueous humor

—is a **plasma-like fluid,** located in the anterior compartment of the eye, and **formed by epithelial cells lining the ciliary processes**.

—is secreted into the posterior chamber of the eye and then flows to the anterior chamber and then into the venous system via the canal of Schlemm.

2. Lens *- from lens placode (surface ectoderm)*

—is a biconvex, **transparent,** flexible structure composed of the lens capsule, subcapsular epithelium, and lens fibers.

a. Lens capsule

—is a thick basal lamina that envelops the entire lens epithelium.

b. Subcapsular epithelium

–is located only on the **anterior surface** of the lens.

–is composed of a single layer of cuboidal cells, which communicate with each other via **gap junctions**. These cells interdigitate with lens fibers, especially at the equator where the epithelial cells become more elongated.

c. Lens fibers

–represent highly differentiated, elongated cells filled with a group of proteins called **crystallins**.

–differentiate from the subcapsular epithelium at the level of the equator. Production of the lens fibers diminishes with increasing age.

–**lack** both nuclei and organelles when mature.

d. Suspensory ligament

–stretches between the lens and the ciliary body, keeping tension on the lens and **enabling it to focus on distant objects**.

3. Vitreous body

–is a **refractile gel** composed mainly of water, collagen, and hyaluronic acid. This gel fills the interior of the globe posterior to the lens.

D. Retina — from cup (neural ectoderm) — .

–is the innermost of the three layers of the eye and is responsible for **photoreception**.

–has a shallow depression in its posterior wall that contains only cones; this avascular region, called the **fovea centralis,** exhibits the greatest visual acuity.

[handwritten left margin: -sensitive (to light)]

–develops initially as the optic vesicle, which subsequently invaginates to form the double-walled **optic cup**. The outer wall gives rise to the outer **pigmented layer** of the retina, and the inner wall gives rise to inner **neural retina,** a complex structure.

[handwritten left margin: Insensitive (2 layer - covers ciliary body + iris - posterior)]

–displays **10 distinct layers** discussed in order from the outermost to the innermost.

1. Retinal pigment epithelium

–is a layer of **columnar cells** firmly attached to **Bruch's membrane**.

a. Structure—pigment epithelial cells

[handwritten left margin: - outer layer of optic cup / PIGMENT layer / inner layer of optic cup / neural layer]

–have **junctional complexes** and **basal invaginations** that contain mitochondria, suggesting the involvement of these cells in ion transport.

–contain smooth endoplasmic reticulum (SER) and many **melanin granules** located apically in cellular processes.

–extend **pigment-filled microvillar processes** that invest the tips of the rods and cones.

b. Function—pigment epithelial cells

[handwritten left margin: single layer epith. cells]

–**esterify vitamin A** and transport it to the rods and cones where it is used in the formation of visual pigment.

–**phagocytose** the shed tips of the outer segments of rods.

–**synthesize melanin,** which absorbs light after the rods and cones have been stimulated.

[handwritten bottom: -Insensitive - 2 cell layers thick - covers posterior aspect of iris + ciliary body / outer, inner - both 1 cell layer / junction - ora serrata - circular ∅ / -sensitive - outer - 1 cell layer / inner - multicell layer]

2. Photoreceptor layer

–consists of **neurons (photoreceptor cells)** referred to as **rods** or **cones**. Their dendrites interdigitate with cells of the pigmented epithelium, and their bases form synapses with cells of the bipolar layer.

a. Rods

–are **sensitive to light of low intensity**.

–have **outer** and **inner segments, a nuclear region,** and a **synaptic region**.

–may synapse with bipolar cells, giving rise to **summation**.

–have a constriction that separates the outer and inner segments and contains an **incomplete cilium** terminating in a basal body within the inner segment.

(1) Outer segments of rods

–consist mainly of hundreds of **flattened membranous disks,** which contain **rhodopsin**.

–eventually shed their disks, which are subsequently phagocytosed by the pigment epithelial cells.

–face the back of the eye; therefore light must pass through all the other retinal layers before reaching the photosensitive region.

(2) Inner segments of rods

–possess mitochondria, glycogen, polyribosomes, and proteins, which migrate to the outer segments to become incorporated into the membranous disks.

(3) Photoreception by rods

–is initiated by the interaction of **light** with **rhodopsin,** which is composed of the integral membrane protein **opsin** bound to **retinal,** the aldehyde form of vitamin A.

(a) The retinal moiety of rhodopsin **absorbs light** in the visible range.

(b) Retinal dissociates from opsin. This reaction, referred to as **bleaching,** permits the diffusion of bound Ca^{2+} **ions** into the cytoplasm of the outer segment of a rod cell.

(c) The excess Ca^{2+} **ions** act to **hyperpolarize** the cell by closing its Na^+ channels, thus preventing Na^+ ions from entering the cell.

(d) These **ionic alterations** in the rod generate electrical activity, which is relayed to other rods via gap junctions.

(e) The dissociated retinal and opsin **reassemble** by an active process in which Müller and pigment epithelial cells also participate.

(f) Ca^{2+} ions are recaptured by the membranous disks, leading to reopening of the Na^+ channels and **reestablishment** of the normal resting membrane potential.

b. Cones

–are **much less numerous than rods**.

–are **sensitive to light of high intensity** and produce **greater visual acuity** than do rods.

−are generally similar in structure to rods and mediate photoreception in the same way with the following exceptions:

(1) The membranous disks in the outer segments of cones are invaginations of the plasma membrane, whereas in rods they are not.

(2) The proteins synthesized in the inner segments of cones are passed to the entire outer segment, whereas in rods they are added only to newly forming disks.

(3) Cones possess **iodopsin** in their disks. This photopigment varies in amount in different cones, making them differentially sensitive to red, green, or blue light.

(4) Each cone synapses with a **single** bipolar neuron, whereas each rod may synapse with several bipolar neurons.

3. External limiting membrane

−is not a true membrane but an area where **zonulae adherentes** (belt desmosomes) are located between the photoreceptor cells and the retinal glial cells (Müller cells).

−also contains microvilli that project from the Müller cells.

4. Outer nuclear layer

−consists primarily of the **nuclei of the rods and cones**.

5. Outer plexiform layer

−contains **axodendritic synapses** between the axons of photoreceptor cells and the dendrites of bipolar and horizontal cells.

−displays **synaptic ribbons** within the rod and cone cells at synaptic sites.

6. Inner nuclear layer

−contains the **cell bodies of bipolar neurons,** horizontal cells, and amacrine cells and the nuclei of Müller cells.

7. Inner plexiform layer

−contains **axodendritic synapses** between the axons of bipolar cells and the dendrites of ganglion cells.

−is the layer where processes of amacrine cells are located.

8. Ganglion cell layer

−contains the **somata of ganglion cells,** which form the final link in the retina's neural chain.

a. Structure—ganglion cells

−are typical neurons that project their axons to a specific region of the retina called the **optic disk**.

−contain **midget, diffuse,** and **stratified ganglion cells**.

b. Function—ganglion cells

−are activated by **hyperpolarization of rods and cones** and generate an action potential, which is transmitted to horizontal and amacrine cells.

−the **action potential** is carried to the visual relay system in the brain.

9. Optic nerve fiber layer

—consists primarily of the **unmyelinated axons** of ganglion cells, which form the fibers of the **optic nerve**. As each fiber pierces the sclera, it acquires a myelin sheath.

10. Inner limiting membrane

—consists of the terminations of Müller cell processes and their basement membranes.

E. Accessory structures of the eye

1. Conjunctiva

—is a **transparent mucous membrane** that lines the eyelids.
—is reflected onto the anterior portion of the orb up to the cornea, where it becomes continuous with the corneal epithelium.
—is a **stratified columnar epithelium** possessing many **goblet cells**.
—is separated by a basal lamina from an underlying lamina propria of loose connective tissue.

2. Eyelids

—are lined internally by conjunctiva and externally by skin that is elastic and covers a supportive framework of **tarsal plates**.
—contain highly modified **sebaceous glands** (meibomian glands), **modified sebaceous glands** (glands of Zeis), and **sweat glands** (glands of Moll).

3. Lacrimal apparatus

a. Lacrimal gland

—is a compound **tubuloalveolar gland** whose secretory units are surrounded by an incomplete layer of **myoepithelial cells**.
—secretes **tears,** which drain via 6–12 ducts into the conjuctival **fornix,** from which the tears flow over the cornea and conjunctiva, keeping them moist. Tears (which contain **lysozyme,** an antibacterial enzyme) then enter the lacrimal puncta and then the lacrimal canaliculi.

b. Lacrimal canaliculi

—are lined by a **stratified squamous epithelium** and unite to form a common canaliculus, which empties into the lacrimal sac.

c. Lacrimal sac

—is lined by a **pseudostratified ciliated columnar epithelium.**

d. Nasolacrimal duct

—is the inferior continuation of the lacrimal sac and also is lined by a **pseudostratified ciliated columnar epithelium.**

IV. Ear (Vestibulocochlear Apparatus)

—consists of three parts: the **external ear,** which receives sound waves; the **middle ear,** through which sound waves are transmitted; and the **internal ear,** where sound waves are transduced into nerve impulses. The vestibular organ, responsible for equilibrium, is also located in the inner ear.

A. External ear

1. Auricle (pinna)

–is composed of irregular plates of **elastic cartilage** covered by **thin skin**.

2. External auditory meatus

–is lined by **skin** containing hair follicles, sebaceous glands, and **ceruminous glands,** which are modified sweat glands that produce **earwax (cerumen)**.

3. Tympanic membrane (eardrum)

–is covered by **skin** on its external surface and by a **simple cuboidal epithelium** on its inner surface.

–possesses **fibroelastic connective tissue** interposed between its two epithelial coverings.

–**transmits sound vibrations** that enter the ear to the ossicles in the middle ear.

B. Middle ear

1. Tympanic cavity

–houses the **ossicles**.

–is connected to the pharynx via the **auditory tube (eustachian tube)**.

–is lined by a **simple squamous epithelium,** which changes to **pseudostratified ciliated columnar epithelium** near its opening to the auditory tube.

–has a **lamina propria** composed of dense connective tissue tightly adherent to the bony wall.

2. Oval and round windows

–are small, **membrane-covered regions devoid of bone** in the medial bony wall of the tympanic cavity.

–separate the middle ear from the bony labyrinth of the inner ear.

3. Ossicles

–include the **malleus, incus,** and **stapes**.

–**transmits movements of the tympanic membrane to the oval window.**

C. Internal ear

1. Bony labyrinth

–houses the membranous labyrinth.

–is filled with **perilymph,** which has a similar ionic composition to that of extracellular fluids but has a very low protein content.

a. Semicircular canals

–house the **semicircular ducts** of the membranous labyrinth.

b. Vestibule

–houses the **saccule** and **utricle**.

c. Cochlea

–winds 2½ times around a bony core (the **modiolus**) containing blood vessels and the spiral ganglion. The **osseous spiral lamina** is the lateral extension of the modiolus.

−has a thickened periosteum forming the **spiral ligament**.

−is subdivided into three spaces: the **scala vestibuli** and **scala tympani,** which are filled with perilymph, and the **scala media,** or cochlear duct, which is filled with endolymph.

2. Membranous labyrinth

−is filled with **endolymph,** which is similar to intracellular fluid and has a low sodium and protein content and a high potassium content.

−possesses various **sensory structures,** which represent **specializations of the epithelial lining**.

a. Saccule and utricle

−are located within the vestibule.

−are sac-like bodies composed of a thin sheath of connective tissue lined by **simple squamous epithelium**.

−each give rise to a duct; the two ducts join, forming the **endolymphatic sac**.

−possess small, specialized regions, called **maculae,** which contain two types of neuroepithelial hair cells, supporting cells, and a gelatinous layer (otolithic membrane) on their free surface.

(1) Vestibular hair cells (Figure 21.1)

−are **neuroepithelial cells** containing many mitochondria and a well-developed Golgi complex.

−possess 50–100 elongated, rigid **stereocilia** (sensory microvilli) arranged in rows and a single cilium (**kinocilium**). These extend from the apical surface of the hair cells into the overlying gelatinous layer.

(a) Type I hair cells

−are **bulbar** in shape and contain a round nucleus in the expanded portion.

−are almost completely surrounded by a **cup-shaped afferent nerve ending**.

(b) Type II hair cells

−are **columnar** and contain a round, basally located nucleus.

−make contact with **small afferent terminals** containing synaptic vesicles.

(2) Supporting cells

−are generally columnar and possess a round, basally located nucleus, many microtubules, and an extensive terminal web.

(3) Otolithic membrane

−is a thick, **gelatinous** (glycoprotein) **layer**.

−contains small calcified particles called **otoliths,** or **otoconia**. These are arranged vertically in the saccule and horizontally in the utricle; they function in the **detection of linear acceleration** (positive or negative) of the head in these two planes.

b. Semicircular ducts

−are continuous with and arise from the utricle.

(1) Ampullae

−are dilated regions of the semicircular ducts located near their junctions with the utricle.

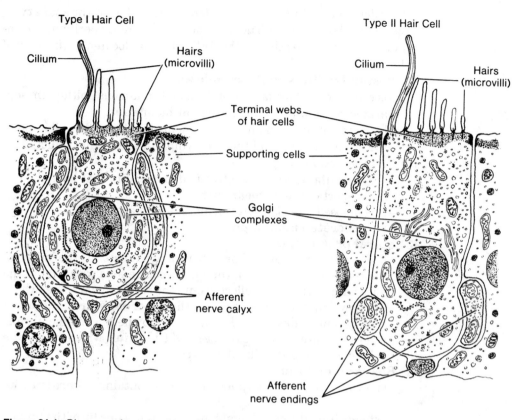

Figure 21.1. Diagrams of type I and type II hair cells located in the maculae of the saccule and utricle. The stereocilia extend into the otolithic membrane, which overlies these cells. (Reprinted from Krause WJ and Cutts JH: *Concise Text of Histology,* 2nd ed. Baltimore, Williams & Wilkins, 1985, p 215.)

(2) Cristae ampullares

- are specialized **sensory regions** within the ampullae of the semicircular ducts.
- are similar to maculae but have a thicker, cone-shaped glycoprotein layer (**cupula**), which does not contain otoliths.
- are so positioned (perpendicular to each other) that they can **detect angular acceleration** of the head along any of the three axes.

c. Endolymphatic duct

- ends in the expanded endolymphatic sac.

d. Endolymphatic sac

- has an epithelial lining containing **electron-dense** columnar cells, which have an irregularly shaped nucleus, and **electron-lucent** columnar cells, which possess long microvilli, many pinocytic vesicles, and vacuoles.
- contains **phagocytic cells** (macrophages, neutrophils) in its lumen.
- may function in **resorption of endolymph**.

e. Cochlear duct

- is a specialized diverticulum of the saccule that contains the **spiral organ of Corti**.

—is bordered above by the **scala vestibuli** and below by the **scala tympani** of the bony cochlea. These scalae, which contain perilymph, communicate with each other at the **helicotrema,** located at the apex of the cochlea.

(1) Vestibular (Reissner's) membrane
  —is composed of **two layers** of flattened squamous epithelium separated by an intervening basement membrane.
  —helps maintain the high ionic gradients between the perilymph in the scala vestibuli and the endolymph in the cochlear duct.

(2) Stria vascularis
  —is a **vascularized** pseudostratified epithelium that lines the lateral aspect of the cochlear duct.
  —is composed of basal, intermediate, and marginal cells.
  —may **secrete endolymph**.

(3) Spiral prominence
  —is an epithelium-covered protuberance that extends the length of the cochlear duct. This epithelium is continuous with that of the stria vascularis and is reflected onto the basilar membrane, where it follows an indentation to form the **external spiral sulcus**.
  —cells become cuboidal and continue onto the basilar membrane where they are known as the **cells of Claudius,** which overlie the polyhedral-shaped **cells of Boettcher**.

(4) Basilar membrane
  —is a thick layer of amorphous material containing **keratin-like fibers**.
  —extends from the spiral ligament to the tympanic lip of the limbus spiralis.
  —has two zones: the medial **zona arcuata** and the lateral **zona pectinata**.

(5) Tectorial membrane
  —**makes contact with the processes of the hair cells.**
  —is secreted by the interdental cells of the spiral sulcus.

(6) Spiral organ of Corti
  —lies upon both parts of the basilar membrane.
  —displays the **inner tunnel of Corti** and the **outer tunnel (space of Nuel),** which communicate with each other via intercellular spaces.
  —is composed of **hair cells** and various **supporting cells**.

(a) Hair cells
  —are **neuroepithelial cells** containing a round, basally located nucleus, which is surrounded by many mitochondria.
  —possess many long, stiff **stereocilia** (arranged in a W-shaped formation) on their free surface, as well as a **basal body** (but no kinocilium).
  —are divided into two types: **inner hair cells,** which are organized in a **single row** along the entire length of the cochlear duct and receive numerous afferent synaptic terminals on their basal surface; and **outer hair cells,** which are organized in **three or more rows** and ensconced within a cup-shaped afferent nerve ending where synaptic contacts are made.

–function in the **reception of sound** and can respond to different sound frequencies.

(b) Inner and outer pillar cells

–are intimately associated with each other and both rest on the basilar membrane.

–enclose and support the inner tunnel of Corti, which lies between the inner and outer pillar cells.

–possess a wide base and have elongated processes, which contain microtubules, intermediate filaments, and microfilaments (actin).

(c) Inner and outer phalangeal cells

–are intimately associated with the inner and outer hair cells, respectively.

–support the slender nerve fibers that form synapses with the hair cells.

(d) Cells of Hensen and border cells

–delineate the inner and outer borders of the spiral organ of Corti.

3. Auditory function of the inner ear

a. Movement of the stapes at the oval window causes disturbances in the perilymph, which cause deflection of the **basilar membrane.** (The oscillations set in motion at the oval window are dissipated at the secondary tympanic membrane covering the round window of the cochlea.)

b. Large areas of the basilar membrane vibrate at many frequencies. However, optimal vibrations are detected at only specific areas. Sound waves of low frequency are detected farther away from the oval window.

c. The **pillar cells** attached to the basilar membrane move laterally in response to this deflection, in turn causing a lateral shearing of the stereocilia of the sensory hair cells of the organ of Corti against the tectorial membrane.

d. Movement of the stereocilia is transduced into electrical impulses that travel via the **cochlear nerve** to the brain.

4. Vestibular function of the inner ear

a. Change in the position of the head causes a flow of the endolymph in the semicircular ducts (**circular movement**) or in the saccules and utricles (**linear movement**).

b. Movement of the endolymph in the semicircular ducts displaces the cupula overlying the **cristae ampullares,** causing bending of the stereocilia of the sensory hair cells.

c. Movement of the endolymph in the saccules and utricles displaces the **otoliths.** This deformation is transmitted to the **maculae** via the overlying gelatinous layer, causing bending of the stereocilia of the sensory hair cells.

d. In both cases, **movement of the stereocillia** is transduced into **electrical impulses,** which are transmitted to the brain via **vestibular nerve fibers**.

V. Clinical Considerations

A. Glaucoma

—is a condition of abnormally **high intraocular pressure**.

—is caused by obstructions that prevent drainage of aqueous humor from the eye.

1. Acute glaucoma

—is relatively rare but leads to blindness within a few days if not treated.

2. Chronic glaucoma

—is much more common than acute glaucoma.

—may be associated with few symptoms except for a gradual loss of peripheral vision.

—usually can be controlled with eye drops.

B. Cataract

—is an **opacity of the lens** resulting from the accumulation of pigment or other substances in the lens fibers.

—is often associated with **aging**.

—leads to a **gradual loss of vision if untreated**.

C. Stye (hordeolum)

—is a localized, inflammatory **infection of a sebaceous gland of the eyelid,** frequently caused by staphylococcal bacteria.

—can involve the meibomian glands or glands of Moll.

D. Detachment of the retina

—occurs when the neural and pigmented retinae become separated from each other.

—can be treated successfully by **laser surgery**.

E. Conductive hearing loss

—results from a defect in the conduction of sound waves in the external or middle ear.

—may be caused by **otitis media,** a common inflammation of the middle ear; **obstruction** by a foreign body; or **otosclerosis** of the middle ear.

F. Nerve deafness

—results from a **lesion** in any of the nerves transmitting impulses from the spiral organ of Corti to the brain.

—may be caused by disease, exposure to drugs, or prolonged exposure to loud noises.

Review Test

Directions: Each of the numbered items or incomplete statements in this section is followed by answers or by completions of the statement. Select the **one** lettered answer or completion that is **best** in each case.

1. Meissner's corpuscles are found in all of the following EXCEPT

(A) hairless skin
(B) lips
(C) nipples
(D) serous membranes

2. Which of the following statements regarding aqueous humor is FALSE?

(A) It is produced by ciliary processes
(B) It flows from the posterior chamber into the anterior chamber
(C) It exits via the canal of Schlemm
(D) It becomes the vitreous body

3. Which of the following statements about the cornea is FALSE?

(A) It represents the anterior portion of the tunica vasculosa
(B) It is the anterior transparent portion of the sclera
(C) It is composed of five layers
(D) It is rich in nerve endings

4. Which of the following statements regarding the choroid is FALSE?

(A) It is avascular
(B) It is part of the pigmented layer of the eye
(C) It is loosley attached to the tunica fibrosa
(D) It contains many melanocytes

5. Which of the following statements concerning the membranous labyrinth is FALSE?

(A) It contains perilymph
(B) It includes the saccule
(C) It is housed within the bony labyrinth
(D) It includes the utricle

6. Communication of the scala vestibuli and scala tympani occurs at the

(A) round window
(B) oval window
(C) helicotrema
(D) endolymphatic sac

7. The bony ossicles of the middle ear cavity are arranged in a series bridging the tympanic cavity beginning at the tympanic membrane and ending at the

(A) endolymphatic duct
(B) round window
(C) helicotrema
(D) oval window

8. Which of the following cells in the inner ear are involved in detecting movements of the head?

(A) Hair cells in the maculae
(B) Outer pillar cells
(C) Inner pillar cells
(D) Cells of Hensen
(E) Hair cells in the organ of Corti

9. Rods and cones form synapses with which of the following cells?

(A) Amacrine
(B) Bipolar
(C) Ganglion
(D) Müller

Directions: The group of items in this section consists of lettered options followed by a set of numbered items. For each item, select the **one** lettered option that is most closely associated with it. Each lettered option may be selected once, more than once, or not at all.

Questions 10–14

Match each structure below with the corresponding letter in the photomicrograph of the cochlea.

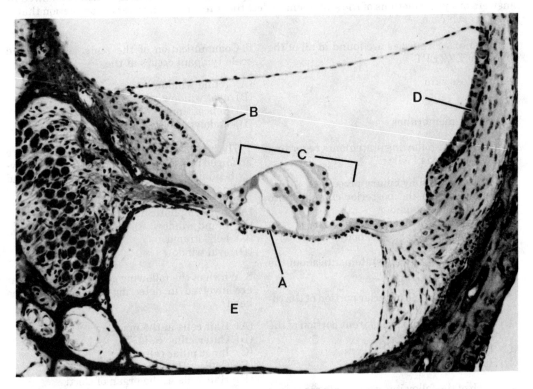

Reprinted from Gartner LP and Hiatt JL: *Atlas of Histology.* Baltimore, Williams & Wilkins, 1987, p 295.

10. Organ of Corti

11. Basilar membrane

12. Scala tympani

13. Tectorial membrane

14. Stria vascularis

Questions 15–19

Match each numbered structure below with the corresponding letter in the photomicrograph of the retina.

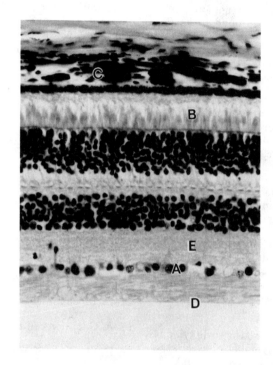

Reprinted from Gartner LP and Hiatt JL: *Atlas of Histology*. Baltimore, Williams & Wilkins, 1987, p 291.

15. Choroid

16. Layer of rods and cones

17. Inner limiting membrane

18. Ganglion cell layer

19. Inner plexiform layer

Answers and Explanations

1–D. Meissner's corpuscles are not found in serous membranes; they are found in hairless skin, lips, and nipples.

2–D. The vitreous body is a refractile gel filling the chamber of the globe posterior to the lens; it is not related to the aqueous humor.

3–A. The cornea is the anterior portion of the tunica fibrosa, the outer covering of the eye, and not the tunica vasculosa. It is composed of five distinct layers and contains many sensory nerve endings.

4–A. The choroid is the vascular, pigmented layer of the eye containing many melanocytes, which impart dark pigment to the eye.

5–A. The membranous labyrinth, which includes the semicircular ducts, saccule, utricle, and cochlear duct, is filled with endolymph. Perilymph is the fluid contained within the bony labyrinth, which lies outside the membranous labyrinth.

6–C. The scala vestibuli and the scala tympani are in reality one perilymphatic space separated by the cochlear duct (scala media). The scala vestibuli and tympani communicate with each other at the helicotrema.

7–D. The bony ossicles of the middle ear cavity articulate in a series from the tympanic membrane to the oval window.

8–A. Neuroepithelial hair cells in the maculae of the saccule and the utricle detect linear movement of the head. These cells are connected to the vestibular portion of the acoustic nerve.

9–B. Rods and cones synapse with bipolar cells and horizontal cells.

10–C. Organ of Corti.

11–A. Basilar membrane.

12–E. Scala tympani.

13–B. Tectorial membrane.

14–D. Stria vascularis.

15–C. Choroid.

16–B. Layer of rods and cones.

17–D. Inner limiting membrane.

18–A. Ganglion cell layer.

19–E. Inner plexiform layer.

Comprehensive Examination

Directions: Each of the numbered items or incomplete statements in this section is followed by answers or by completions of the statement. Select the **one** lettered answer or completion that is **best** in each case.

1. Which of the following statements concerning RNA synthesis is FALSE?

(A) DNA acts as a template for synthesis of RNA
(B) Synthesis of rRNA, mRNA, and tRNA is catalyzed by three different RNA polymerases, I, II, and III, respectively
(C) In the synthesis of mRNA, exons are excised and the introns are spliced together, yielding mRNPs
(D) Protein moieties are removed from mRNPS as they leave the nucleus, yielding functional mRNAs
(E) Transcription of DNA into rRNA occurs in the nucleolus

2. Osteoporosis is a condition characterized by a decrease in bone mass. It commonly occurs with advancing age and may result from any of the following EXCEPT

(A) diminished secretion of growth hormone
(B) immobility
(C) diminished estrogen production
(D) calcium deficiency

3. RNA contains all of the following nucleotides EXCEPT

(A) thymine
(B) adenine
(C) uracil
(D) cytosine
(E) guanine

4. The centroacinar cells of the pancreas secrete

(A) an alkaline, enyzme-poor fluid
(B) pancreatic digestive enzymes
(C) secretin
(D) cholecystokinin
(E) glucagon

5. The zona fasciculata of the adrenal cortex synthesizes and secretes

(A) mineralocorticoids
(B) glucagon
(C) epinephrine
(D) aldosterone
(E) none of the above

6. Which of the following statements about bony joints is FALSE?

(A) Long bones generally are united by diarthroses
(B) Diarthroses are surrounded by a two-layered capsule
(C) Synarthroses are classified as synovial joints
(D) Type A cells of the synovial membrane are phagocytic
(E) Type B cells of the synovial membrane secrete synovial fluid

7. Which of the following characteristics can be used to distinguish neutrophils and basophils histologically?

(A) Size of specific granules
(B) Shape of the nucleus
(C) Number of azurophilic granules
(D) The presence or absence of peroxidase
(E) All of the above

8. A long-time user of smokeless tobacco noticed several whitish, thickened, painless patches on the lining of his cheeks. The most probable diagnosis is

(A) aphthous ulcers
(B) adenocarcinoma
(C) keloids
(D) oral leukoplakia
(E) epidermolysis bullosa

9. Primordial follicles of the ovary possess

(A) a layer of cuboidal follicular cells
(B) an oocyte arrested in prophase of meiosis I
(C) an oocyte arrested in metaphase of meiosis II
(D) a well-defined theca interna and theca externa
(E) a thick zona pellucida

10. Which of the following statements regarding the membranous labyrinth of the inner ear is TRUE?

(A) It contains the saccule and utricle
(B) Maculae contain neuroepithelial cells, which possess numerous stereocilia and a single kinocilium
(C) Cristae ampullares in the semicircular canals detect angular acceleration of the head
(D) The otolithic membrane contains small calcified particles
(E) All of the above

11. Which of the following statements concerning euchromatin is TRUE?

(A) It represents about 90% of the chromatin
(B) It appears as basophilic clumps of nucleoprotein in the light microscope
(C) It is concentrated near the periphery of the nucleus
(D) It is transcriptionally active
(E) None of the above

12. Intercalated disks function in all of the following EXCEPT

(A) end-to-end attachments between cardiac muscle cells
(B) intercellular movement of large protein molecules
(C) ionic coupling of cardiac muscle cells
(D) anchoring of actin filaments

13. Release of thyroid hormones from the follicular cells of the thyroid gland depends on thyroid-stimulating hormone (TSH). TSH stimulation involves all of the following EXCEPT

(A) activation of a G protein
(B) binding of TSH to receptors on the basal plasma membrane
(C) formation of cAMP catalyzed by adenyl cyclase
(D) change in cell shape from columnar to flattened
(E) endocytosis of colloid droplets

14. Which of the following statements related to development of the tooth crown is FALSE?

(A) The enamel organ is derived from epithelium
(B) The dental papilla is derived from ectomesenchyme
(C) The four-layered enamel organ appears during the cap stage
(D) Dentin and enamel are formed during the appositional stage
(E) None of the above

15. Which of the following statements concerning liver sinusoids is FALSE?

(A) They are surrounded by a well-developed basal lamina
(B) Their lining includes Kupffer cells
(C) They are lined by fenestrated endothelial cells
(D) They deliver blood to the central vein
(E) None of the above

16. Which of the following statements concerning events during the G_1 phase of the cell cycle is FALSE?

(A) Daughter cells are restored to normal volume and size
(B) Cells failing to reach the restriction point proceed to the S phase
(C) "Trigger proteins" are synthesized, enabling the cell to reach the restriction point
(D) Cells remain in the G_1 phase for a few hours to several days
(E) None of the above

17. A young college student experienced nausea, vomiting, visual disorders, and muscular paralysis after eating canned tuna fish. The probable diagnosis is botulism, caused by ingestion of the *Clostridium botulinum* toxin. The physiologic effect of this toxin is to

(A) inactivate acetylcholinesterase
(B) bind to and thus inactivate acetylcholine receptors at myoneural junctions
(C) prevent release of calcium ions from the sarcoplasmic reticulum, thus inhibiting muscle contraction
(D) inhibit release of acetylcholine from presynaptic membranes
(E) inhibit hydrolysis of ATP during the contraction cycle

18. Which of the following statements concerning veins is FALSE?

(A) The tunica adventitia of large and medium veins is thicker than the tunica media
(B) Some large veins possess cardiac muscle cells in their walls
(C) Veins have thicker walls than the corresponding arteries
(D) Some veins contain valves
(E) None of the above

19. Which of the following statements concerning the adult tooth is FALSE?

(A) Enamel is continuously elaborated to compensate for the decrease in tooth length due to abrasion
(B) The pulp is a gelatinous connective tissue
(C) The matrix of cementum possesses collagen
(D) Odontoblastic processes are located in dentinal tubules
(E) Cementum is continuously elaborated to compensate for the decrease in tooth length due to abrasion

20. Which of the following statements about the nucleosome is FALSE?

(A) It is the structural unit of chromatin packing
(B) Histones form the nucleosome core around which linker DNA is wound
(C) Groups of nucleosomes are packaged together to form the structural unit of the chromosome
(D) It consists of an RNA molecule and two copies each of four different histones
(E) None of the above

21. A high-school student complains of fatigue and a sore throat and has swollen, tender lymph nodes and a fever. Blood tests reveal an increased white blood cell count with many atypical lymphocytes; the number and appearance of the erythrocytes are normal. This student is likely to be suffering from

(A) AIDS
(B) pernicious anemia
(C) infectious mononucleosis
(D) Hodgkin's disease
(E) factor VIII deficiency

22. Which of the following statements about the gallbladder is TRUE?

(A) The gallbladder dilutes bile
(B) Bile enters the gallbladder via the common bile duct
(C) Bile leaves the gallbladder via the cystic duct
(D) The gallbladder is lined by a simple squamous epithelium
(E) Secretin stimulates the wall of the gallbladder to contract, forcing bile from its lumen

23. Which of the following statements regarding the retina is FALSE?

(A) The outer segments of cones possess numerous membranous disks containing rhodopsin
(B) The nuclei of rods and cones are located in the outer nuclear layer
(C) The axons of the ganglion cells are projected to the optic disk and become the fibers constituting the optic nerve
(D) The outer plexiform layer contains synapses between the axons of photoreceptor cells and bipolar cells
(E) None of the above

24. Which of the following statements concerning mitochondria is FALSE?

(A) They change from the condensed to the orthodox form in response to the uncoupling of oxidation from phosphorylation
(B) They may have originated as intracellular parasites
(C) They possess the enzymes of the Krebs cycle in their matrix
(D) They contain elementary particles on their cristae
(E) They possess circular DNA

25. Which of the following components is present in muscular arteries but absent from elastic arteries?

(A) Fenestrated membranes
(B) Vasa vasorum
(C) Factor VIII
(D) A thick, complete internal elastic lamina
(E) Smooth muscle cells

26. Which of the following statements concerning the layers of the cornea is FALSE?

(A) Bowman's membrane, a noncellular layer composed of interlacing collagen fibers, provides form, stability, and strength to the cornea
(B) Part of the cornea is lined by a stratified squamous nonkeratinized epithelium
(C) Descemet's membrane, a thick basal lamina, separates the corneal stroma from the endothelium
(D) The corneal endothelium, a simple squamous epithelium, lines the anterior of the cornea
(E) None of the above

27. Epiphyseal plates are composed of five histologically distinct zones. Which of the following statements about these zones is FALSE?

(A) Zone of reserve, located on the epiphyseal side of plates, contains inactive chondrocytes
(B) Cartilage becomes calcified and chondrocytes die in the zone of calcification
(C) Osteoblasts enlarge and mature into osteocytes in the zone of hypertrophy and maturation
(D) Zone of proliferation contains isogenous chondrocytes
(E) All of the above

28. Which of the following statements concerning the pancreas is TRUE?

(A) Islets of Langerhans secrete enzymes
(B) It possesses mucous acinar cells
(C) The endocrine pancreas has more beta cells than delta cells
(D) Its alpha cells secrete insulin
(E) None of the above

29. The most common cause of peptic ulcer disease is excess secretion of hydrochloric acid (HCl) in the stomach. All of the following either stimulate or inhibit gastric secretion EXCEPT

(A) urogastrone
(B) cholecystokinin
(C) gastrin
(D) somatostatin
(E) vagus nerve stimulation

30. Which of the following statements about the tunica vasculosa of the eye is FALSE?

(A) The choroid layer is a highly pigmented layer containing many melanocytes
(B) Bruch's membrane separates the retina from the choroid
(C) The outer, pigmented layer of ciliary processes is responsible for secretion of aqueous humor
(D) The lens is anchored in place via suspensory ligaments, which arise from the ciliary processes
(E) None of the above

31. A deficiency or an excess of which of the following vitamins results in short stature?

(A) Vitamin A
(B) Vitamin C
(C) Vitamin D
(D) Vitamin K

32. The intercellular spaces in the stratum spinosum of the epidermis contain lipid-containing sheets that are impermeable to water. This material is released from

(A) keratohyalin granules
(B) Langerhans' cells
(C) membrane-coating granules
(D) sebaceous glands
(E) melanosomes

33. Which of the following statements about early basophilic erythroblasts is FALSE?

(A) Their nucleus has a coarse chromatin network
(B) They possess nucleoli
(C) They can undergo mitosis
(D) Their cytoplasm contains ferritin

34. Luminal surface modifications of the duodenum and jejunum include all of the following EXCEPT

(A) microvilli
(B) rugae
(C) plicae circulares
(D) villi
(E) valves of Kerckring

35. Blood coagulation involves a cascade of reactions that occur in two interrelated pathways, the extrinsic and intrinsic. All of the following are associated with the intrinsic pathway of blood coagulation EXCEPT

(A) conversion of fibrinogen to fibrin
(B) platelet aggregation
(C) release of tissue thromboplastin
(D) von Willebrand's factor
(E) calcium

36. A 25-year-old woman complains about a frequently recurring, painful lesion on her upper lip that exudes a clear fluid. She is probably suffering from

(A) oral leukoplakia
(B) herpetic stomatitis
(C) aphthous ulcer
(D) bullous pemphigoid

37. Which of the following statements concerning eosinophils is FALSE?

(A) They have antiparasitic activity
(B) They possess receptors for IgE on their plasma membrane
(C) Their concentration in the blood is much lower than that of neutrophils
(D) They have a bilobed nucleus
(E) They possess enzymes that inactivate histamine and leukotriene C

38. Which of the following statements concerning the functions of the skin is TRUE?

(A) Infrared radiation, necessary for synthesis of vitamin D, is absorbed by the skin
(B) The skin provides protection against desiccation
(C) The skin contains temperature receptors and plays a role in regulating body temperature
(D) Melanin is synthesized by melanocytes, which are located in the dermis
(E) All of the above

39. A person with diabetes insipidus will have all of the following signs or symptoms EXCEPT

(A) hypotonic urine
(B) polyuria
(C) proteinuria and hematuria
(D) dehydration
(E) polydipsia

40. Stratified squamous keratinized epithelium is always present in the

(A) rectum
(B) esophagus
(C) pyloric stomach
(D) jejunum
(E) anus

41. Which of the following properties is exhibited in all three types of cartilage?

(A) Is involved in bone formation
(B) Possesses type II collagen
(C) Possesses type I collagen
(D) Grows interstitially and appositionally
(E) Has an identifiable perichondrium

42. Which of the following is a receptor for fine touch?

(A) Pacinian corpuscle
(B) Crista ampullaris
(C) Ruffini's ending
(D) Krause's end-bulb
(E) None of the above

43. Which of the following statements concerning the spiral organ of Corti is FALSE?

(A) Both inner and outer phalangeal cells are intimately associated with inner and outer pillar cells
(B) Basilar membrane vibrations produced by disturbances in the perilymph are detected by the hair cells
(C) Both inner and outer pillar cells rest on the basilar membrane
(D) The tips of the hair cells are embedded in the tectorial membrane
(E) None of the above

44. A premature infant was found to have labored breathing, which is eventually alleviated by administration of glucocorticoids. The most probable diagnosis is

(A) immotile cilia syndrome
(B) emphysema
(C) hyaline membrane disease
(D) asthma
(E) chronic bronchitis

45. Which of the following statements concerning the cribriform plate is TRUE?

(A) It is the inner layer of the alveolar bone
(B) It lacks Sharpey's fibers
(C) It is also known as the spongiosa
(D) It is composed of cancellous bone
(E) All of the above

46. All of the following substances are synthesized in the hypothalamus EXCEPT

(A) oxytocin
(B) antidiuretic hormone (ADH)
(C) somatotropin
(D) neurophysin

Directions: Each group of items in this section consists of lettered options followed by a set of numbered items. For each item, select the **one** lettered option that is most closely associated with it. Each lettered option may be selected once, more than once, or not at all.

Questions 47–49

Match each description below with the corresponding glycoprotein.

(A) Chondronectin
(B) Plasma fibronectin
(C) Osteonectin
(D) Matrix fibronectin

47. A circulating protein in the blood that functions in wound healing

48. An adhesive glycoprotein that forms fibrils in the extracellular matrix

49. A multifunctional protein that attaches chondrocytes to type II collagen

Questions 50–54

Although fibroblasts are the predominant cells present in connective tissue, many other cell types also are found in connective tissue. Match each description below with the most appropriate type of cell.

(A) Pericytes
(B) Macrophages
(C) T lymphocytes
(D) Plasma cells
(E) B lymphocytes
(F) Foreign-body giant cells
(G) Mast cells
(H) Eosinophils

50. Antibody-manufacturing cells

51. Principal phagocytes of connective tissue

52. Cells that can bind IgE antibodies and mediate immediate hypersensitivity reactions

53. Cells responsible for initiating cell-mediated immune responses

54. Pluripotential cells located primarily along capillaries

Questions 55–59

The plasma membrane is a complex structure that functions in membrane transport processes and cell-to-cell communication. Match each description below with the most appropriate component of the plasma membrane.

(A) Phospholipid
(B) Glycocalyx
(C) Carrier protein
(D) Band 3 protein
(E) G protein
(F) K^+ leak channel

55. Functions in activation of secondary messenger systems

56. Is primarily responsible for establishing the potential difference across the plasma membrane

57. Is an amphipathic molecule

58. Is a carbohydrate-containing covering associated with the outer leaflet of the plasma membrane

59. Functions in antiport transport

Questions 60–64

For each description below, select the most appropriate molecule associated with the immune response.

(A) CD4
(B) CD8
(C) Interleukin 1
(D) Interleukin 2
(E) Perforin
(F) Interferon γ

60. Is a surface marker on T suppressor cells

61. Is a surface marker on T helper cells

62. Mediates lysis of tumor cells

63. Is released by macrophages and stimulates activated T helper cells

64. Stimulates activation of natural killer (NK) cells

Questions 65–69

Match each description below with the most appropriate type of cell.

(A) Sertoli cells
(B) Primary spermatocytes
(C) Secondary spermatocytes
(D) Interstitial cells of Leydig
(E) Spermatids

65. Replicate their DNA during the S phase of the cell cycle and undergo meiosis

66. Form a temporary cylinder of microtubules called the manchette

67. Possess receptors for luteinizing hormone (LH) and produce testosterone in response to binding of LH

68. Are responsible for formation of the blood–testis barrier

69. Synthesize androgen-binding protein (ABP) when stimulated by follicle-stimulating hormone (FSH)

Questions 70–73

Match each description below with the most appropriate cytoplasmic organelle.

(A) Rough endoplasmic reticulum
(B) Smooth endoplasmic reticulum
(C) Mitochondrion
(D) Annulate lamellae
(E) Lysosomes
(F) Polysomes

70. Possesses mixed function oxidases that detoxify phenobarbital and other drugs

71. Contains ribophorins

72. Are parallel stacks of membranes that resemble the nuclear envelope

73. Are the sites where hemoglobin and other cytosolic proteins are synthesized

Questions 74–78

Various enzymes and hormones are secreted by the cells lining the digestive tract. Match each physiologic activity with the appropriate hormone or enzyme.

(A) Gastrin
(B) Somatostatin
(C) Motilin
(D) Pepsinogen
(E) Lysozyme
(F) Urogastrone

74. Stimulates parietal cells to secrete HCl

75. Is produced by Brunner's glands and inhibits HCl secretion by parietal cells

76. Functions as an antibacterial agent

77. Stimulates contraction of smooth muscle in the wall of the digestive tract

78. Inhibits secretion by nearby enteroendocrine cells

Questions 79–80

Match each description below with the appropriate amino acid.

(A) Lysine
(B) Isodesmosine
(C) Hydroxyproline
(D) Arginine

79. Is partly responsible for the elasticity of elastin

80. Cross-links elastin molecules to form an extensive network

Questions 81–82

Match each characteristic with the condition that it most often causes.

(A) Simple goiter
(B) Exophthalmic goiter
(C) Graves' disease
(D) Addison's disease

81. Inadequate amounts of iodine in the diet

82. Destruction of the adrenal cortex

Questions 83–85

For each function, select the muscle protein that performs it.

(A) Troponin C
(B) Globular head (S1 fragment) of myosin
(C) Myoglobin
(D) Actin

83. Hydrolyzes ATP

84. Binds oxygen

85. Binds calcium ions

Questions 86–87

Match each description with the disease to which it applies.

(A) Epidermolysis bullosa
(B) Basal cell carcinoma
(C) Malignant melanoma

86. A rare form of skin cancer that may be fatal

87. A hereditary skin disease characterized by blister formation following minor trauma

Questions 88–91

Match each description with the most appropriate structure of the respiratory system.

(A) Trachea
(B) Nasopharynx
(C) Terminal bronchiole
(D) Alveolar duct
(E) Intrapulmonary bronchi

88. Is the region involved in patients with adenoids

89. Possesses C-shaped rings of hyaline cartilage

90. Contains smooth muscle at openings into alveoli

91. Is lined by an epithelium containing ciliated cells and Clara cells

Questions 92–95

For each characteristic, select the condition most closely associated with it.

(A) Endometriosis
(B) Cervical cancer
(C) Ectopic tubal pregnancy
(D) Breast cancer

92. Is associated with rupture of the oviduct and hemorrhaging into the peritoneal cavity

93. May be classified as lobular carcinoma

94. May be detected by a Pap smear

95. Commonly results in hemorrhaging into the peritoneal cavity dependent on the stage of the menstrual cycle

Questions 96–99

Match each process with the substance associated with it.

(A) Gastrin
(B) Enzyme associated with the glycocalyx of the intestinal striated border
(C) Lipase
(D) Chylomicron
(E) Gastric intrinsic factor

96. Absorption of vitamin B_{12}

97. Digestion of carbohydrates

98. Digestion of proteins

99. Transport of triglycerides into lacteals

Questions 100–104

Match each description with the appropriate lettered structure in the photomicrograph of the basal surface of an epithelial cell.

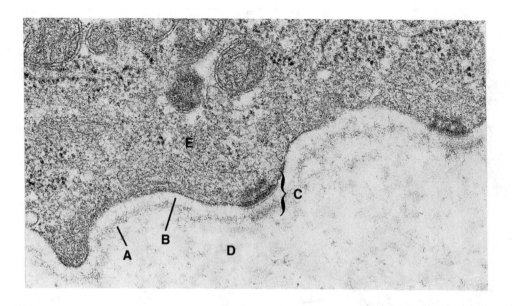

100. Contains heparan sulfate

101. Contains type IV collagen

102. Acts as a self-antigen in the autoimmune disease bullous pemphigoid

103. Is the reticular layer of the basement membrane

104. Attaches epithelial cell to basal lamina

Questions 105–108

Match each description with the appropriate lettered structure in the micrograph of a section of the renal medulla.

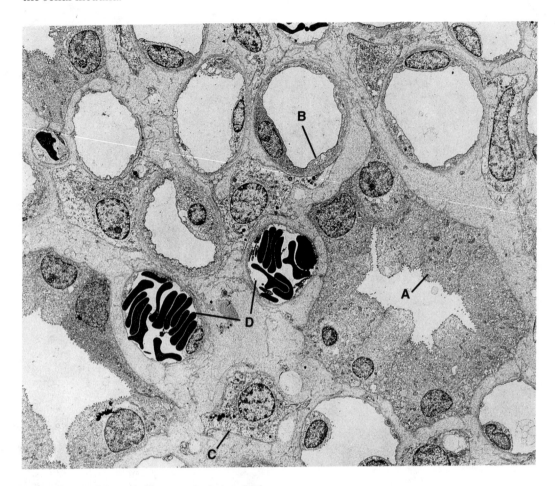

105. Portion of the uriniferous tubule in which the filtrate becomes hypertonic to blood plasma

106. Portion of the uriniferous tubule whose epithelial lining cells are responsive to antidiuretic hormone

107. Small capillaries that extend deep into the medulla

108. Site where prostaglandins and other vasodepressor substances are produced

Questions 109–113

Match each description with the appropriate lettered cell component in the electron micrograph of a cell.

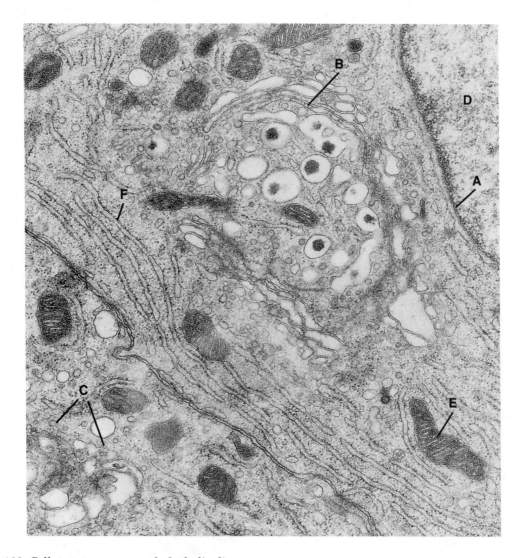

109. Cell structure composed of tubulin dimers

110. Site where sulfation of secretory proteins may occur

111. A layer composed of three intermediate filament proteins

112. Cell component containing transcriptionally active DNA

113. Structure in which ATP is produced by two different mechanisms

Questions 114–118

Match each structure below with the appropriate letter in the electron micrograph of a peripheral nerve.

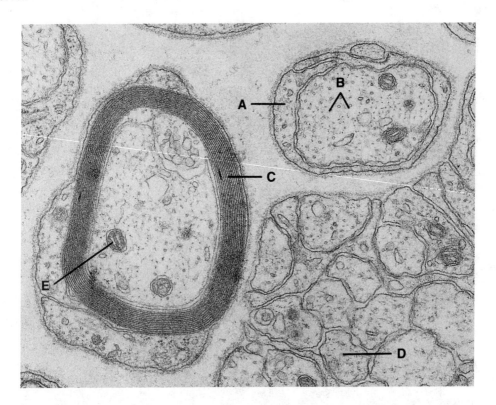

114. Neurotubule

115. Unmyelinated nerve fiber

116. Schwann cell

117. Mitochondrion

118. Myelin

Answers and Explanations

1–C. Both exons and introns are transcribed to form a primary transcript. After associating with protein as an hnRNP, the introns are excised and the exons spliced to form a mRNP.

2–D. Although patients with osteoporosis exhibit decreased bone mass, their bones have a normal ratio of mineral to matrix. Calcium deficiency in adults can lead to osteomalacia, which is characterized by softened bones with a low ratio of mineral to matrix.

3–A. In DNA, thymine replaces uracil as one of the pyrimidines. RNA contains adenine and guanine (purines) and cytosine and uracil (pyrimidines).

4–A. Pancreatic centroacinar cells form the initial segment of the intercalated duct and are part of the exocrine pancreas. They secrete an alkaline, enzyme-poor fluid when stimulated by secretin. Pancreatic digestive enzymes are synthesized by the acinar cells of the exocrine pancreas; their release is stimulated by cholecystokinin. Glucagon is produced in the endocrine pancreas (islets of Langerhans).

5–E. The zona fasciculata, the largest region of the adrenal cortex, produces glucocorticoids (cortisol and corticosterone). The zona glomerulosa produces mineralocorticoids, primarily aldosterone. Epinephrine is produced in the adrenal medulla. Glucagon is produced in the pancreas, not in the adrenal gland.

6–C. Synarthroses, which are immovable joints, are composed of connective tissue, cartilage, or bone and lack a synovial membrane. Diarthroses, or synovial joints, permit maximum movement and generally unite long bones. The internal layer of their two-layered capsule is composed of phagocytic type A cells and type B cells that secrete synovial fluid into the articular cavity.

7–E. Neutrophils have a nucleus with three or four lobes, many azurophilic granules, and small specific granules that lack peroxidase. In contrast, basophils have an S-shaped nucleus, few azurophilic granules, and large specific granules that contain peroxidase.

8–D. Oral leukoplakia, which results from epithelial hyperkeratosis, is usually of unknown etiology but often is associated with the use of smokeless tobacco. Although the characteristic painless lesions are benign, they may transform into squamous cell carcinoma. Aphthous ulcers are painful lesions of the oral mucosa that are surrounded by a red border. Adenocarcinoma is a form of cancer arising in glandular tissue. Keloids are swellings in the skin that arise from increased collagen formation in hyperplastic scar tissue. Epidermolysis bullosa is a group of hereditary skin diseases characterized by blister formation following minor trauma.

9–B. A primordial follicle is composed of a flattened layer of follicular cells surrounding a primary oocyte, which is arrested in prophase of meiosis I. Well-defined thecal layers and a thick zona pellucida are found in growing follicles. A graafian (mature) follicle possesses a secondary oocyte, which becomes arrested in metaphase of meiosis II.

10–E. All of the statements are true. Linear acceleration of the head is detected by the neuroepithelial hair cells of the maculae, which are specialized regions of the saccule and utricle.

11–D. Euchromatin, the transcriptionally active form of chromatin, represents only about 10% of the chromatin. In the light microscope, it appears as a light-staining, dispersed region of the nucleus.

12–B. Large protein molecules cannot move across intercalated disks (the step-like junctional complexes present in cardiac muscle). These junctional structures possess three specializations: desmosomes, which provide end-to-end attachment of cardiac muscle cells; fasciae adherentes, to which actin filaments attach; and gap junctions, which permit intercellular movement of small molecules and ions (ionic coupling).

13–D. Flattened, squamous cells are characteristic of an unstimulated, inactive thyroid gland. TSH binds to G-protein–linked receptors on the basal surface of follicular cells. Under TSH stimulation, thyroid follicular cells become columnar and form pseudopods, which engulf colloid. Lysosomal enzymes split thyroxine (T_4) and triiodothyronine (T_3) from thyroglobulin; the hormones are then released basally.

14–C. The four-layered enamel organ appears during the bell stage of crown formation. It develops by elaboration of a fourth layer in the three-layered enamel organ that appears in the cap stage.

15–A. Liver sinusoids are vessels that lack a basal lamina. They arise at the periphery of the classic liver lobule, receiving blood from branches of the portal vein and delivering it to the central vein.

16–B. During the G_1 phase, cells that fail to reach the restriction point proceed to the G_0 state, becoming quiescent (resting) cells. In some cases (e.g., peripheral lymphocytes), the cell cycle is only temporarily suspended, and the resting cells can reenter the cell cycle and begin to divide again. In the case of differentiated cells such as cardiac muscle cells and neurons, the G_0 state is permanent. For cells that are mitotically active, the G_1 phase is the period during which cell growth and protein synthesis occur, restoring cell volume and size. A cell remains in this phase (from a few hours to several days) until the synthesis of "trigger proteins" enables it to reach the restriction point and proceed to the S phase.

17–D. The toxin from *Clostridium botulinum* inhibits the release of acetylcholine, the neurotransmitter at myoneural junctions. As a result, motor nerve impulses cannot be transmitted across the junction and muscle cells are not stimulated to contract.

18–C. Veins have thinner walls and larger, more irregular lumina than the corresponding arteries. The large veins entering the heart possess cardiac muscle in the tunica adventitia for a short distance. Valves are present in some veins to prevent retrograde movement of blood. The tunica adventitia is the thickest, most prominent layer in all veins.

19–A. The enamel of the adult tooth is acellular and thus new enamel matrix cannot be produced. The decrease in tooth length resulting from abrasion of the surface enamel is offset by elaboration of cementum, which possesses matrix-producing cells (cementoblasts).

20–D. The nucleosome, the structural unit of chromatin packing, does not contain RNA. In extended chromatin, two copies each of histones H2A, H2B, H3, and H4 form the nucleosome core around which a DNA molecule is wound. Condensed chromatin contains an additional histone, H1, that wraps around groups of nucleosomes, forming the structural unit of the chromosome.

21–C. Only infectious mononucleosis is characterized by all the signs and symptoms indicated. AIDS is associated with a decreased lymphocyte count (particularly of T helper cells). Pernicious anemia is associated with a decreased red blood cell count. Hodgkin's disease is associated with fatigue and enlarged lymph nodes, but the nodes are not painful and the presence of Reed-Sternberg cells is diagnostic of this disease. Factor VIII deficiency, a coagulation disorder, is not associated with any of the indicated signs and symptoms.

22–C. The gallbladder, which concentrates and stores bile, is lined by a simple columnar epithelium. Cholecystokinin stimulates contraction of the gallbladder wall, forcing bile from the lumen into the cystic duct; this joins the common hepatic duct to form the common bile duct, which delivers bile to the duodenum.

23–A. Rhodopsin is the visual pigment present in rods. Cones contain a related, but different, pigment called iodopsin.

24–A. Uncoupling of oxidation from phosphorylation induces mitochondria to change from the orthodox to the condensed form (not the reverse). Condensed mitochondria are often present in brown fat cells, which produce heat rather than ATP.

25–D. Muscular (distributing) arteries have a thick, complete internal elastic lamina in the tunica intima, whereas elastic (conducting) arteries have an incomplete internal elastic lamina. Both types of arteries have vasa vasorum, factor VIII, and smooth muscle cells in their walls. Muscular arteries possess numerous layers of muscle cells in the tunica media, but elastic arteries do not. Only elastic arteries possess fenestrated (elastic) membranes in the tunica media in which smooth muscle cells are dispersed.

26–D. The corneal endothelium is a simple squamous epithelium that lines the posterior (not the anterior) of the cornea.

27–C. The zone of hypertrophy is the region of the epiphyseal plate where chondrocytes are greatly enlarged. Bone-forming cells are not present in this zone.

28–C. In the endocrine pancreas, beta cells account for about 70% of the secretory cells; alpha cells, about 20%; and delta cells, less than 5%. Polypeptide hormones are synthesized by and released from the islets of Langerhans (endocrine pancreas). The exocrine pancreas possesses serous (not mucous) acinar cells. Insulin is produced by beta cells.

29–B. Cholecystokinin, a hormone produced in the small intestine, stimulates release of pancreatic enzymes. Urogastrone, gastrin, and somatostatin are hormones, produced in the stomach and/or small intestine, that directly or indirectly influence secretion of gastric HCl, as does stimulation of the vagus nerve.

30–C. Aqueous humor is secreted by the inner, nonpigmented layer of the ciliary processes.

31–A. A deficiency of vitamin A inhibits bone formation and growth, while an excess stimulates ossification of the epiphyseal plates, thus leading to premature closure of the plates. Both conditions result in short stature. A deficiency of vitamin D reduces calcium absorption from the small intestine and results in soft bones, whereas an excess of vitamin D stimulates bone resorption. A deficiency of vitamin C results in poor bone growth and fracture repair. Vitamin K plays no role in bone formation.

32–C. Membrane-coating granules are present in keratinocytes in the stratum spinosum (and stratum granulosum). The contents of these granules are released into the intercellular spaces to help "waterproof" the skin. Keratocytes in the stratum granulosum also possess keratohyalin granules; these contain proteins that bind keratin filaments together.

33–A. Basophilic erythroblasts are relatively undifferentiated cells; therefore, their nucleus possesses a fine, not coarse, chromatin network.

34–B. Rugae are surface modifications of the empty stomach. They are longitudinally directed folds of the mucosa and submucosa, which disappear when the stomach is distended with food.

35–C. The extrinsic pathway is initiated by the release of tissue thromboplastin following trauma to extravascular tissue. Platelet aggregation is promoted by von Willebrand's factor, which is associated with the intrinsic pathway only. Calcium is required in both pathways, and the final reaction—the conversion of fibrinogen to fibrin—is the same in both.

36–B. Herpetic stomatitis is characterized by painful fever blisters on the lips or near the nostrils. These blisters exude a clear fluid. Aphthous ulcers (canker sores) do not exude fluid. Bullous pemphigoid is an autoimmune disease marked by chronic, generalized blisters in the skin.

37–B. Basophils and mast cells, not eosinophils, have surface receptors for IgE. Eosinophils constitute 2%–4% of the leukocytes in the blood, whereas neutrophils constitute 60%–70%.

38–C. The skin, which consists of the epidermis and dermis, is important in the regulation of body temperature and contains temperature receptors in the dermis. Ultraviolet (not infrared) radiation absorbed by the skin is necessary for the synthesis of vitamin D. Protection against desiccation is provided by the hypodermis, a layer of loose connective tissue that binds skin to underlying tissues; the hypodermis is not considered part of the skin. Melanocytes are located in the deepest layer of the epidermis (stratum basale).

39–C. Diabetes insipidus is caused by the inability of the body to produce adequate amounts of antidiuretic hormone. This hormone is necessary for reabsorption of water from the urine in the collecting tubules of the kidney. Patients with this disorder produce large volumes of dilute urine (polyuria) and exhibit polydipsia and dehydration. Type I (juvenile-onset) diabetes mellitus also is characterized by polyuria, but the urine is not dilute. Hematuria and proteinuria are associated with glomerulonephritis.

40–E. The anus is lined by stratified squamous keratinized epithelium. The rectum, jejunum, and pyloric stomach are lined by simple columnar epithelium. The esophagus is lined by stratified squamous (nonkeratinized) epithelium.

41–D. Hyaline cartilage, elastic cartilage, and fibrocartilage all exhibit both interstitial and appositional growth. Hyaline and elastic cartilage have type II collagen in their matrix and are surrounded by a perichondrium, whereas fibrocartilage has type I collagen and lacks an identifiable perichondrium. Only hyaline cartilage is involved in endochondral bone formation.

42–E. Meissner's corpuscles, located in the papillary layer of the dermis, are fine-touch receptors. Pacinian corpuscles perceive pressure, touch, and vibration; they are located in the dermis, hypodermis, and connective tissue of mesenteries and joints. Cristae ampullares are special regions of the semicircular canals that detect circular movements of the head. Ruffini's endings, located in the dermis and joints, function in pressure and touch perception. Krause's end-bulbs are cold and pressure receptors located in the dermis.

43–A. Inner and outer phalangeal cells are among the supporting cells of the organ of Corti. They are intimately associated with the hair cells and support the nerve fibers that synapse with the hair cells.

44–C. Hyaline membrane disease, which results from inadequate amounts of pulmonary surfactant, is characterized by labored breathing and typically is observed in premature infants. Glucocorticoids stimulate synthesis of surfactant and can correct the condition.

45–A. The cribriform plate, the inner layer of the alveolar bone, is composed of compact bone. It is attached to the principal fiber groups of the periodontal ligament via Sharpey's fibers. The outer layer of the alveolar bone is the cortical plate. The spongiosa is the region of cancellous (spongy) bone enclosed between the cortical and cribriform plates.

46–C. Somatotropin (growth hormone) is synthesized by cells called somatotrophs, which are acidophils located in the pars distalis of the anterior lobe of the pituitary gland. Oxytocin and antidiuretic hormone (also called vasopressin) are produced in the hypothalamus and transported to the pars nervosa of the pituitary. Neurophysin, a binding protein, aids in this transport.

47–B. Plasma fibronectin functions in wound healing, blood clotting, and phagocytosis of material from the blood.

48–D. Matrix fibronectin mediates cell adhesion to the extracellular matrix by binding to fibronectin receptors on the plasma membrane.

49–A. Chondronectin has binding sites for collagen, proteoglycans, and cell-surface receptors.

50–D. Plasma cells, which arise from antigen-activated B lymphocytes, produce antibodies and thus are directly responsible for humoral-mediated immunity.

51–B. Macrophages, the principal phagocytes of connective tissue, remove large particulate matter and assist in the immune response by acting as antigen-presenting cells.

52–G. Mast cells (and basophils) have receptors for IgE antibodies on their surface. These cells release histamine, heparin, leukotriene C (slow-reacting substance of anaphylaxis), and eosinophil chemotactic factor, whose effects constitute immediate hypersensitivity reactions.

53–C. T lymphocytes initiate cell-mediated immune responses.

54–A. Pericytes are smaller than fibroblasts and are located along capillaries. When necessary, they assume the pluripotential role of embryonic mesenchymal cells.

55–E. G proteins are membrane proteins that are linked to certain cell-surface receptors. Upon binding of a signaling molecule to the receptor, the G protein functions as a signal transducer by activating a secondary messenger system that leads to a cellular response.

56–F. K^+ leak channels are ion channels that are responsible for establishing a potential difference across the plasma membrane.

57–A. The term "amphipathic" refers to molecules, such as phospholipids, that possess both hydrophobic (nonpolar) and hydrophilic (polar) properties. The plasma membrane contains two phospholipid layers (leaflets) with the hydrophobic tails of the molecules projecting into the interior of the membrane and the hydrophilic heads facing outward.

58–B. The glycocalyx (cell coat) is associated with the outer leaflet of the plasma membrane. It is composed of proteoglycans, which possess polysaccharide side chains.

59–C. Membrane carrier proteins are highly folded transmembrane proteins that undergo reversible conformational alterations, resulting in transport of specific molecules across the membrane. The Na^+–K^+ pump is a carrier protein that mediates antiport transport, which is the transport of two molecules concurrently in opposite directions.

60–B. T suppressor cells and T cytotoxic cells have CD8 marker molecules on their surface.

61–A. T helper cells have CD4 marker molecules on their surface.

62–E. Perforin, which is released by cytotoxic T cells, mediates lysis of tumor cells and virus-infected cells.

63–C. Interleukin 1, which is produced by macrophages, stimulates activated T helper cells. In turn, activated T helper cells produce interleukin 2 and other cytokines involved in the immune response.

64–F. Interferon γ (macrophage-activating factor) stimulates activation of NK cells and macrophages, thereby increasing their cytotoxic and/or phagocytic activity.

65–B. Primary spermatocytes undergo the first meiotic division following DNA replication in the S phase. The resulting secondary spermatocytes undergo the second meiotic division, without an intervening S phase, forming spermatids.

66–E. During spermiogenesis, the manchette is formed. This temporary structure aids in elongation of the spermatid.

67–D. Interstitial cells of Leydig produce testosterone when they are stimulated by luteinizing hormone.

68–A. Sertoli cells are columnar cells that extend from the basal lamina to the lumen of the seminiferous tubules. Adjacent Sertoli cells form basal tight junctions, which are responsible for the blood–testis barrier, thus protecting the developing sperm cells from autoimmune reactions.

69–A. Sertoli cells also produce androgen-binding protein, which binds testosterone and maintains it at a high level in the seminiferous tubules.

70–B. Smooth endoplasmic reticulum possesses mixed function oxidases that detoxify phenobarbital and certain other drugs.

71–A. The membrane of the rough endoplasmic reticulum contains ribophorins, which are receptors that bind the large ribosome subunit.

72–D. Annulate lamellae are parallel stacks of membranes that resemble the nuclear envelope. They are present in the cytoplasm of rapidly growing cells, but their function remains obscure.

73–F. Cytosolic proteins are synthesized on polysomes (polyribosomes). In contrast, secretory proteins are synthesized at the rough endoplasmic reticulum.

74–A. Gastrin, a paracrine hormone secreted in the pylorus and duodenum, stimulates HCl secretion by parietal cells in the gastric glands.

75–F. Urogastrone, produced by Brunner's glands in the duodenum, inhibits gastric HCl secretion and enhances division of epithelial cells.

76–E. Lysozyme, manufactured by Paneth cells in the crypts of Lieberkühn, is an enzyme that has antibacterial activity.

77–C. Motilin, a paracrine hormone secreted by cells in the small intestine, increases gut motility by stimulating smooth muscle contraction.

78–B. Somatostatin, produced by enteroendocrine cells in the pylorus and duodenum, inhibits secretion by nearby enteroendocrine cells.

79–B. Isodesmosine and desmosine are two unusual amino acids in elastin that are responsible for its elasticity.

80–A. Lysine cross-links elastin molecules forming a network. Fibrillin is a glycoprotein that organizes elastin into fibers.

81–A. Simple goiter is an enlargement of the thyroid gland resulting from inadequate dietary iodine (less than 10 μg/day). It is common where the food supply is low in iodine.

82–D. Addison's disease is most commonly caused by an autoimmunity that destroys the adrenal cortex. As a result, inadequate amounts of glucocorticoids and mineralocorticoids are produced. Unless these are replaced by steroid therapy, the disease is fatal.

83–B. The globular head of the myosin molecule has ATPase activity, but interaction with actin is required for the noncovalently bound reaction products (ADP and P_i) to be released. This ATPase activity is retained by the S1 fragment resulting from digestion of myosin with proteases.

84–C. Myoglobin is a sarcoplasmic protein that, like hemoglobin, can bind and store oxygen. The myoglobin content of red (slow) muscle fibers is higher than that of white (fast) muscle fibers.

85–A. Troponin C is one of the three subunits of troponin, which along with tropomyosin, binds to actin (thin) filaments in skeletal muscle. Binding of calcium ions by troponin C results in unmasking of the myosin-binding sites on thin filaments.

86–C. Malignant melanoma, a relatively rare form of skin cancer, arises from melanocytes. It is aggressive and invasive. Surgery and chemotherapy usually are necessary for successful treatment of this cancer.

87–A. Epidermolysis bullosa is a group of hereditary skin diseases characterized by the separation of the layers in skin with consequent blister formation.

88–B. The nasopharynx is the site of the pharyngeal tonsil; when enlarged and infected, this tonsil is known as adenoids.

89–A. The trachea and extrapulmonary (primary) bronchi have walls supported by C-shaped hyaline cartilages (C-rings), whose open ends face posteriorly.

90–D. The alveolar duct has alveoli whose openings are rimmed by sphincters of smooth muscle. Alveoli more distal than these have only elastic and reticular fibers in their walls.

91–C. Terminal bronchioles are lined by a simple cuboidal epithelium containing ciliated cells and Clara cells. Clara cells can divide and regenerate both cell types.

92–C. An ectopic tubal pregnancy occurs when the embryo implants in the wall of the oviduct (rather than in the uterus). Because the oviduct cannot support the developing embryo, the duct eventually bursts, causing hemorrhaging into the peritoneal cavity.

93–D. Breast cancer that originates from the epithelium lining the terminal ductules of the mammary gland is classified as lobular carcinoma.

94–B. Abnormal cells associated with cervical cancer are revealed in a Pap smear, providing a simple method for the early detection of this cancer.

95–A. Endometriosis is a condition in which uterine endometrial tissue is located in the pelvic peritoneal cavity. The misplaced endometrial tissue undergoes cyclic hormone-induced changes, including menstrual breakdown and bleeding.

96–E. Gastric intrinsic factor, which is produced by parietal cells in the gastric glands, is necessary for absorption of vitamin B_{12} in the ileum.

97–B. Disaccharidases located in the glycocalyx of the striated border hydrolyze disaccharides to monosaccharides.

98–A. Digestion of proteins begins with the action of pepsin in the stomach, forming a mixture of polypeptides. Activation of pepsinogen to pepsin only occurs at a low pH. Thus stimulation of gastric HCl secretion by gastrin facilitates digestion of proteins.

99–D. After free fatty acids and monoglycerides, in micelles, enter the surface absorptive cells of the small intestine, they are reesterified to form triglycerides. These are complexed with proteins, forming chylomicrons, which are released from the lateral cell membrane and enter lacteals in the lamina propria.

100–B. The lamina lucida (rara), the electron-lucent layer of the basal lamina, contains heparan sulfate.

101–A. The lamina densa, the electron-dense layer of the basal lamina, contains type IV collagen.

102–C. In bullous pemphigoid antibodies are formed against hemidesmosomes. As a result, the epithelium separates from the underlying connective tissue causing blisters to form.

103–D. The basement membrane comprises the lamina lucida, lamina densa, and underlying reticular lamina.

104–C. Hemidesmosomes function to attach epithelial cells to the basal lamina.

105–B. In the descending thin limb of the loop of Henle, the filtrate loses water and gains sodium ions, thus becoming hypertonic relative to blood plasma.

106–A. In the presence of antidiuretic hormone (vasopressin), the epithelial cells of the collecting tubules become permeable to water. In the absence of ADH, these cells are impermeable to water.

107–D. The vasa recta are long, thin vessels that follow a straight path into the medulla and then loop back toward the corticomedullary boundary. These vessels passively exchange sodium and water with the loops of Henle.

108–C. Interstitial cells of the renal medulla manufacture vasodepressor substances, including prostaglandins.

109–C. Microtubules have a rigid wall composed of 13 protofilament strands, each of which consists of a linear arrangement of α- and β-tubulin dimers.

110–B. Several Golgi cisternae (saccules) arranged in a stack make up part of the Golgi complex. The various steps in the processing of secretory proteins occur in distinct cisternae within the stack.

111–A. The nuclear lamina is composed of lamins A, B, C, which are intermediate filament proteins. The nuclear lamina faces the nuclear chromatin and lies adjacent to the inner nuclear membrane. It helps organize the nuclear envelope and perinuclear chromatin.

112–D. Euchromatin is composed of histones, acidic proteins, and transcriptionally active DNA. It appears as electron-lucent regions dispersed in the nucleus. In contrast heterochromatin, containing transcriptionally inactive DNA, appears as electron-dense clumps and is concentrated mostly at the periphery of the nucleus.

113–E. Mitochondria produce ATP via the Krebs cycle and a chemiosmotic coupling mechanism. The Krebs-cycle enzymes are located in the mitochondrial matrix (except for succinate dehydrogenase). The chemiosmotic mechanism involves enzyme complexes of the electron transport chain and elementary particles that are located in membranes of the cristae.

114–B. Neurotubule.

115–D. Unmyelinated nerve fiber.

116–A. Schwann cell.

117–E. Mitochondrion.

118–C. Myelin.

Index

Note: Page numbers in *italic* denote illustrations; those followed by (t) denote tables; those followed by Q indicate questions; and those followed by E indicate explanations.